Nano-Society
Pushing the Boundaries of Technology

RSC Nanoscience & Nanotechnology

Series Editors:
Professor Paul O'Brien, *University of Manchester, UK*
Professor Sir Harry Kroto FRS, *University of Sussex, UK*
Professor Harold Craighead, *Cornell University, USA*

Titles in the Series:
1: Nanotubes and Nanowires
2: Fullerenes
3: Nanocharacterisation
4: Atom Resolved Surface Reactions: Nanocatalysis
5: Biomimetic Nanoceramics in Clinical Use: From Materials to Applications
6: Nanofluidics: Nanoscience and Nanotechnology
7: Bionanodesign: Following Nature's Touch
8: Nano-Society: Pushing the Boundaries of Technology

Visit our website at www.rsc.org/nanoscience

How to obtain future titles on publication:
A standing order plan is available for this series. A standing order will bring delivery of each new volume immediately on publication.

For further information please contact:
Sales and Customer Care, Royal Society of Chemistry, Thomas Graham House, Science Park, Milton Road, Cambridge, CB4 0WF, UK
Telephone: +44 (0)1223 432360, Fax: +44 (0)1223 420247,
Email: sales@rsc.org
Visit our website at http://www.rsc.org/Shop/Books/

Nano-Society
Pushing the Boundaries of Technology

Michael Berger
Nanowerk LLC

RSC Publishing

RSC Nanoscience & Nanotechnology No. 8

ISBN: 978-1-84755-883-1
ISSN: 1757-7136

A catalogue record for this book is available from the British Library

Published by The Royal Society of Chemistry,
Thomas Graham House, Science Park, Milton Road,
Cambridge CB4 0WF, UK

Registered Charity Number 207890

For further information see our web site at www.rsc.org

Preface

This book is a journey through the world of nanotechnology research and development that takes a very personal look at how nanotechnologies are created today. You will read about 122 research projects that are taking place in laboratories around the world and you will meet the scientists who developed the theories, conducted the experiments, and built the new materials and devices that will each take us one step further into our nanotechnology future.

Chemists and biologists have always dealt with naturally occurring nanoparticles: think molecules or viruses. Toxicologists have dealt with nanoparticles that are the result of modern human life, such as carbon particles in combustion engine exhaust. Without being aware of it, tire manufacturers used nanoparticles (carbon black) to improve the performance of tires as early as the 1920s. Medieval artists (unknowingly) used gold nanoparticles to achieve the bright red colors in church windows. You might even say that we are surrounded by, and made of, nanomaterials—atoms and molecules are nanoscale objects, after all. So why all the fuss about "nano" now?

The ongoing quest for miniaturization has resulted in tools like the atomic force microscope and the scanning tunneling microscope. Combined with refined processes such as electron beam lithography, these instruments allow researchers to deliberately manipulate and manufacture nanostructures; something they couldn't do before. Engineered nanomaterials, either by way of a top-down approach (a bulk material is reduced in size to nanoscale patterns) or a bottom-up approach (larger structures are built or grown atom by atom or molecule by molecule), go beyond just a further step in miniaturization. They have broken a size barrier below which quantization of energy for the electrons in solids becomes relevant. The so-called *quantum size effect* describes the physics of electron properties in solids with great reductions in particle size. This effect does not come into play when we go from macro to micro dimensions. However, it becomes dominant when the lower nanometer size range is

RSC Nanoscience and Nanotechnology No. 8
Nano-Society: Pushing the Boundaries of Technology
By Michael Berger
© Michael Berger 2009
Published by the Royal Society of Chemistry, www.rsc.org

reached. Materials reduced to the nanoscale can suddenly show very different properties compared to what they show on a macroscale. For instance, opaque substances become transparent (copper); inert materials become catalysts (platinum); stable materials turn combustible (aluminium); solids turn into liquids at room temperature (gold); insulators become conductors (silicon).

A second important aspect of the nanoscale is that the smaller a nanoparticle gets, the larger its relative surface area becomes. Its electronic structure changes dramatically. Both effects lead to greatly improved catalytic activity but can also lead to aggressive chemical reactivity.

The fascination with nanotechnology stems from the unique quantum and surface phenomena exhibited by matter at the nanoscale, making possible novel materials and revolutionary applications.

What we call nanotechnology is not an industry; nor is it a single technology or a single field of research. It consists of sets of enabling technologies applicable to many traditional industries (therefore it is more appropriate to speak of *nanotechnologies* in the plural).

This book is a scientific journey that shines a spotlight on some of the scientists who are pushing the boundaries of technology. It illustrates examples of their work and how they are advancing scientific knowledge and technological capabilities one tiny step at a time. Some stories are more like an introduction to nanotechnology, some are about understanding current developments, and some are advanced technical discussions of leading-edge research. Reading this book will shatter the monolithic term "nanotechnology" into the myriad of facets that it really is.

Major technology shifts don't happen overnight; and rarely are they the result of a single breakthrough discovery. Nowhere is this more true than for the vast set of capabilities that we have come to simply call *nanotechnology*. Rather than standing on the shoulders of a few intellectual giants, nanotechnologies get created by tens of thousands of researchers and scientists working on minute and sometimes arcane aspects of their fields of expertise in areas as diverse as medicine, fuel cells, solar energy, water filtration, food formulations, pathogen detection, airplane design, solid state lighting, coatings or ever smaller transistors for electronic devices; they come from different science backgrounds; live in different parts of the world; work for different organizations (government labs, industry labs, universities, private research facilities) and follow their own set of rules—they get papers reviewed and published; achieve scientific recognition from their peers; struggle to get funding for new ideas; look to make that breakthrough discovery that leads to the ultimate resumé item, a Nobel prize; are pushed by their funders to secure patent rights and commercialize new discoveries.

The collection of stories in this book barely scratches the surface of the vast and growing body of research that leads us into the nanotechnology age. The selection presented here is not meant to rank some labs and scientists higher than others, nor to imply that the work introduced in this book is more important or valuable than all the work that is not covered. The intention is to give the interested reader an idea of the incredibly diverse aspects that make up nanotechnology research and development, the results of which will bring about a new era of industrial and medical technologies.

Nanoscience and nanotechnology research is a truly multidisciplinary and international effort. Each of the following chapters is based on a particular scientific paper that has been published in a peer-reviewed journal and, while each story revolves around one or two scientists who were interviewed for this book, many, if not most, of the scientific accomplishments covered here are the result of collaborative efforts by several scientists and research groups, often from different organizations and from different countries.

Our world of tomorrow will be based upon ubiquitous and pervasive nanotechnologies as fundamental enabling technologies for human activities. This nano-society will be the result of thousands upon thousands of untold stories like the ones found in this book.

Contents

RSC Nanoscience and Nanotechnology No. 8
Nano-Society: Pushing the Boundaries of Technology
By Michael Berger
© Michael Berger 2009
Published by the Royal Society of Chemistry, www.rsc.org

PART I:
THE NANOTECHNOLOGIST'S TOOLBOX

CHAPTER 1

How Can You 'See', 'Feel' or 'Hear' Something so Incredibly Tiny?

In 1981, the scanning tunneling microscope (STM) was invented, followed 4 years later by the atomic force microscope (AFM)—and that's when nanoscience and nanotechnology really started to take off. Various forms of scanning probe microscopes (SPM) based on these discoveries are essential for many areas of today's research. Conventional optics cannot resolve objects measuring tens of nanometers or less because the visible wavelength of light is roughly between 400 and 750 nm. With scanning probe techniques, all of a sudden the nanoworld became accessible to scientists in many countries, and these instruments have been the workhorses of nanoscience and nanotechnology research ever since.

"Today these methods are still making a tremendous impact on many disciplines that range from fundamental physics and chemistry through information technology, quantum computing, spintronics and molecular electronics, and all the way to life sciences," Christoph Gerber and Hans Peter Lang from the National Competence Center for Research in Nanoscale Science at the University of Basel in Switzerland write in an article in *Nature Nanotechnology*.[1] "Indeed, some 4000 AFM-related papers were published in 2006 alone, bringing the total to 22 000 since it was invented, and the STM has inspired a total of 14 000 papers. There are also at least 500 patents related to the various forms of SPM. Commercialization of the technology started in earnest at the end of the 1980s, and approximately 10 000 commercial systems have been sold so far to customers in areas as diverse as fundamental research, the car industry

---

[1] C. Gerber and H. P. Lang, How the doors to the nanoworld were opened, *Nat. Nanotechnol.*, 2006, **1**, 3–5.

RSC Nanoscience and Nanotechnology No. 8
Nano-Society: Pushing the Boundaries of Technology
By Michael Berger
© Michael Berger 2009
Published by the Royal Society of Chemistry, www.rsc.org

and even the fashion industry. There are also a significant number of home-built systems in operation. Today some 30–40 companies are involved in manufacturing SPM and related instruments, with an annual worldwide turnover of \$250–300 million. Moreover, the market of SPMs is predicted to double over the next 5 years."

Unless they work in a state-of-the-art laboratory equipped with multimillion-dollar high-tech instruments, most people find it impossible to visualize nanoscale objects. The overused description that one nanometer is 50 000–100 000 times smaller than the diameter of a hair isn't really helpful either. Even scientists who work with the latest electron microscope techniques on a daily basis, and who have brought us all these amazing images from the nanoworld, often find it difficult, if not impossible, to make the mental connection between what they see with their own eyes and what the read-outs on their AFM show them. In this respect, a nanoscientist peering into the nanorealm isn't that different from an astronomer looking at the farthest reaches of the observable universe—the scales, be it nanometers or light years, overwhelm our brain's capacity for visualization.

While our five senses do a reasonably good job at representing the world around us on a macro scale, we have no existing intuitive representation of the nanoworld, ruled by laws entirely foreign to our experience. This is where molecules mingle to create proteins; where you wouldn't recognize water as a liquid; and where minute morphological changes would reveal how much 'solid' things such as the ground or houses are constantly vibrating and moving. Therefore, before we delve into the world of nanoscale probing and imaging, our first story is about an idea that could result in tools to explore the boundaries between the nanoscopic and the macroscopic worlds—touching nanoscale water, shaking hands with bacteria, crushing a virus between your fingers, playing nano-Lego. For scientists, it could also lead to a new generation of professional lab tools that allow nanoscale manipulation with precise control of tool interaction with nano-objects.

1.1   Getting all Touchy-Feely

Our sense of touch connects us to the world around us and it is an integral part of how we experience things, both physically and emotionally. In the virtual world of remote-control robots, scientific models, or computer games, users generally lack tactile, or haptic,[2] feedback, which either makes delicate manipulative tasks difficult or keeps the subject purely visual and often inscrutable (such as an electron microscope image of a nanoscale object). The desire for natural and intuitive human machine interaction has led to the inclusion of haptics in human–machine interfaces. The user is able to control inputs to the system through hand movements and in turn receives feedback through tactile stimulation in the hands. Sophisticated, state-of-the-art haptic

---

[2] From the Greek word *haphe*, pertaining to the sense of touch.

user-interface software is capable of adding interactive, realistic virtual touch capabilities to human–computer interactions. Among the uses are medical applications, remote vehicle or robotic control, military applications, and video games. Users are said to feel realistic weight, shape, texture, dimension, dynamics, and force effects. Applying the use of real-time virtual reality and multisensory user interface to nanoscience, scientists in France have begun to open up the otherwise only scientifically described reality of the nanoworld to a nonscientific public.

"A central challenge is how we can put our hands on scientifically explored parts of reality that cannot be reached by our senses and whose rules are completely foreign to our representation of reality," Joël Chevrier tells us. "Since science is full of abstract descriptions, it is hard to represent it in an easy way. But thanks to computer sciences and robotics we now have the necessary tools to use human senses to explore, in real time, model worlds as they are described by science, or even true reality when coupling these multisensory interfaces to real nanosensors and nanoactuators."

Chevrier, a professor at the Université Joseph Fourier in Grenoble, France, together with his collaborators hopes to open up a completely new field for our perception. This new 'playground'—using haptic, vision, and sound interfaces—is the world we live in; but explored at scales entirely foreign to everything we experience around us.

"In the nanoworld simulacrum that we have begun to build, object identification will be based on the intrinsic physical and chemical properties of the probed entities, on their interactions with sensors, and on the final choices made in building a multisensory interface so that these objects become coherent elements of the human sphere of action and perception," says Chevrier.

In other words, we might be able to touch, feel, and interact with the nanoworld which otherwise is not open to our direct experience. Chevrier hopes that this will be a major step in helping nonscientists understand nanosciences and nanotechnologies. The scientifically described part of our reality—much of what mathematics, physics, or chemistry is about—is usually inaccessible to people not trained in these subjects, *i.e.* to most of us. Opening up this part of reality to everybody could go a long way in creating interest in science education and science careers, and help a better-informed public to lead a more objective discussion on the pros and cons of nanotechnologies.

Rather than using the abstract descriptions and experiments of a classical science education, the French team has begun to use real-time virtual reality combined with a multisensory human–machine interface to allow the direct perception of and interaction with the nanoworld.

"One way to develop this extension of the sphere where our senses are efficient can be based on nanosensors and nanoactuators," explains Chevrier. "Another approach is to use virtual environments which can bring the nanoworld to us through real-time multisensory interfaces. This can dramatically enhance the possibilities for easy exploration of remote realities foreign to our senses and can trigger spontaneous motivation in users, similar to what we observe in video game players."

Chevrier and his team have built a virtual AFM and coupled it to an advanced haptic interface as well as a sonification and visualization system. The resulting instrument allows its user to experience contact of a surface at the nanoscale. About 10 000 people used this demonstrator during three exhibitions in Grenoble, Paris, and Geneva.

A central part of this concept is not a new idea. It actually goes back to the earliest days of experimental science: Galileo's use of a telescope to observe the Moon and coming to the immediate conclusion that the Moon is Earth-like. As Galileo immediately emphasized, this dramatic change in the human representation of the universe is caused by direct use of senses technically extended by an instrument, and not by *a posteriori* rational demonstration.

"Our proposal can be seen as a revival of this famous tale," says Chevrier. "There is a major difference, however. Two points can illustrate the need for new approaches in implementing the nanoscale in virtual environments:

(1) As we gradually approach the nanoscale, continuous description no longer stands and the molecular, discontinuous structure of matter is revealed. Atomic scale is a radical rupture with our common experience that is based on the objective existence of isolated continuous objects.

(2) Can we manage to 'see' and 'touch' an electron, a particle that has a mass and an electric charge but has no classical spatial extension in the sense of a material sphere, although it is at the root of the stability of matter? In fact, seeing or touching an electron has no intrinsic meaning. Electron-based objects can, however, be created and our interaction with these unusual objects defined."

Almost all scientific data today is represented visually. That's why we have all these amazing electron microscope images and artists' impressions of nanoscale objects. That's also why most people can't really get a grip (literally) on scientific discoveries unless they result in a better TV set or more stain-resistant shirts. Enriching the visual component with interactive tactile and sound aspects, and wrapping the whole thing into a virtual reality environment, will give us a much richer and more real experience of these objects.

At the Center for Cognitive Ubiquitous Computing (CUbiC) at Arizona State University in the USA they have developed some interesting haptic visualization schemes. Many object features are easy to invoke in human memory and are presented through tactile cueing. There are, however, some features that are not primary haptic features but may contribute to further knowledge of the object. One example is the weight of the object. At CUbiC they have developed a haptic visualization scheme for the presentation of weight. In this scheme, a user is able to bounce the virtual object off an imaginary surface. When the object hits back, it generates a vibrotactile stimulation analogous to its weight.

Even if technology will one day offer us sophisticated tools to explore the nanoworld with our senses, the question is whether we will be able to really grasp it. Imagine an atom. Chances are you are seeing a Nagaoka (Figure 1.1).

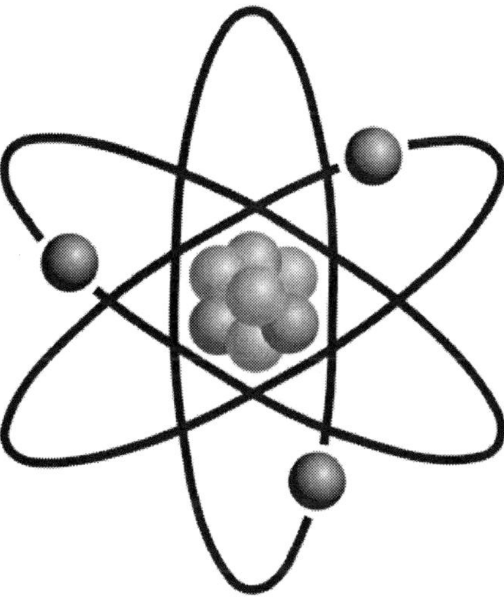

Figure 1.1　The classic atomic model created by Hantaro Nagaoka (Source: Wiki Commons)

In 1904, a Japanese physicist named Hantaro Nagaoka created the classic atom image with planet-like electrons orbiting around a nucleus.

This is the picture that many people have in mind—cute, but wrong. Reality at the atomic scale is much, much weirder: atoms are mostly empty space and the solid world we experience around us is an illusion. Timothy Ferris has described this nicely in his book *Coming of Age in the Milky Way*:[3]

"A bar of gold, though it looks solid, is composed almost entirely of empty space. The nucleus of each of its atoms is so small that if one atom were enlarged a million billion times, until its outer electron shell was as big as greater Los Angeles, its nucleus would still be only about the size of a compact car parked downtown. The electron shells would be zones of insubstantial lightning, each a mile or so thick, separated by many miles of space. Nor, to return to the old classical metaphor, does a cue ball strike a billiard ball. Rather, the negatively charged fields of the two balls repel each other; on the subatomic scale, the billiard balls are as spacious as galaxies, and were it not for their electrical charges they could, like galaxies, pass right through each other unscathed."

So, while your 'reality' tells you that you are sitting in your chair right now as you are reading this, reality at the subatomic level means that you are not really in contact with your chair—thanks to the repulsion of the chair's electrons and your own, you are actually floating on it at a height of a fraction of a nano-meter. The point is that, even if we might have the tools one day to truly

---

[3] T. Ferris, *Coming of Age in the Milky Way*, Harper Perennial, New York, 2003.

experience the nanoworld, its rules are so foreign to our human experience that we might not be able to comprehend it anyway.

Of course, this first instrument built by Chevrier's team in Grenoble is more Galileo telescope than Hubble space observatory. But it is an interesting beginning that one day might result in virtual worlds that will allow us to go all weird at the nanoscale.

> *Featured scientist:* Joël Chevrier
> *Organization:* Université Joseph Fourier, Grenoble, France
> *Relevant publication:* Implementation of perception and action at
>   nanoscale. *Proceedings of ENACTIVE/07*, 4th International
>   Conference on Enactive Interfaces, Grenoble, France, 19–22
>   November 2007.

1.2   Feeling our Way Through the Nanoworld

Let's come back to the AFM mentioned at the beginning and look at an example of how scientists continuously work to improve these instruments. Since the nanoscale world is accessible only with specialized—and often very expensive—tools, the ongoing improvement of these instruments, and the development of new ones, is a crucial aspect of continuous progress in nanosciences and nanotechnologies. Contrary to much of the hype surrounding the field, a large part of 'nanotechnology' today is about developing new tools, techniques, and applications to explore and understand phenomena at the nanoscale.

"Children begin to learn by seeing, hearing, tasting and, above all, by touching. In a very similar approach, we are currently learning to orient ourselves in the nanoworld by 'feeling' materials—not with our fingers, but with microscopes that allow us to probe these materials with atomic resolution."[4]

The ability of researchers to engineer novel materials that possess superior electronic, thermal, magnetic, and mechanical properties depends on tools that can identify and characterize material components and their spatial arrangement at the nanoscale. Equally important, understanding structure–function relationships in biological systems also demands tools that can probe structural properties with molecular resolution: AFMs are the most widely used tools to image matter at the nanoscale.

The operating principle of an AFM is based on an atomically sharp tip, placed at the end of a flexible cantilever beam, that is brought into physical contact with the surface of a sample. The cantilever beam deflects in proportion to the force of interaction. Scanning across the surface, the sharp tip follows the bumps and grooves formed by the atoms on the surface. A topography of the

---

[4] R. W. Stark, Atomic force microscopy: getting a feeling for the nanoworld, *Nat. Nanotechnol.* 2007, **2**, 461–462.

surface can be generated by monitoring the deflections of the flexible cantilever beam.

Because of its mechanical operation, the AFM can in principle also perform nanomechanical measurements. This aspect of the AFM has been explored by researchers over the past two decades. However, current state-of-the-art techniques are very slow—it takes about a second for the AFM tip to approach, push into, and retract from the surface of a material—and rather large forces are applied during the measurement process that damage the tip and the sample.

Researchers at Harvard and Stanford universities have developed a specially designed AFM cantilever tip, the *torsional harmonic cantilever* (THC), which offers orders-of-magnitude improvements in temporal resolution, spatial resolution, indentation, and mechanical loading compared to conventional tools. With high operating speed, increased force sensitivity, and excellent lateral resolution, this tool facilitates practical mapping of nanomechanical properties.

Notwithstanding all the advances that have been made in the field of AFM, Ozgur Sahin from the Nanomechanical Sensing Group at the Rowland Institute at Harvard University in the USA says that so far a technique that can quantitatively map mechanical properties in detail with nanoscale resolution is missing. "Mechanical properties of matter are largely determined by the nature of chemical bonds and their spatial organization in the material," he explains. "Furthermore, materials used in everyday life exhibit a huge variation in their mechanical properties. Diamond, for example, is almost a million times stiffer than rubbery materials. The spectrum of mechanical properties of materials spans the range between these two extremes. These observations tell us that there is a lot of information in mechanical properties of materials."

Sahin, together with collaborators from Stanford University, led by Olav Solgaard, and Veeco Instruments, describes the design of a special cantilever tip that allows the material properties of a surface to be determined and mapped in detail with nanoscale spatial resolution. "In order to create a high-speed and sensitive nanomechanical measurement tool, we have started from the most commonly used AFM technique called the tapping mode,"[5] explains Sahin. "The primary advantage of this technique is that it protects the tip and the sample during the imaging process and minimizes the interaction forces.

"For our goal of performing mechanical measurements, tapping mode also provides a unique opportunity because the sharp tip is moving back and forth against the surface and feels the variation of force during the interaction. If one can detect those forces varying with tip–sample distance, one can perform a clear and detailed mechanical analysis."

Unfortunately, there are major difficulties in measuring the forces between the tip and the sample. These forces change at a rate much faster than the vibration of the cantilever, therefore the force-sensing cantilever cannot

---

[5] In this mode, the tip is vibrated perpendicular to the sample surface to avoid gouging the specimen as the tip is scanned laterally.

respond to them. Indeed, there is a wealth of publications in the literature working on the nonlinear dynamics of tapping cantilevers that seek indirect ways to measure these forces. "In a way, our work stands on the shoulders of these pioneers, because they have reached a very good understanding of the complicated cantilever dynamics in AFMs," says Sahin. "Nevertheless, we have taken a different approach by engineering the force-sensing cantilever to measure the interaction forces directly."

The AFM cantilever has many vibration modes, each of which can act as an independent force sensor. The rapidly changing forces demand the use of a fast mode with high-resonance-frequency. The problem with high resonance frequency modes is that they are stiff and do not bend easily to give a good signal.

"What we have noticed is that torsional vibration modes allow good signal levels and they have high enough resonance frequencies," says Sahin. "Unfortunately, tip sample forces do not excite torsional oscillations because the conventional cantilevers have their tips on the center line. Therefore, we designed cantilevers that have their tips off-centered. When this cantilever hits the surface, tip–sample forces generate a torque that bends the cantilever torsionally. Torsional vibrations can be detected in a commercial AFM system simultaneously with the vertical vibrations."

When this cantilever is operated in conventional tapping-mode—touching the surface ever so lightly some 50 000 times per second (50 kHz)—the torsional vibrations can be simultaneously detected and translated into a time-varying tip–sample force waveform which contains detailed information about the mechanical properties of the sample. Figure 1.2

"In principle, the speed of these measurements is limited by the oscillation frequency of the cantilever," says Sahin. "At the moment, we are not fully benefiting from the speed enhancement. However, it is still more than a factor of 1000 times faster than conventional mechanical measurements, yet it is much gentler to the sample. Improved speed enables mapping mechanical properties across a surface with nanometer resolution. I believe that in the near future we will see mechanical measurements performed within a microsecond. This will open up a new window to study time dependent phenomena at the nanoscale, such as protein folding and chemical reactions in general."

Sahin and his collaborators have demonstrated their technique on a 50 nm thick polymer composite (PMMA—a clear, hard plastic also known under its trade name Plexiglas) that has two components with submicron features. These components have different thermal characteristics. Sahin explains that as the PMMA is heated up, it goes through a glass transition at around 100 °C where the hard, brittle polymer becomes rubbery and compliant. "When you heat up a polymer composite, each component has a different glass transition temperature and therefore a bulk measurement does not tell us how individual components behave. The high-resolution mechanical measurement technique we presented allowed us to observe the changes in the two components of the polymer independently. This kind of ability is crucial for the development of any material system with multiple components."

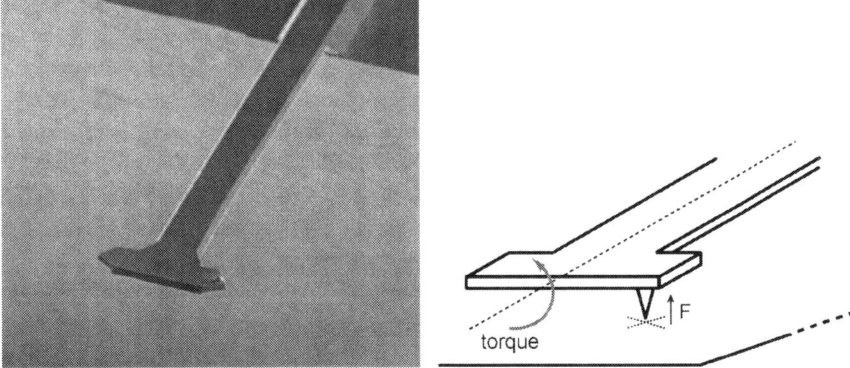

Figure 1.2 Design of the torsional harmonic cantilever with an off-axis tip. Left: A SEM image of a torsional harmonic cantilever. Right: An illustration of the THC interacting with the surface. The offset position of the tip results in a torque around the axis of the cantilever. (Image and illustration: Ozgur Sahin)

An intriguing application with significant potential lies in using mechanical sensing to detect biomolecular reactions. When molecules bind, their mechanical properties change dramatically and in a specific way. Sahin says that a mechanical sensing technique such as theirs could lead to label-free, yet specific, detection with single-molecule resolution.

Developing this technique further, Sahin sees two distinct paths. "One major direction will be towards biomolecular detection, and the other is material analysis," he says, "but there are a number of challenges. AFM in general is difficult to work with in liquids. If this can be resolved, mechanical sensing in liquids will provide exciting opportunities in the study of biological systems and biomolecules."

Featured scientist: Ozgur Sahin
Organization: Rowland Institute at Harvard, Cambridge, MA, USA
Relevant publication: Ozgur Sahin, Sergei Magonov, Chanmin Su, Calvin F. Quate, Olav Solgaard: An atomic force microscope tip designed to measure time-varying nanomechanical forces. *Nat. Nanotechnol.*, **2**, 507–514.

1.3 Real-time Single-molecule Imaging of an Entire Chemical Reaction

In its more than 25 years of existence, STM has predominantly brought us extremely detailed images of matter at the molecular and atomic level. An

STM—not to be confused with a scanning electron microscope (SEM)[6]—is a nonoptical microscope that scans an electrical probe over a surface to be imaged to detect a weak electric current flowing between the tip and the surface. The STM allows scientists to visualize regions of high electron density and hence infer the position of individual atoms and molecules on the surface of a lattice.

Researchers have advanced this technique to a point where they can use STM to perform real-time single-molecule imaging of an entire chemical reaction. Many chemical reactions are catalyzed by metal complexes, and insight into their mechanisms is essential for the design of future catalysts. Applying the STM approach to studying chemical reactions in a dynamic environment can provide valuable information about reaction mechanisms and rates as well as catalyst activity and stability.

"A couple of years ago, we took up the idea of imaging all the steps of a chemical reaction in an STM," says Hans Elemans. "Some examples of imaging chemical reactions have been described before, but only under extreme conditions—for example low temperature, ultra-high vacuum—and we wanted to make it work in an environment familiar to chemists, such as the interface of a liquid and a solid." Elemans, a member of the Nolte Group for Supramolecular Chemistry at Radboud University in Nijmegen, The Netherlands, together with members from his group and collaborators from the University of Sydney and the Institute for Molecules and Materials at Radboud University, are the first team to use an STM to image a complete catalytic reaction at the interface of a liquid and a solid by following the catalyst molecules, when they are in action, at the single-molecule level.

As the catalysts for their chemical reactions the researchers used manganese porphyrins. These are dye molecules, whose naturally occurring magnesium analogues are involved in processes such as light-harvesting in plants. In their experiments, manganese porphyrin was added to a liquid-cell STM set-up and the researchers were able to observe the formation of extended monolayer domains at the liquid–solid interface, with the porphyrins adsorbed face-on to the surface in regular patterns.

"The reaction that we investigated led to several unexpected new insights," explains Elemans. "First, the gold surface we used appeared to activate the adsorbed manganese porphyrin catalysts to react with molecular oxygen. When the catalysts are in their dissolved state in a solution, this reaction does not occur. The second surprise was that each oxygen molecule appeared to have a preference to oxidize two adjacent catalysts. This perfectly illustrates that STM is a unique technique, because it measures so locally, *i.e.* at the single-molecule level, that such a process can be identified. A final surprise was that the surface-bound catalysts were extremely stable. In similar reactions in solution, manganese porphyrin catalysts cluster together and oxidize themselves. At the surface, they remain well-separated from each other and many turnovers per single catalyst could be observed."

---

[6]A SEM uses a finely focused electron beam to scan the surface of a sample.

The researchers' motivation for this work was that they wanted to develop an additional technique to study a chemical reaction. Nearly all conventional techniques are ensemble techniques and measure the average behavior of millions of molecules at a particular time. STM measures the behavior of single molecules, and as such it might be used in the future as an additional analytic technique to the conventional ones.

Elemans points out that there are drawbacks to the new technique. One of the problems still is that STM is slow. Scanning takes seconds, and if the processes to be studied are faster than that they might not be captured by the instrument. "For that reason, we chose an oxidation reaction that was known to be very slow, so we were sure that we were looking at real-time processes," he says. "However, it limits the use of this technique for studying many other, faster reactions. Solving the problem relies on the development of faster STM equipment, which is in fact currently ongoing."

Being able to gain a highly detailed insight into chemical reaction mechanisms might one day enable chemical engineers to improve the reaction, and, on an industrial scale, maybe improve entire chemical synthesis and catalysis processes. At present this is only a vision, but the possibilities are almost endless.

"This is only the first chemical reaction investigated with STM at a liquid–solid interface, and there might be thousands of other reactions that can be investigated in a similar manner," says Elemans. "All of these might provide new insights. Our group is investigating other types of reactions with similar porphyrin catalysts. A particular challenge is to study cascade reactions in this way. These are reactions in which two or more different catalysts are bound to the surface, and in which the product of a catalytic reaction at a catalyst A is the starting material of the reaction at catalyst B. It is of great interest to see how such reactions are coupled in time and space, at the single-molecule level."

Featured scientist: Hans Elemans
Organization: Cluster for Molecular Chemistry, Radboud University, Nijmegen, The Netherlands
Relevant publication: Bas Hulsken, Richard Van Hameren, Jan W. Gerritsen, Tony Khoury, Pall Thordarson, Maxwell J. Crossley, Alan E. Rowan, Roeland J. M. Nolte, Johannes A. A. W. Elemans, Sylvia Speller: Real-time single-molecule imaging of oxidation catalysis at a liquid–solid interface. *Nat. Nanotechnol.,* **2**, 285–289.

1.4　Sucking Spaghetti Through Nanopores

There is a significant and growing need across the research and medical communities for low-cost, high-throughput DNA separation and quantification techniques. The isolation of DNA is a prerequisite for many molecular biology techniques and experiments. Although single-molecule techniques such as Elemans' STM technique afford extremely high sensitivity, so far such experiments have remained within the confines of academic and research laboratories. The primary reasons for this relate to throughput speed, detection efficiencies, and analysis times. For example, in a conventional solution-based single-molecule detection experiment, one can only detect approximately 10 000 molecules per minute, or one molecule every 6 ms. Although this may sound a lot, consider that a small drop of water (~ 5 ml) contains $\sim 1.67 \times 10^{23}$ molecules. At this speed you would need over 100 trillion years to detect all the water molecules in that single drop. Using a novel nanopore array developed by researchers in the UK, expect to be able to detect up to 1 million molecules simultaneously in the same 6 ms time window (and bringing the timeframe for analyzing the molecules in a single water drop down to some 60 billion years— about five times the estimated age of the universe).

The above example exaggerates a bit, of course. Compared to a water molecule, which is very small (it consists of only three atoms with an overall diameter of less than 0.3 nm), a DNA molecule is very large. In the real world you would never analyze a drop of DNA but much, much smaller quantities; even then, existing methods are considered to be unacceptably slow. Although it is now several years since the completion of the rough draft of the human genome was announced in 2003, much effort is still focused on identifying genes responsible for specific biological functions or diseases, and determining the DNA sequence bearing this information.

"In recent years, the creation of nanochannels or nanopores in thin membranes has attracted much interest because of the potential to isolate and sense single DNA molecules while they translocate through the highly confined channels," Joshua Edel, a lecturer in micro- and nanotechnology at Imperial College London, explains. "Nanopores for such applications have already been fabricated, but in all studies to date the detection of translocation events is performed electrically by measuring the ionic current."[7]

Edel's group, together with collaborators from Drexel University, presented proof-of-concept studies that describe a novel approach for optically detecting DNA translocation events through an array of solid-state nanopores which allows for ultra-high-throughput parallel detection at the single-molecule level.

"The single-molecule studies performed by Guillaume Chansin, a PhD student within my research group, are very exciting and novel from a technological perspective," says Edel. "Firstly, this work is the first demonstration of using fluorescence detection to monitor translocation events within a nanopore array. Secondly, this is the first true demonstration of an approach leading towards high-throughput single-molecule detection confined within nanofluidic structures."

This technique functions by electrokinetically driving DNA strands through submicron-sized holes (in this case 300 nm) on an aluminium/silicon nitride membrane. During the translocation process, the molecules are confined to the walls of the nanofluidic channels, allowing 100% detection efficiency.

---

[7] What this means is that molecules translocating through a nanopore will momentarily perturb the ionic current, with the duration of the perturbation and the magnitude of the current blockade providing more detailed information about molecular shape and structure.

Importantly, the opaque aluminium layer acts as an optical barrier between the illuminated region and the analyte reservoir. In these conditions, high-contrast imaging of single-molecule events can be performed.

Edel notes that the majority of work to date using nanopores utilizes ionic blockage currents to monitor translocation events. "Unfortunately, simply measuring blockage currents can only be performed in a single pore," he says. "One of our motivations for using optical detection was to ensure we can probe multiple holes simultaneously, allowing for true high-throughput detection. Our results indicate that it is possible to obtain high spatial resolution DNA analysis while independently controlling the applied voltage that drives the molecules into the nanopore. A critical feature of the generic approach is the possibility of parallelizing molecular analysis by probing an entire array of nanopores under uniform illumination."

According to Edel, there are plenty of potential applications including DNA sequencing, fragment sizing, sieving, separations, and rare event diagnostics. "For example, using analytical technologies that exist today it is essentially impossible to detect a single DNA strand within a standard blood sample (of a few millilitres) within a reasonable time frame. The technology we describe can potentially allow for such detection to be performed both rapidly and efficiently."

It seems that nanopore research is an exciting and growing field. Expect optical detection to play a dominant role in the future of this technology.

> *Featured scientist:* Joshua Edel
> *Organization:* Imperial College London, UK
> *Relevant publication:* Guillaume A. T. Chansin, Rafael Mulero, Jongin Hong, Min Jun Kim, Andrew J. deMello, Joshua B. Edel: Single-molecule spectroscopy using nanoporous membranes. *Nano Lett.*, **7**, 2901–2906.

1.5 Nanoprobes and Biosensors

During the past two decades, biosensors have been developed for environmental, industrial, and biomedical diagnostics. The application of nanotechnology to biosensor design and fabrication promises to revolutionize diagnostics and therapy at the molecular and cellular levels. The convergence of nanotechnology, biology, and photonics opens the possibility of detecting and manipulating atoms and molecules using a new class of fiberoptic biosensing and imaging nanodevices. These nanoprobes and nanosensors have the potential for a wide variety of medical uses at the cellular level.

The potential for monitoring *in vivo* biological processes within single living cells, *e.g.* the capacity to sense individual chemical species in specific locations within a cell, will greatly improve our understanding of cellular function, thereby revolutionizing cell biology. Over the past few years, nanoprobes have already demonstrated the capability to perform biologically relevant measurements inside single living cells.

A fiberoptic nanosensor is a nanoscale probe that basically consists of a biologically or chemically sensitive layer that is covalently attached to an optical transducer. Biological sensing elements can be either a biological molecular species (*e.g.* an antibody, an enzyme, a protein, or a nucleic acid) or a living biological system (*e.g.* cells, tissue, or whole organisms) that uses a biochemical mechanism for recognition. In the case of a receptor-based nanosensor, an interaction between the immobilized receptor and its substrate—the molecule it binds to—produces a perturbation that the optical transducer converts to an electrical signal via laser-induced fluorescence.

Tuan Vo-Dinh, professor of biomedical engineering and chemistry and director of the Fitzpatrick Institute for Photonics at Duke University's Pratt School of Engineering, in the USA, has devoted extensive research and development to the development of a variety of fiberoptic chemical nanosensors and nanobiosensors.

Vo-Dinh explains that preparing fiberoptic nanosensors is fairly straightforward if one has the right tools and good practice. "Using the so-called 'heat and pull' method, a micron-scale diameter silica optical fiber is placed in a commercially available puller that heats the fiber using a carbon dioxide laser and then pulls the fiber to the desired thickness, usually between 20 and 100 nm in diameter. The pulled fiber is then cut in half, yielding two nanoscale fiber tips. Subsequently, vapor deposition is used to deposit a thin layer of silver, aluminium, or gold on the side walls of the tip, followed by a two-step chemical treatment of the tip that provides covalent attachment points for the biosensor molecules."

The need for fast and specific assays has induced researchers to explore alternative optical detection technologies for diagnostic applications. In addition to the nanobisensor technology, Vo-Dinh and co-workers have developed a new type of nanoprobe using Raman and surface enhanced Raman spectroscopy (SERS) detection.

"Because of the inherently small Raman cross-section, Raman spectroscopy has not been widely used in the past for trace analysis," says Vo-Dinh. "Nevertheless, there has been a renewed interest in Raman techniques as a result of the discovery of the SERS effect." In SERS, the Raman effect is found to be greatly enhanced when it is close to a rough metal surface consisting of gold or silver nanoparticles, as a result of surface plasmon resonance. In recent years it has been demonstrated that detection of single molecules with SERS is possible.

Vo-Dinh points out that a significant advantage of nanobiosensors for cell monitoring is the minimal invasiveness of the technique. This makes optical nanobiosensors promising tools for dynamic analyses of proteins in biochemical pathways within single living cells.

"These optical nanoprobes provide a new method in cell-based assays offering highly miniaturized nanoscale devices that make cell-based analysis accessible at the single-cell level," says Vo-Dinh. "Future applications of optical nanoprobes could include multianalyte detection and analysis of protein-protein interactions and similar analyses of other proteins involved in cellular biochemical pathways."

> *Featured scientist:* Tuan Vo-Dinh
> *Organization:* Fitzpatrick Institute for Photonics, Duke University's
>   Pratt School of Engineering, Durham, NC, USA
> *Relevant publication:* Tuan Vo-Dinh, Paul Kasili, Musundi Wabuyele:
>   Nanoprobes and nanobiosensors for monitoring and imaging
>   individual living cells. *Nanomed. Nanotechnol. Biol. Med.* **2,** 22–30.

1.6 Nanomechanical Holography can Reveal Nanoparticles inside Intact Cells

Even less invasive than these optical nanoprobes is a novel ultrasonic holography technique that provides a noninvasive way of looking inside a cell.

Nanomaterial-based drug delivery and nanotoxicology are two of the areas that require sophisticated methods and techniques for characterizing, testing, and imaging nanoparticulate matter inside the body. In particular, the potential risk factors of certain nanomaterials have become a topic of heated discussion. Most, if not all, toxicological studies on nanoparticles rely on current methods, practices, and terminology as gained and applied in the analysis of micro- and ultrafine particles and mineral fibers. The development of novel imaging techniques that can visualize local populations of nanoparticles at nanometer resolution within the structures of cells—without destroying or damaging the cells—is therefore important.

Researchers in the United States have demonstrated that at ultrasonic frequencies, intracellular nanomaterial causes sufficient wave scattering that a probe outside the cell can respond to it.

"The novelty of our findings lies in the fact that it provides an alternative way of studying a cell under ambient conditions, *i.e.* without placing it in a vacuum, coating it with a metal, bombarding it with electrons, or inserting other molecules, as is the case with other techniques such as electron microscopy or fluorescent tagging," explains Ali Passian, a researcher with the Nanoscale Science and Devices Group at the Oak Ridge National Laboratory (ORNL) in Tennessee, USA.

"The use of nanomaterials is becoming ubiquitous and there is therefore a pressing need to understand how engineered nanomaterials interact with biological species," Passian says. "Health effects and environmental factors are currently of major importance in our group at ORNL and a lot of our research resources have been focused on tackling the associated problems."

He points out that with this novel imaging technique, scientists do not need to cut up the cell or inject artificial light-emitting molecules to find out whether or not a certain type of nanomaterial is present inside it. This avoids altering the intracellular configuration when attempting to pinpoint the nanoparticles.

The relatively new technique known as *scanning near field ultrasound holography* (SNFUH) is a revolutionary approach which provides noninvasive nanoscale imaging capabilities for deeply buried and embedded structures. It offers the ability to acquire simultaneous topography and holography information with nanoscale image resolution. SNFUH synergistically integrates three disparate approaches: a unique combination of SPM platform (which enjoys excellent lateral and vertical resolution) coupled to microscale ultrasound source and detection (which facilitates 'looking' deeper into structures, section by section) and a novel holography approach (to enhance phase resolution and phase coupling in imaging).

Applying these techniques to biological structures, it becomes possible to image soft samples and probe structures that are below their surfaces. For instance, if a cell is oscillated at megahertz frequencies using a piezoelectric crystal, the ultrasonic waves traveling through the oscillating cell structure may weakly drive an AFM cantilever that is in contact with the cell surface, as long as the elastic properties of the cell can support a propagation mode in the ultrasonic spectrum.

In their work, Passian and his team explored the viability of SNFUH as a technique to probe cellular uptake of nanoparticles. Specifically, they tried to determine the cellular fate of single-walled carbon nanohorns (carbon nanotubes aggregates having conical tips are referred to as carbon nanohorns) using a mouse model to detect and visualize particles within lavage cells and blood.

"We found that the nanoparticles cause sufficient phase change for it to be measured with an external probe," he describes the research findings. "Bear in mind that the nanoparticles were not artificially placed within the cell but got there as a result of exposing a living mouse to carbon nanohorns and then sacrificing the mouse a few days later and preparing the sample cells."

"The sizes of these particles are statistically consistent with the size distribution (70–110 nm) established from analyzing several AFM images of the nanohorn solution, indicating that individual nanoparticles, rather than larger aggregates, were taken up by the macrophages," explains Passian. "The contrast measures the phase of the local tip–cell surface coupling and originates from the difference in elasticity and density between the nanohorns and the cell."

This work clearly demonstrates that specific problems that require subsurface knowledge can be tackled using this technique. This will mostly benefit research areas such as nanotoxicological investigations, drug delivery, and pharmaceutical work where, so far, most studies can only target the surface of samples and suffer from the lack of probes that can, with nanoscale resolution, provide information on what may be within a sample.

Passian explains the team's next challenge: "Our next step is to enable our approach to visualize the interior of a cell in its natural milieu, that is, in a fluid."

Featured scientist: Ali Passian
Organization: Nanoscale Science and Devices Group, Oak Ridge
   National Laboratory, Oak Ridge, TN, USA
Relevant publication: Laurene Tetard, Ali Passian, Katherine T.
   Venmar, Rachel M. Lynch, Brynn H. Voy, Gajendra Shekhawat,
   Vinayak P. Dravid, Thomas Thundat: Imaging nanoparticles in cells
   by nanomechanical holography. *Nat. Nanotechnol.,* **3**, 501–505.

1.7 Direct Imaging Technology Reaches the Nanoscale

Magnetic resonance imaging (MRI) is a powerful imaging technology that serves as a noninvasive method to render images of the inside of an object. During an MRI scan, the nuclei of an object's hydrogen atoms align with the scanner's powerful, uniform magnetic field. Pulses of radio waves are then sent into the scanner that knock these hydrogen nuclei out of their normal position. After the radio wave pulsing stops, the nuclei realign back into their proper position. During this realignment process, the nuclei send out signals. These signals are captured by the computer system that analyzes and translates them into an image of the object being scanned.

MRI is primarily used in medical imaging to demonstrate pathological or physiological alterations of living tissues. It also has uses outside the medical field, for instance as a nondestructive testing method to characterize the quality of products such as produce and timber. Conventional MRI usually operates at the scale of millimetres to micrometres—3 μm at best—which is good enough for the mostly medical diagnostic purposes it is used for. Researchers have now shown that the imaging of nuclear spins using magnetic resonance, the basis for MRI, can be pushed to sub-100 nm resolution, into the nanoscale realm. They demonstrated that, using an emerging technique based on force detection, they can image nuclear spins with a sensitivity that is 60 000 times better than MRI. The resolution is about 30 times better than the most advanced conventional MRI imaging. By improving this technique, researchers will be able to push deeper into the nanorange and approach the capability needed for direct 3-D imaging of individual macromolecules.

As MRI increasingly finds applications in material and biological sciences, scientists are trying to find ways to overcome its sensitivity limitations and push its resolution into the nanorange and, ultimately, the atomic scale. *Magnetic resonance force microscopy* (MRFM) seems to be the solution.

"Our MRFM method measures a very tiny magnetic force between the nuclei in the sample and a nearby nanomagnet," explains John Mamin. "Many common types of atoms, including hydrogen, fluorine, phosphorus, copper, and aluminium, have nuclei that are very weakly magnetic. Sometimes this basic magnetic property is referred to as the nuclear 'spin'. MRI typically uses the magnetism of hydrogen nuclei contained in water. In our work, we use the magnetism of fluorine nuclei contained in a small sample of calcium fluoride, though we can also use hydrogen." Mamin is a member of an IBM team at the company's Almaden Research center in California that succeeded, for the first time, in pushing NMR imaging below the 100 nm threshold.

MRFM uses a magnetic tip and an ultrasensitive cantilever to sense the magnetic force generated between the tip and spins in a sample. Unlike the permanent magnet tips previously used for MRFM detection of electron spin resonance, the tips used by the IBM researchers are based on a thin film of magnetic material that has a high magnetic moment, but is magnetically soft. "The tips we developed are compatible with sample-on-cantilever experiments," says Mamin. "In such an experiment, the sample is placed on the distal end of an ultra-sensitive cantilever situated above the magnetic tip. By choosing the sample-on-cantilever configuration, rather than tip-on-cantilever, we eliminated the magnetic damping that occurs when a soft magnetic tip vibrates in the presence of an external magnetic field."

The ability to image a single molecule with atomic resolution would have a huge impact in structural biology, in particular the area of protein structure determination—an important and famously difficult problem—and would aid the development of new drugs. It also could revolutionize the study of materials, ranging from pharmaceuticals to integrated circuits, for which a detailed understanding of the atomic structure is essential. Knowing the exact location of specific atoms within nanoelectronic structures, for example, would enhance designers' insight into manufacture and performance.

The IBM team is motivated by the prospect of developing such a tool—the 'holy grail' of molecular imaging—an instrument that is able to perform direct, 3-D imaging of complex structures such as molecules with atomic resolution. "Our detection limit of about 1000 nuclear spins is a significant step toward the goal of detecting a single nuclear spin," says Mamin. "The best conventional MRI today requires at least 100 million nuclear spins to obtain a detectable signal. If we can achieve single spin detection in the future, this would open up the possibility of taking 3-D MRI images of the atoms in a molecule. Obviously, there is still considerable progress needed to extend our capability to the single spin level."

"The main issues are related to the very tiny magnetic forces that must be detected. Better nanomagnetic tips are needed to generate stronger force signals, and noise reduction will also be key. There is still a lot of room for innovation in these areas."

Featured scientist: John Mamin
Organization: IBM Almaden Research Center, San Jose, CA, USA
Relevant publication: H. J. Mamin, M. Poggio, C. L. Degen, D. Rugar: Nuclear magnetic resonance imaging with 90-nm resolution. *Nat. Nanotechnol.*, **2**, 301–306.

1.8 My corona! Novel Spectroscopic Technique could Revolutionize Chemical Analysis

Laser-based analytical techniques such as Raman spectroscopy, fluorescence spectroscopy or the state-of-the-art *laser-induced breakdown spectroscopy* (LIBS) are highly sophisticated techniques for analyzing minute amounts of matter with regard to its structure, elemental composition, and other chemical properties. LIBS has been shown to be capable of analyzing extremely small samples with high sensitivity—nanolitre volumes with levels of detection in water of one part per million. LIBS works by focusing short laser pulses on to the surface of a sample to create a hot plasma with temperatures of $10\,000$–$20\,000\,°C$. The plasma emits radiation that allows the observation of the characteristic atomic emission lines of the elements. The downside is that LIBS is complicated by the need for multiple laser pulses to generate a sufficiently hot plasma and the need for focusing and switching a powerful laser, requiring relatively large and expensive instruments.

Research coming out of Drexel University has shown that light emitted from a new form of cold plasma in liquid—field emission generated, highly nonequilibrium, and high energy density—permits optical emission spectroscopy (OES) analysis of the elemental composition of solutions within nanoseconds from femtolitre (10^{-15} L) volumes.

"We were able to generate, for the first time, a nonthermal corona discharge in liquid around electrodes with ultrasharp tips and around nanowires," says Yury Gogotsi. "We have demonstrated that plasmas created with 50 nm probe tips or carbon nanotubes (CNTs)—what we have termed *nanoscale corona discharge probes* (NCDPs)—dispersed in solution allow simultaneous chemical analysis of multiple dissolved elements within nanoseconds. Time-resolved OES of NCDPs demonstrates narrow spectral lines that prove very useful for simple yet sensitive multi-elemental analysis, thus opening new possibilities in chemical detection, environmental monitoring, medicine, and many other applications."

Gogotsi, a professor in the Department of Materials Science and Engineering, heads the A.J. Drexel Nanotechnology Institute at Drexel University, and has worked with colleagues Gary Friedman from Drexel's Department of Electrical and Computer Engineering; Alexander Gutsol (currently with Chevron); and Alexander Fridman (director of the A.J. Drexel Plasma Institute).

Gogotsi says that the OES method proposed by the Drexel scientists can be applied for ultra-fast, time-resolved, multi-elemental analysis of liquid in microfluidic reactors, living biological systems, or environmental sensors and for diagnostics of femtolitre volumes with $1\,\mu m$ or better spatial resolution. "Using this method, we have detected one part-per-million concentrations of sodium, calcium, and other elements in aqueous solutions."

In comparison to LIBS, the researchers found that the observed spectra are of better quality, have significantly smaller analytical volumes, and are accomplished using drastically simpler, smaller, and less expensive equipment and materials. Furthermore, OES can be performed remotely, using nanorods and nanotubes dispersed in fluid.

The team describes how in a typical experiment, a tungsten wire with a tip sharpened to less than 50 nm radius was used to generate the corona discharge. Negative corona discharges at the tips have been demonstrated in all cases. The pulsed voltage source provides 2–30 kV pulses 10–500 ns in duration at approximately 30 Hz repetition rate, achieving negative corona for a 50 nm radius tip with as little as 3 kV.

"Considering existing theories of discharge initiation in liquids and our current experiments, there are several factors that contribute to the reasons why the NCDP is different from the streamer coronas previously observed in liquid and specifically why the two initial stages of the negative corona are observed," Gogotsi explains. "Our study is the first that simultaneously combines short rise-time voltages, nanosecond-duration pulses, high temporal resolution emission spectra, and most importantly, nanoscale tips."

The Drexel team is hopeful that the OES nanoscale probes may open a new era in micro/nanoscale chemical, environmental, and biological sensing and detection techniques.

"Just as the discovery of AFM changed the world of microscopy, the OES nanoscale probes may change the world of chemical analysis, replacing the large and expensive instruments that are used for elemental analysis or measurements of cation concentration in thousands of labs worldwide by simple, portable, and very inexpensive tools that also add analytical capabilities not available today, *e.g.* fast simultaneous quantitative analysis of multiple cations in solution," says Gogotsi.

With regard to the nonthermal plasma in liquid, nanoscale corona discharge OES is presented as only the first of many potential applications for this newly discovered tool; applications in nanopatterning and surface functionalization as well as tools for cellular surgery are readily conceivable. Gogotsi and his colleagues expect this research to affect a broad spectrum of fields ranging from pharmaceuticals and biomedicine to nanotechnology and fundamental plasma chemistry.

Featured scientist: Yury Gogotsi
Organization: A.J. Drexel Nanotechnology Institute, Philadelphia, PA (USA)
Relevant publication: David Staack, Alexander Fridman, Alexander Gutsol, Yury Gogotsi, Gary Friedman: Nanoscale corona discharge in liquids, enabling nanosecond optical emission spectroscopy. *Angew. Chem., Int. Ed.*, **47**, 8020–8024.

1.9   Listening to the Music of Molecules

As we have seen, detecting the presence of a given substance at the molecular level, down to a single molecule, remains a considerable challenge for many nanosensor applications that range from nanobiotechnology research to environmental monitoring and antiterrorist or military applications.

Currently, chemical functionalization techniques are used to specify what a nanoscale detector will sense. For biological molecules, this might mean developing an antibody–antigen (*i.e.* lock and key) pair, or an alternative synthetically generated ligand. For gases, it is much more challenging to develop the right 'glue' that sticks a given gas—and only that gas—to a substrate. Thus, for many gas-sensing applications, appropriate functionalization may not even be possible.

The advantage of spectroscopic techniques—measuring and interpreting electromagnetic spectra arising from either emission or absorption of radiant energy by various substances—such as Raman, infrared, and NMR spectroscopy is that they are label-free, *i.e.* they require no preconditioning in order to identify a given analyte. They are also highly selective, capable of distinguishing species that are chemically or functionally very similar. On the downside, spectroscopic methods face enormous challenges in measuring dilute concentrations of an analyte and generally involve the use of large, expensive equipment.

"We have been working on ways to overcome the functionalization bottleneck in sensing and, instead of trying to see a molecule by using photons or electrons—as in optical spectroscopy or electron microscopy—we have been using vibrational energy exchange to in effect 'listen' to the vibrations of the molecule," says Jeffrey Grossman. "The concept is much like bringing a set of nano tuning forks up to a molecule and seeing which ones become excited. Those would form a chord of 'notes' that are unique to that particular molecule. Thus, the molecule can be identified."

In his work, Grossman, who leads the Computational Nanoscience Group at the University of California, Berkeley and is Executive Director of Berkeley's Center of Integrated Nanomechanical Systems, has been taking advantage of the unique manner in which vibrational energy transfers between nanoscale objects, with applications spanning chemical, biological, radiation, and even acoustic sensing.

"The scientific core of our work is aimed at utilizing the unique way in which mechanical energy—in other words, heat—is exchanged at the nanoscale" Grossman explains. "Specifically, we have shown that if one nanoscale object vibrates at the same frequency as another, 'in resonance', then it is possible for these two objects to exchange heat extremely efficiently. Conversely, if they are not vibrating at the same frequency then the flow of heat is blocked and little or no energy is exchanged."

In essence, what Grossman and his group have done is to demonstrate that one can take advantage of this nanomechanical exchange of energy for detection or characterization of an unknown molecule type. They have termed their novel chemical detection technique *nanomechanical resonance spectroscopy* (NRS). NRS basically employs an array of nanomechanical resonators that are used to directly interrogate a heated ('thermally excited') analyte's vibrational frequencies.

They laid some of the groundwork for the theory behind nanomechanical energy exchange, which allowed them to demonstrate this effect. The proposed

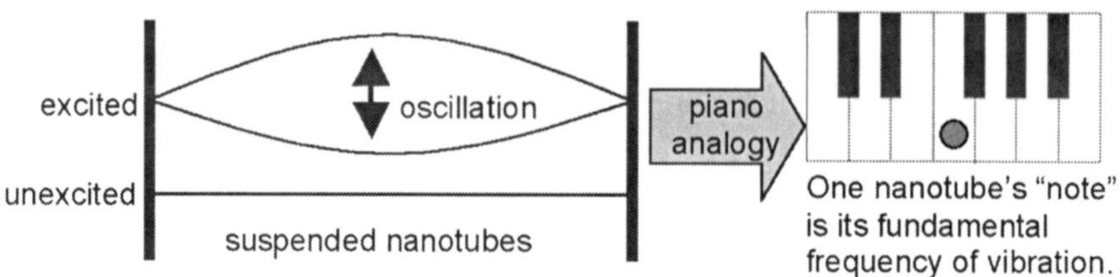

Figure 1.3   Left: Illustration of nanotube or nanowire when its fundamental mode of
oscillation is unexcited (bottom) or excited (top). Right: analogy with a
piano—the frequency of oscillation of the nanotube or nanowire can be
thought of as a note on a piano. (Image: Dr J. Grossman, UC Berkeley)

NRS sensor could 'listen to the music' a molecule makes without needing to
adhere the molecule to a surface, thereby allowing for continuous measurements with little or no cleaning, resetting, or degradation of the sensor. The
result would be a nanodetection system that could detect many different species
without functionalization chemistry steps. The NRS has become possible with
the ability to synthesize nanoscale objects that resonate at the same frequencies
as the natural vibrations of molecules. Figure 1.3

The work in Grossman's group is part of an emerging field of study of frequency-dependent thermal phenomena in nanoscale systems, *i.e.* recognizing
and exploiting the wave nature of heat. In these small systems the macroscopic
concept of temperature (the time-averaged thermal energy) is insufficient to
describe how a system will behave.

Grossman points out that one must also take into account how the thermal
energy is distributed in frequency; this distribution can influence the way in
which heat is transported, or blocked, as well as how heat effects other
important properties such as electron transport.

"There are other researchers who are studying the importance of the frequency dependence of heat in nanostructures, for example in thermoelectric
materials, or in thermally rectifying materials," he says. "However, our work is
unique in proposing an application that depends solely on frequency-dependent
phenomena and also by utilizing these phenomena for label-free detection."

"One could envision several NRS device set-ups" says Grossman. "For
example, a series of CNTs of differing length (or radius) suspended over a
trench, similar to the strings in a harp. Alternatively one could pass analytes
through holes in graphene membranes or over substrates coated in fullerenes of
different radius. The design of a practical NRS is more limited by capabilities
for detecting the excitation of a vibrational mode than the ability to fabricate
nanoscale devices."

The number of applications for a sensitive, label-free detection system is
quite extensive. It could be extremely useful in areas such as medicine, home-
land security, environmental monitoring, and clean energy and water.

Research in this area of thermal phenomena and heat flow at the nanoscale is
just beginning. "With the development of theory for describing heat transport
in nanoscale systems and advances in nanoscale fabrication and characterization techniques, we are now well equipped to study these phenomena," says

Grossman. "As a result we should look forward to the development of more applications that exploit the wave nature of heat."

> *Featured scientist:* Jeffrey Grossman
> *Organization:* Center of Integrated Nanomechanical Systems (COINS), University of California, Berkeley, CA, USA
> *Relevant publication:* P. Alex Greaney, Jeffrey C. Grossman: Nanomechanical resonance spectroscopy: a novel route to ultrasensitive label-free detection. *Nano Lett.*, **8**, 2648–2652.

1.10 Nano-scales Weigh Atoms with Carbon Nanotubes

The two principal types of components common to most electromechanical systems, irrespective of scale, are mechanical elements and transducers. The mechanical element either deflects or vibrates in response to an applied force. Depending on their type, the mechanical elements can be used to sense static or time-varying forces. The transducers in microelectromechanical systems (MEMS) and nanoelectromechanical systems (NEMS) convert mechanical energy into electrical or optical signals and *vice versa*. For example, MEMS are used as accelerometers in modern automobile airbags where they sense deceleration and initiate the inflation of the airbag if the force is beyond a programmed threshold.

NEMS devices have two particular attributes—minuscule mass and high quality factor (Q)—that provide them with unprecedented potential for mass sensing down to zeptogram (10^{-21} gram, zg) resolution. A Spanish team has demonstrated an ultra-sensitive CNT-based mass sensor in which they measured chromium atoms with a mass resolution of only 1.4 zg. For comparison, the best sensitivity achieved before was 7 zg using resonators microfabricated in silicon.

Adrian Bachtold, whose group at the Centre Investigacions Nanociencia Nanotecnologia (CSIC-ICN) in Barcelona developed this ultra-low-mass sensor, describes the device: "The sensor consists of a device based on a CNT that is suspended and clamped at the extremities. The nanotube acts like a guitar string: when actuated, it oscillates at specific frequencies. When atoms or molecules are deposited on to the nanotube, the mass of the oscillating nanotube increases and the frequency decreases. In other words, the reduction of the velocity of the nanotube motion is directly related to the mass of the deposited atoms or molecules." Figure 1.4

Unlike the silicon often used in NEMS, CNTs are chemically inert and do not suffer from the surface roughness inherent to lithographically patterned NEMS. This is very important in order to obtain high-Q resonance quality factors. CNTs are also the stiffest material known and have low density, so that frequencies are expected to be very high, more than 1 GHz. One of the goals of Bachtold's group is to fabricate nanotube resonators with resonance frequencies and Q factors as high as possible.

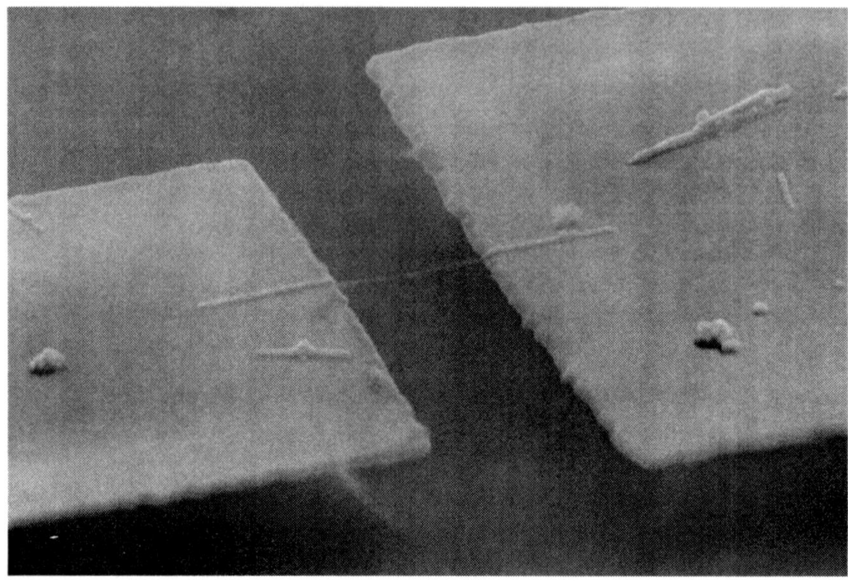

Figure 1.4 SEM image of the nanotube resonator. (Image: Dr A. Bachtold, CSIC-ICN)

The novelty of the Spanish team's work is the use of a CNT as the oscillating element. The mass of a nanotube is ultra low, typically a few attograms (10^{-18} g), so even a tiny amount of atoms deposited on to the nanotube makes up a significant fraction of the total mass.

Although CNT resonators have proved to be excellent mass sensors in laboratory set-ups, and might one day even provide compact alternatives to current mass spectrometers, the way they are currently fabricated lacks the ability to precisely control positioning that would allow large-scale industrial type applications.

Bachtold's team fabricated their nanotube resonators by growing single-walled CNTs by chemical-vapor deposition on a highly doped silicon substrate, and then using electron-beam lithography to connect a CNT nanotube to two chromium/gold electrodes. Wet etching and subsequent annealing was performed to suspend the CNT and remove impurities.

A necessary feature required for real-world applications of CNT 'scales' is the ability to be reset. According to Bachtold, his team has done that successfully and repeatedly without any loss in sensitivity.

"This process is accomplished by applying a few microamperes of current through the nanotube for several minutes," he says. "Adsorbed atoms are removed by heating and/or electromigration. As a result, the resonance frequency is reset to its initial value, and the nanotube resonator is ready for new sensing measurements."

The team has been focusing on improving their measurement set-up and they hope to ultimately achieve a resolution of 0.001 zg (or 1 yoctogram, 10^{-24}

gram)—the mass of one atomic nucleus. "Such a mass resolution would open new perspectives for mass spectrometry," says Bachtold. "It would be possible to weigh large molecules with subatomic precision. Individual atoms or molecules could also be placed on the nanotube in order to probe the variation of their mass. Chemical reactions in organic or biological molecules could then be monitored in real time, as well as nuclear reactions in individual atoms."

Featured scientist: Adrian Bachtold
Organization: Centre Investigacions Nanociencia Nanotecnologia
 (CSIC-ICN), Barcelona, Spain
Relevant publication: B. Lassagne, D. Garcia-Sanchez, A. Aguasca,
 A. Bachtold: Ultrasensitive mass sensing with a nanotube
 electromechanical resonator. *Nano Lett.,* **8**, 3735–3738.

CHAPTER 2

Building Nanostructures—from the Top Down and from the Bottom Up

Michelangelo was a 'top-down' artist. He took one huge, raw block of Carrara marble and after years of chiseling away produced a spectacular masterpiece like David. In the process, he reduced the original block of marble to half its original volume and left the other half as waste. The nanotechnology equivalent of this is lithography and other top-down methods where you start by taking a block of material and remove the bits and pieces you don't want until you get the shape and size you do want. In the process you expend (quite a lot of) energy, use (sometimes very toxic) chemicals, produce (often quite a bit of) waste, need a lot of patience (these processes are relatively slow)—and frequently the results are quite unique and not easily replicable.

Bottom-up methods are much more elegant and efficient. Take Lego blocks. Just pick the shape and sizes you need, and—piece by piece—build more or less anything you want with them. Replace your hands with a (tiny) machine, or some other assembly process, and the Lego bricks with atoms or molecules, and you have molecular assembly. Unfortunately, this analogy is too simplistic.

Self-Assembly

To make things a bit more complicated, there are two fundamentally different ways of fabricating things from the bottom up. This has caused a lot of confusion about the terminology used in nanotechnology.

One bottom-up method is Nature's way: self-assembly. Self-organizing processes are common throughout Nature and involve components at scales

RSC Nanoscience and Nanotechnology No. 8
Nano-Society: Pushing the Boundaries of Technology
By Michael Berger
© Michael Berger 2009
Published by the Royal Society of Chemistry, www.rsc.org

ranging from the molecular (*e.g.* protein folding) to the planetary (*e.g.* weather systems) and even beyond (*e.g.* galaxies). The key to using self-assembly as a controlled and directed fabrication process lies in designing the components that are required to self-assemble into desired patterns and functions. Self-assembly reflects information coded—as shape, surface properties, charge, polarizability, magnetic dipole, mass, *etc.*—in individual components; these characteristics determine the interactions among them.

On a very small scale you wouldn't even speak of 'self-assembly' but rather of 'chemical synthesis'—the processes chemists have refined over many years. However, the stability of covalent bonds enables the synthesis of almost arbitrary configurations of only up to 1000 atoms. Larger molecules, molecular aggregates, and forms of organized matter more extensive than molecules cannot be synthesized bond by bond. Self-assembly is one strategy for organizing matter on these larger scales.

This technique can be summarized as follows: we have no clue why certain atoms and molecules self-assemble the way they do, but once we can initiate and control the process we can use it to build structures from the bottom up, atom by atom.

Self-assembly has become an especially important concept in nanotechnology. As miniaturization reaches the nanoscale, conventional manufacturing technologies fail because it has not yet been possible to build machinery that assembles nanoscale components into functional devices. Until robotic assemblers capable of nanofabrication can be built, self-assembly, together with chemical synthesis, will be the necessary technology to develop for bottom-up fabrication.

Exploiting Nature's self-assembly tricks is real science, and it's happening in labs already. As a matter of fact, it's actually about to lead to real-world products. One example is IBM's announcement in 2007 of the first-ever application of self-assembling nanotechnology to conventional chip manufacturing, borrowing a process from Nature to build the next generation of computer chips.[1]

Self-assembly is also the reason why nanotechnologies have had such a profound impact on the chemical industry. One example is the huge range of polymers used for industrial products (think plastics). Chemists are using molecules' tendency to self-align to design molecular structures with specific properties. Once you know how certain nanoparticles behave and what properties they possess, you can use this knowledge to deliberately create structures with desired properties. This is a much more efficient way than the cement-mixer chemistry of old where you mix compounds in a more or less arbitrary way based on best guesses and see what materials you get, and then try figure out what you could do with them.

---

[1] Read the IBM press releases here: http://www-03.ibm.com/press/us/en/presskit/21463.wss

Molecular Assembly

The other way of doing bottom-up nanotechnology is our human way: molecular assembly. Although it sounds not unlike self-assembly, it is a very different concept. This is the vision that proponents of revolutionary nanotechnology put forward: molecular assembly as a factory concept, assembly lines and all, just scaled down to the nano level. The notion of 'self-assembly' becomes relevant in this context with regard to 'self-replicating' nanomachines, *i.e.* machines that self-assemble themselves; but this is very different from the type of self-assembly found in Nature.

There is one very big catch, though: today, universal molecular assembly is a vision; in a scientific sense it is not even a theoretical concept yet.

With our current technical capabilities, the most advanced bottom-up nanotechnologies are a combination of chemical synthesis and self-assembly. But they already allow us to perform atomically precise manufacturing on a modest scale and this will lead to vastly improved materials, much more efficient manufacturing processes, and entirely new medical procedures.

2.1 Applying a 250 Year Old Discovery to Nanofabrication

Back in 1756, the German physicist Johann Gottlob Leidenfrost published a manuscript titled *De Aquae Communis Nonnullis Qualitatibus Tractatus* ('A tract about some qualities of common water') in which he described a phenomenon in which a liquid, in near contact with a mass significantly hotter than its boiling point, produces an insulating vapor layer which keeps that liquid from boiling rapidly. This effect came to be called the *Leidenfrost effect* and the associated temperature point the *Leidenfrost temperature*.

An everyday example of this can be seen in your own kitchen. Sprinkle a drop of water into a hot frying pan. If the pan's temperature is at or above the Leidenfrost temperature, the water skitters across the metal and takes longer to evaporate than it would in a pan that is hot, but at a temperature below the Leidenfrost point. Researchers in Germany have used this effect for a novel, template-free synthesis and patterning method of nanostructures.

While nanoscale self-assembly is one of the core concepts in Nature, scientists are only beginning to scratch the surface of the potential that self-assembly holds for materials engineering. The list of self-assembly fabrication methods for the formation of nanocluster structures seems to get longer and longer but they all feature different timescales, complexities, and versatilities. The one thing they usually have in common is the requirement for two steps—the formation of nanoparticles from precursors in the liquid, solid, or gas phases employing either chemical or physical deposition processes and, in a subsequent step, the organization of these particles into useful structures.

"Wet-chemical strategies utilizing fluid mechanics appear to be the simplest and most effective way of template-free, self-assembling nanostructuring," explains Rainer Adelung. "However, structuring of nanocluster arrays or wire-like morphologies from a droplet still faces certain challenges."

Adelung, a researcher at the Department of Multicomponent Materials at the University of Kiel, Germany, tells the story of his colleague Mady Elbahri experimenting with new routes for zinc oxide synthesis and discovering a novel way for a droplet-disposition-based nanostructuring technique. "Inspired by kitchen experiments with his wife Julia, he discovered that it might be possible to ignore the typical temperature limit of 100 °C for water-based synthesis. He found that at temperatures of 250 °C, well above the boiling point, it is possible to use water droplets that contain a small amount of chemicals for nanostructuring—thanks to the Leidenfrost effect."

What the researchers in Kiel subsequently developed was a droplet-deposition-based, template-free, and rapid (only a few seconds) approach for fabricating nanostructures without the use of any surfactant. "Our general set-up can be understood as a kind of *anti-lotus effect*," says Adelung. "The well-known lotus effect is the removal of dust particles from the surface of a lotus leaf by gathering them into a droplet that is moving over the surface, thus cleaning it. The effect is based on the ability of certain surfaces to form spherical droplets with contact angles near 180° (a property called *super-hydrophobicity*), enabling the incorporation of surface particles as well as a reduction in friction. In contrast to this, our work makes use of an anti-lotus effect, in which the droplet delivers material while moving over the surface."

It seems that, similar to room temperature droplets on superhydrophobic surfaces, a Leidenfrost droplet can move with greatly reduced friction over an arbitrary surface such as plain silicon and deposit nanoparticles. These can be nanoparticles already dispersed in the water droplet or formed in the overheated steam from a mix of chemicals. In this way, reactions leading to nanoparticle synthesis can be performed that do not occur at room temperature (a method the Kiel researchers have come to call the *Elbahri synthesis*). Combining this method with a top-down approach enables the formation of regular patterns. A grid with a regular pattern of openings is enough to force the self-organized structures into a reproducible shape.

Adelung points out that the random deposition of a solution drop would be totally ineffective in fathering any technologically useful nanoscale structures such as arrangements of well-separated patterns prepared from monodisperse particles. "In order to deposit material from a droplet in an organized manner, a deeper insight into the underlying mechanism of the fluid drop is necessary," he says.

The researchers prepared an extremely dilute solution of the desired material powder in water. Then they gently placed a droplet from this solution on a substrate, such as a silicon wafer, that was maintained at a temperature of 230 °C, which is the estimated Leidenfrost temperature for any suspension droplet. They allowed the drop to stay on the hot substrate for approximately 5 s before the substrate was tilted 30° to the horizontal to exploit gravity for the propagation of the loaded droplet across the substrate surface. When a water droplet loaded with silver nanoparticles was subjected to this procedure, nanowires or cluster chains were successfully produced.

"Apart from straight, parallel lines in series, nanoparticles can also be arranged in concentric circles with our Leidenfrost structuring," says Adelung. "We attribute this patterning to a different and interesting phenomenon that also occurs at the Leidenfrost temperature, under slightly modified fluid dynamics, where droplet impact is emphasized. Interfacial, viscous, and capillary forces are the three basic forces governing the behavior of a droplet on a surface. At temperatures below the Leidenfrost temperature, we observed a spreading behavior that results in the formation of a disk like thin film upon impact."

> *Featured scientist:* Rainer Adelung
> *Organization:* Department of Multicomponent Materials, University of
>   Kiel, Germany
> *Relevant publication:* M. Elbahri, D. Paretkar, K. Hirmas, S. Jebril,
>   R. Adelung: Anti-lotus effect for nanostructuring at the Leidenfrost
>   temperature. *Adv. Mater. (Weinheim, Ger.)*, **19**, 1262–1266.

2.2   Single Atom Manipulation on a 3-D Surface

In recent years, the manipulation of single atoms and molecules has been a major advance in the application of the scanning tunneling microscope (STM). The main appeal of STM manipulation is the ability to access, control, and modify the interactions between the tip and the adsorbate, a few ångströms apart. So far, however, atom manipulation using an STM or an AFM tip has been restricted to flat surfaces. Manipulation of atoms on a rough terrain requires much more precise control at the atomic scale.

Saw-Wai Hla, a professor at the Nanoscale and Quantum Phenomena Institute at Ohio University, and his team have managed extraction and manipulation of individual silver atoms on 3-D silver nanoclusters. This was the first demonstration that individual atoms can be repeatedly pulled out from a silver cluster on a silver surface using a STM tip. It is also the first atom manipulation work done on a 3-D surface. While it certainly is not a commercial production technique, this achievement furthers the fundamental understanding of the interaction between atoms, and it is an atom production technique that can be used to extract the atoms for atomistic construction.

Hla explains how the atoms can be extracted from the cluster: "We first dip the STM tip into the single-crystal silver surface. This coats the tip with silver. Then by gently touching the tip to the surface at a flat surface area, some of the silver from the tip is transferred to the surface as a cluster."

After achieving a silver cluster on the surface, the researchers take a 3-D STM image of the cluster. Protruding parts of the cluster are chosen as the ideal target zones for atom extraction. To extract the atoms from the cluster, the tip is first approached very closely to the cluster (within 0.6 ångström or 0.06 nm). "Just bringing the tip very close to the cluster results in loosening of the top atom inside the cluster," says Hla. "When we move the tip laterally across the cluster surface, the loose atom follows the tip. Now we have extracted just one atom."

The atomistic details of the atom extraction mechanism are explained by means of statistical analyses and theoretical modeling, which reveal that merely locating the STM tip within the required proximity of the nanocluster greatly reduces the extraction barrier, facilitating repeated removal of the top atoms from the cluster.

Even though the atomistic dynamics of atom extraction can be understood from the already established knowledge of manipulation signals, the environment that the extracted atom faces during the process is clearly different from atomic manipulation on a flat surface. In particular, this atom extraction involves pulling out the top atom from a protruding part of a cluster, and then moving it over rough terrain on a 3-D cluster surface. The energy required to extract the atom from the cluster is the energy barrier to move the atom from its original location to the next site within the cluster. In the absence of the tip, the energy barrier for the atom to diffuse over the step edge is 300 meV, which is much higher than the 35 meV barrier for a silver atom diffusion on a flat Ag(111) terrace. These barriers are altered by the presence of the tip, as the latter drastically modifies the energy landscape of the system. The variation of the tip height has a dramatic effect on the potential energy of the cluster and of the extracted atom. It appears that locating the tip in close proximity to the cluster is sufficient to extract the top atom by overcoming the binding of the atoms within the cluster.

Hla sees this work as just the beginning: "We are continuing our investigations of atomic level interactions to get a deeper understanding of how atoms bind to form matter."

Featured scientist: Saw-Wai Hla
Organization: Ohio University, Athens, OH (USA)
Relevant publication: A. Deshpande, H. Yildirim, A. Kara, D. P. Acharya, J. Vaughn, T. S. Rahman, S.-W. Hla: Atom-by-atom extraction using scanning tunneling microscope tip-cluster interaction. *Phys. Rev. Lett.*, **98**, 028304.

2.3 Native Protein Nanolithography that can Write, Read, and Erase

Proteins are very specific about which other proteins or biochemicals they will interact with, and are therefore very useful for biosensing applications. If a malignant cancer develops in the human body, the cancer cells produce certain types of proteins. Identifying such proteins enables early detection of cancer. One of the goals of nanobiotechnology is to develop protein chips that respond sensitively to a very tiny amount of specific proteins in order to enable such early diagnosis. A protein that is known to bind to a protein produced by a cancer cell could be attached to a biochip. If this particular cancer cell protein were present in a sample passed over the chip, it would bind to the protein on the chip, causing a detectable change in the electrical signal passing through the chip. This change in the electrical signal would be registered by the device, confirming the presence of the protein in the sample.

While this sounds very promising for the future of diagnostic systems, with the promise of protein chips capable of single-molecule resolution, the controlled assembly of proteins into bioactive nanostructures is still a key challenge in nanobiotechnology. Researchers in Germany have taken a further step towards overcoming this problem by developing a native protein nanolithography technique that allows for the nanostructured assembly of even fragile proteins.

Existing nanolithographic techniques are operated under vacuum or ambient atmosphere conditions—settings that are incompatible with most biological molecules. Oxidation, dehydration, and organic solvents impair these delicate entities. Being able to perform direct patterning with proteins has therefore been the exception, and maintaining their bioactivity is only possible in case of rather stable proteins such as antibodies. For this reason, biochemically interesting proteins such as receptors from signal transduction pathways or large macromolecular complexes have not yet been included in protein chip fabrication.

"We recognized a technological gap in this field of protein nanopatterning and aimed to develop a process that protects biomolecules from non-physiological conditions," Ali Tinazli explains. "The technique we came up with, *native protein nanolithography* (NPNL), has a very straightforward process control and it has been transferred easily between different instruments and labs. Standard AFM equipment is quite sufficient for this purpose."

Tinazli's work describes a nanolithography technique that permits rapid writing, reading, and erasing of protein arrays in a versatile manner. Developed by a group of researchers from the Cellular Biochemistry lab of Robert Tampé at the Goethe University of Frankfurt, where Tinazli is a postdoc, and the Max Planck Institute of Biochemistry near Munich, the corresponding protein chip platform is suitable for any His-tagged proteins (His is an amino acid motif in proteins used to detect protein–protein interactions).

The German researchers based their novel and easy-to-use fabrication technique for protein nanoarrays on metal-chelating self-assembled monolayers combined with AFM-based nanolithography. Affinity-captured proteins are mechanically displaced in a special patterning mode of AFM. For lateral structuring of multiple protein assemblies, these proteins are replaced simultaneously or sequentially by different His-tagged proteins.

What is so special about this AFM technique is that it permits instant switching between imaging and replacement of immobilized proteins in their native state without any change of the set-up or the tip. In NPNL, the first self-assembled protein layer acts as a biocompatible and ductile patterning material (the protein resist). Affinity-captured proteins can be replaced by applying the AFM tip in contact oscillation mode (a term the researchers coined for a vibrational mode where the AFM tip is in constant contact with the sample), and the generated structures can be erased and refilled with different proteins in a uniform and functional manner.

"The novelty of our approach is that it allows the fabrication of bioactive protein nanoarrays down to a resolution of 50 nm in a fast and versatile manner without the need of vacuum or ambient atmosphere conditions—but under conditions of a physiological solution," says Tinazli. "NPNL serves requirements in nanobiotechnology, where physiological ambient conditions, such as in aqueous solution, are highly desirable for ensuring the preservation of biological functionality during and after array fabrication." This new technique makes it possible to fabricate rewritable protein nanoarrays, allowing a previously unobtainable flexibility in nanolithography and experimentation in biosensing and single-molecule studies.

The researchers also demonstrated that this erase-and-write (*i.e.* 'displacement and replacement') technique allows a more complex lateral organization of protein assemblies in multiplexed arrays, consisting of a series of different proteins and protein complexes in a unique orientation. Tinazli points out that this higher degree of multiplexity of protein nanoarrays will be key in the next steps the researchers are planning to take. "We will turn our future efforts to the fabrication of nanocatalytic centers and highly multiplexed protein assembly lines," he says.

Specific applications of this work will facilitate the design of protein arrays for biosensors, single molecule studies, and medical diagnostics. Applications in the biosensor field in particular, in combination with nano-optical biosensing elements, will provide new research tools allowing insights into intermolecular interactions.

Featured scientist: Ali Tinazli
Organization: Biocenter, Johann-Wolfgang-Goethe-Universität
   Frankfurt am Main (Germany)
Relevant publication: Ali Tinazli, Jacob Piehler, Mirjam Beuttler,
   Reinhard Guckenberger, Robert Tampé: Native protein
   nanolithography that can write, read and erase. *Nat. Nanotechnol.*,
   2, 220–225.

2.4 Playing Rubik's Cube with Nanoparticles

In order to exploit the unique properties of nanoscale materials for advanced applications it is often necessary to assemble nanoparticles into arrays with specific architectures. The interaction among the nanoparticles, or effects arising from their assembled larger structure, could result in interesting optical, magnetic, or catalytic properties that researchers and engineers could then exploit for new materials and applications. In recent years, there has been much interest in colloidal crystals—which are examples of periodic nanoparticle arrays—as photonic crystals, templates for photonic crystals, sensors, optical and electro-optical devices, and as model systems to study crystallization processes. The success of many of these potential applications is currently limited by scientists' ability to control the structure of colloidal crystals.

Normally, crystallization of uniform colloids produces face-centered cubic (fcc) or hexagonal close packing.[2] A few other colloidal crystal structures have recently been reported, but they either require careful balance of electrostatic interactions between colloidal particles, or rely on directing nanoparticles on a lithographic pattern that then dictates the geometry of a few layers in a thin film.

A completely different and novel approach to colloidal crystallization has been developed by scientists at the University of Michigan—an approach that results in simple cubic colloidal crystals extending over many unit cells in three dimensions. Simple cubic packing is quite rare, even in atomic structures. Here, it results from combined disassembly and self-reassembly of a template-directed structure in a single reaction step.

The cube is the simplest and most common shape found in crystals. Scientists differentiate between three Bravais lattice shapes which form the cubic system—simple, body-centered cubic (bcc), and fcc. Figure 2.1

Colloidal crystals prepared by natural assembly of uniform spherical particles form close-packed arrays, typically with fcc packing. Since many applications involving colloidal crystals (photonics, optoelectronics, combinatorial screening, *etc.*) depend on the specific geometry of the colloidal crystal and/or on interactions between colloidal particles, a simple and low-cost alternative to current fabrication methods for nanoparticle arrays would be good news for many researchers.

Andreas Stein, a professor in the Department of Chemistry at the University of Minnesota, has reported a novel disassembly synthesis of silica nanocubes and nanospheres. These were formed by replication of void spaces in a colloidal crystal template, *i.e.* a mold composed of close-packed uniform spheres.

"Our approach to preparing periodic arrays of uniform nanoparticles is based on a new concept of combining disassembly (top-down) and self-reassembly (bottom-up) syntheses," he says. "In the reassembled colloidal crystals, simple cubic packing, which is relatively rare even among atomic crystals, extends in three dimensions over a large number of unit cells."

"As we explored syntheses of shaped nanoparticles with other compositions, we discovered a novel self-assembly mechanism," says Stein. "When the

---

[2] In geometry, close packing of spheres is the construction of an infinite regular arrangement of identical spheres so that they take up the greatest possible fraction of an infinite 3-D space (*i.e.* they are packed as densely as possible).

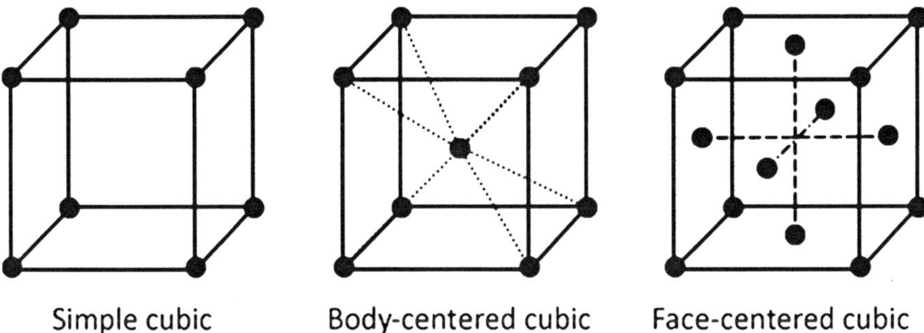

Figure 2.1 The simple cubic system consists of one lattice point on each corner of the cube. Each atom at the lattice points is then shared equally between eight adjacent cubes, and the unit cell therefore contains one atom in total (1/8 * 8). The body-centered cubic (bcc) system has one atom in the center of the unit cell in addition to the eight corner points. It has a contribution from two atoms per cell ((1/8)*8 + 1). The face-centered cubic (fcc) has atoms on the faces of the cube of which each unit cube gets exactly one-half contribution, giving a total of four atoms per unit cell ((1/8 for each corner) * 8 corners + (1/2 for each face) * 6 faces).

disassembly synthesis was applied to titanium oxide–phosphorus oxide compositions, nanocubes could be formed and, surprisingly, these self-reassembled into ordered arrays with simple cubic symmetry. For comparison, if we separated the cubes after their synthesis and tried to reassemble them, they did not form ordered arrays. We believe that self-reassembly occurred because the cubes were placed in the correct positions to reassemble by the original template and because capillary forces from the melting phosphorus oxide component drew the cubes together."

Stein cautions that the order of the simple cubic arrays that his team obtained so far is not yet sufficient for actual photonic crystals, but he hopes that it can be further optimized. "Even though the structures created in our work are far from the perfection necessary for photonic crystals, an adaptation of the self-reassembly method may provide a faster, low-cost alternative to produce simple cubic photonic crystals that are normally prepared by elaborate micromachining, layer-by-layer lithography, holography, and macroporous silicon etching, all expensive and time-consuming methods," he says.

"We had originally been working on monolithic porous silica structures with hierarchical porosity," Stein explains. "This is a mouthful, but it means that we devised methods to create silica with uniform larger pores and uniform smaller pores by combining two types of template. The large pores permit easy transport of guest molecules through the structure and the small pores provide the material with selectivity, like a sieve, and an extremely high surface area for chemical interactions. We discovered the nanocubes in a sample where we had changed one of the components in the synthesis mixture (an acid used to solidify the precursors). The self-reassembly observed for titanium oxide–phosphorus oxide, and now also for two other compositions, was quite unexpected."

It is still difficult to assemble nanoparticles into 3-D arrays with specific symmetries other than close-packed structures. Stein's research provides a new approach to achieve simple cubic packing, adding one more symmetry to a very small list of choices. As this list grows larger over time, it will be easier to manufacture functional devices from nanoparticle arrays. Stein's unique approach to simple cubic arrays of nanoparticles adds to the choices of available geometries, coordination numbers, and packing densities in nanoparticle arrays.

"We plan to confirm the proposed formation mechanism, so that we can apply our method to other compositions and optimize the order of the self-reassembled arrays," Stein explains his group's next steps. "Another challenge for the future will be to obtain other nanoparticle shapes and arrange these into 1-D, 2-D, or 3-D arrays. Hopefully, the particles can self-assemble, but we may have to lend them a hand using various external forces."

> *Featured scientist:* Andreas Stein
> *Organization:* Department of Chemistry, University of Minnesota, Minneapolis, MN (USA)
> *Relevant publication:* Fan Li, Zhiyong Wang, Andreas Stein: Shaping mesoporous silica nanoparticles by disassembly of hierarchically porous structures. *Angew. Chem., Int. Ed.,* **46**, 1885–1888.

2.5 Nanofabrication Based on Inorganic Replicas of Complex Biotemplates

The use of design concepts adapted from Nature is a promising new route to the development of advanced materials. Increasingly, nanotechnology researchers find naturally occurring nanostructures a useful inspiration for overcoming their design and fabrication challenges. Because biological structures are the result of millions of years of evolution, their designs have many unique merits that would be difficult to achieve by a completely artificial simulation.

By replicating the eye of a fruit fly, researchers have demonstrated a highly reliable and low-cost technique for making inorganic replicas of biotemplates for fabricating complex nanostructures with biologically inspired functionality. As Akhlesh Lakhtakia, the Charles G. Binder professor of engineering science and mechanics at Pennsylvania State University, explains it, "The exact imitation of biological structures by using equivalent processes to those used in Nature is extremely difficult since the mechanisms of formation of biological structures are tremendously complex. An alternative approach to fabricating replicas of biological shapes is based on converting templates harvested from a particular species to inorganic materials. This approach could result in a highly reproducible and inexpensive process for the fabrication of complex nanostructures with unique functionalities. This way, structures can be made out of more stable, harder, and high-temperature-tolerant inorganic materials." Lakhtakia cautions, however, that two major problems can arise with this approach. First, there may be no technique available to grow high-fidelity replicas, particularly at the nanoscale; second, most physical or chemical techniques will result in damage or destruction of the original biotemplate.

Lakhtakia, together with Penn State colleagues Carlo Pantano and Raul Jose Martin-Palma, replicated the eye of a fruit fly at the micro- and nanoscales by implementing a novel technique that allows the replication of even curved biotemplates. Pantano is Distinguished Professor of Materials Science and Engineering at Penn State. Martin-Palma is a Professor of Physics at the Autonomous University of Madrid who was visiting Penn State while this work was being conducted.

The scientists were motivated by a very simple fact: "Many structures have evolved in Nature to display interesting and useful properties. The most appealing of these properties are optical, imparting coloration and/or camouflage to the organism. Optical imaging structures such as compound eyes or polarization-sensitive eyes are also attractive. Some of these structures may have properties in the infrared regime, and therefore may not be easily appreciated by casual human observers. If we could easily replicate these and other attractive features—*e.g.* superhydrophobicity of ciliated objects such as lotus leaves—we could exploit them for various technical or scientific purposes." Martin-Palma suggested that, for instance, the development of compound-eye-based miniature cameras and optical sensors could lead to their integration into tight spaces in automobile engineering, credit cards, displays, security and surveillance, and medical technology.

Since the compound eyes of flies are very efficient collectors of light, their replicas could also be used to fabricate solar cell covers and other energy-harvesting structures as well as lenses offering good spatial resolution. Known methods for replication of biological structures include thin-film deposition methods like thermal evaporation, electron-beam evaporation, and sputtering, as well as other chemical methods. Atomic layer deposition (ALD) also has been used to replicate templates; however, it produces replicas that lack spatial precision with increasing thickness. Moreover, ALD is limited to applications with a limited number of materials and suffers from slow growth rate of film and extremely small film thickness.

The three scientists point out that nanoscale and nonplanar features of these biostructures cannot be replicated by most of these known methods. Additionally, the use of biological templates is limited by the use of chemicals and/or high temperatures. Finally, the fidelity of replication of these structures is enormously reduced as the thickness of the deposited film increases.

"Traditional methods are limited to deposition of films that are extremely thin, on the order of a few nanometers, and to a limited selection of materials, thus limiting any practical application," says Lakhtakia. "Furthermore, features oriented close to the normal of the growth direction of the film cannot be precisely replicated by these methods. These conventional methods are therefore not suited for use with biological templates that have curved surfaces."

Lakhtakia, Martin-Palma, and Pantano saw the need to develop a method of replicating intricate features of biotemplates that addresses the deficiencies of the existing methods. In a technique that they call *conformal evaporated film by rotation* (CEFR), the three researchers successfully replicated a biotemplate— the compound eye of a common fruit fly—by rapidly rotating it and implementing the *oblique angle deposition* (OAD) technique. OAD basically combines a typical deposition system with a tilted substrate. "Our method is particularly suited for use with curved templates, including micro- and nanostructured templates," said Martin-Palma. "It creates an actual replica of the template rather than an inverted structure of it, and, moreover, achieves high growth rates, typically about 0.5 μm/min."

The researchers mention that CEFR also may be used to produce replicas from insulating materials, metals, semiconductors, semimetals, polymers, and organic materials. "For example, infrared transparent materials such as chalcogenide glasses can be used to produce infrared micro lenses and visible laser-hardened infrared sensors, as well as photodiodes, solar sensors, solar concentrators, photonic crystals, optical bioprobes, and the like," Pantano said.

> *Featured scientist:* (a) Akhlesh Lakhtakia, (b) Carlo Pantano, (c) Raul Jose Martin-Palma
> *Organizations:* (a) Department of Engineering Science and Mechanics, Penn State University, University Park, PA (USA); (b) Materials Research Institute, Penn State University; (c) Department of Physics, Universidad Autónoma de Madrid (Spain)
> *Relevant publication:* R. J. Martín-Palma, C. G. Pantano, A. Lakhtakia: Replication of fly eyes by the conformal-evaporated-film-by-rotation technique. *Nanotechnology,* **19**, 355704.

2.6   Viruses as Building Blocks for Engineers

Geneticists regularly use viruses as vectors to introduce genes into cells that they are studying. Viruses are also the most common carrier vehicles in gene therapy. Having been genetically altered to carry normal human DNA, they deliver the therapeutic genes to the patient's target cells. These viruses infect cells, deposit their DNA payloads, and take over the cells' machinery to produce the desired proteins.

Current trends in nanotechnology promise to take virus technology in an entirely new direction. From the viewpoint of a materials scientist, viruses can be regarded as organic nanoparticles. Their surface carries specific tools designed to cross the barriers of their host cells. The size and shape of viruses, and the number and nature of the functional groups on their surface, are precisely defined. As such, viruses are commonly used in materials science as scaffolds for covalently linked surface modifications. The powerful techniques developed by biologists are becoming the basis of engineering approaches towards nanomaterials, opening a wide range of applications far beyond biology and medicine.

A virus is a nanoscale particle, ranging in size from 20 to 300 nm, that can infect the cells of a biological organism. Viruses cannot reproduce on their own. They replicate themselves only by infecting a host cell. At the most basic level, viruses consist of genetic material contained within a protective protein coat. For scientists, a particular quality of viruses is that they can be tailored by directed evolution. The properties of the virus can be readily modified by changing the underlying construction plan—the nucleic acid sequence of the viral genome. It is also possible to produce viral chimeras that carry proteins of different viral origins.

"Viruses can be multiplied in appropriate tissue and cell cultures, but they do not have any metabolic activity of their own," says Edwin Donath, a professor at the Institute of Medical Physics and Biophysics at the University of Leipzig in Germany. "This, in principle, allows the use of virus particles as durable building blocks for composite materials. In combination with powerful molecular biology approaches, all these features provide novel and far-reaching possibilities for the production and engineering of hybrid composite materials from these nanoparticles."

"Instead of chemically engineering functions into a composite material, it may be more convenient to take advantage of nanoparticles as carriers of the desired properties," adds Martin Fischlechner, a scientist in Donath's group. "These nanoparticles can then be used as building blocks for the fabrication of a composite material with the required qualities. Engineered viruses may fulfill the role of the nanoparticles and once a convenient and general strategy to attach them to an interface is found, the set-up can be standardized. The fabrication of a large variety of functionalized surfaces becomes possible by bringing together the potential of viruses for combinatorial surface display and a general strategy for surface attachment."

Donath and Fischlechner describe two basic approaches to modifying viral nanoparticles to obtain desired functionality: chemical modification and genetic engineering. "By using the various techniques available for covalent coupling of molecules on to proteins, viruses can be used as a platform for spatially defined chemical surface modifications," says Fischlechner. "The advantage of using viruses instead of synthesized nanoparticles is the defined number and orientation of the accessible functional groups on their surface. This provides the possibility to arrange different chemical functionalities with nanoscale precision."

The concept of utilizing viruses as a synthesis platform can thus be seen as a nanoparticle-based 3-D equivalent of the nanotechnology methods used for arranging molecules in two dimensions. Viruses have already been employed as scaffolds for metallization or for the growth of minerals, resulting in metallized or mineralized building blocks. They can also be used as nanocages for the entrapment of substances. "However, the true power of using viral systems originates from their unique quality as nanocomposites that carry all the necessary information for the production of their components within their host cell," says Donath. "This allows the display of specific polypeptides on the viral surface by engineering the viral nucleic acid sequence inside. The most straightforward engineering concept aimed at modification of the viral surface is thus to modify the genetic code of the viral genome itself. If successful, the product would be a genetically modified virus that is still able to replicate."

The two scientists explain how viruses represent composite nanoparticles that offer many degrees of freedom to design their surface properties. This can be achieved either by techniques based on chemical conjugation or by genetic engineering methods.

Donath believes that in the near future we are likely to see a lot of developments in virus technologies based on genetic code design. "Genetic engineering has a clear analogy to software production, in that pieces of code can be multiplied at low cost, put together in an artificial genome, and subsequently processed in the cellular machinery. Repositories for standard genetic building blocks ready for customization for special needs are about to be established."

He concludes that recent developments in microfluidics will certainly contribute to the parallel production of surface-engineered viruses at low cost and with a great diversity of functions: "Although this field is still in its infancy, automatic parallel production of viral building blocks seems possible when recent progress ranging from chip-based DNA manipulation to cell-culture techniques is considered".

> *Featured scientists:* Edwin Donath, Martin Fischlechner
> *Organization:* Institute of Medical Physics and Biophysics, University
>   of Leipzig (Germany)
> *Relevant publication:* Martin Fischlechner, Edwin Donath: Viruses as
>   building blocks for materials and devices. *Angew. Chem., Int. Ed.,*
>   **46**, 3184–3193.

2.7   Nanoassembly with Living Materials

Most nanotechnologists, even if they manage to self-assemble functional nanodevices, still operate exclusively at the nanoscale. Bridging the gap between the nanoworld and the macroworld has proved to be a huge hurdle. In a novel approach that merges materials chemistry, biology, and medicine, researchers in Germany have used living bacteria to show that self-assembly of functional materials and living systems is possible through a chemically programmed construction.

"We set out to realize the first step toward the exchange of specific information between synthetic systems and bacteria," says Luisa de Cola. "We have functionalized biocompatible artificial nanocontainers (the synthetic mineral zeolite L) and attached them to nonpathogenic bacteria (*E. coli*). The functional systems (the zeolite) can be filled with different molecules rendering them fluorescent or active for further actions. Because of the particularly defined geometrical arrangement of the zeolite and bacteria, we are also able to self-organize two bacteria by using the nanocontainer as a junction."

de Cola, a professor in the Physikalische Institut and CeNTech (Center for Nanotechnology) at the University of Münster in Germany, and her research group are working with zeolites by loading their 1-D channels with suitable small molecules. The hybrid systems thus produced are suitably designed to play an important role as photo- and electroresponsive materials, with interesting applications in nanophotonics and nanoelectronics. In combination with living systems, they could also be used for molecular *in vivo* imaging and drug delivery experiments.

de Cola's research was triggered by her close cooperation with Professor Gion Calzaferri at the University of Berne in Switzerland. Calzaferri's group is working with zeolite and develops highly organized dye–zeolite materials for nanosensors and photoelectronic devices. "I was convinced that we could push the field in a different direction and we started a collaboration which is very successful," says de Cola. "One idea was to move into the nanomedicine field with these objects and we started with bacteria assembly."

Zeolite L is a crystalline aluminosilicate, a biocompatible material—basically a little piece of transparent volcanic rock. It contains 1-D channels running through the whole crystal; a single crystal with a diameter of 550 nm consists of about 80 000 parallel channels. The German research team has been trying to use these as nanocontainers for *in vivo* applications. "I can foresee several applications in nanomedicine but also in other material fields," says de Cola. She points out that the use of such materials rather than more conventional nanoparticles and quantum dots is appealing because of their tunable size, ranging from 30 nm up to several microns, but also their shape, optical transparency, biocompatibility, and of course their porous character and easy covalent functionalization. "We have shown that we can assembly them into rod-type structures because of their selective channel entrance functionalization."

To construct their linear zeolite—*E. coli* assembly, de Cola's team decided to explore an electrostatic-type binding between the negatively charged outer cell membrane of the bacterium and a positively charged zeolite L crystal. "Interestingly, we are able to position the charge only at the entrance of the channels, so that the entire crystal retains its own character on the surface," says de Cola. "To realize the construction, we first loaded 1 µm long zeolite L with the green luminescent organic dye pyronine by means of an ion-exchange procedure. We then functionalized the channel entrances of the zeolite with thousands of amino derivatives, which are protonated, leading to the desired positively charged systems. Amino-functionalized zeolite crystals and bacteria in an estimated 1:1 ratio were then incubated together for 1 hour at 37 °C in phosphate-buffered saline (PBS) solution."

The resulting geometrically linear assembly stimulated a perhaps obvious question: "Can we assemble living systems by using the nanocontainers as a junction?" asked de Cola. To achieve this goal, the researchers changed the estimated ratio between the cells and the zeolite L by using a large excess of bacteria. "After mixing, we observe, under the microscope, a linear structure that does not correspond to a single cell," says de Cola. "An accurate analysis of the assemblies proved that the ratio between the zeolite and the bacteria is now 1:2. At this stage, we do not know if the zeolite can play an active role in the communication of the two linked systems. In fact, we wish to stress that our zeolite, used as a connector, is not only biocompatible and stable but also functional and modular."

This research opens the possibility of arranging living cells and nanomaterials in various combinations and having them self-assemble into complex structures. de Cola even believes that an exchange of specific information between the zeolite and/or the bacteria is possible. For instance, substances stored inside the zeolite's nanochannels could be transferred to the living cells; conversely, substances discharged from the bacteria could be captured by the crystals and analyzed. "It will be fascinating to explore the consequences of this," says de Cola.

Featured scientist: Luisa de Cola
Organization: CeNTech (Center for Nanotechnology), University
   of Münster, Germany.
Relevant publication: Zoran Popovi, Matthias Otter, Gion Calzaferri,
   Luisa de Cola: Self-Assembling living systems with functional
   nanomaterials. *Angew. Chem., Int. Ed.,* **46**, 6188–6191.

CHAPTER 3

Techniques and Concepts for the Nanotechnologist

Until the 20th century, a single craftsman or team of craftsmen would create each part of an industrial product individually and assemble the parts together into a single item, making changes to them so that they would fit together and work together—the so-called 'English system' of manufacture. Then Henry Ford came along and in 1907–08 developed the assembly line for his Model T automobile. This innovation revolutionized not only industry but also society, because it allowed mass production of industrial goods at much lower cost than before. At its core, an assembly line is a manufacturing process in which interchangeable parts are added to a product in a sequential manner to create a finished product. Nanotechnology techniques today are pretty much where the industrial world was before Ford's assembly line—a domain of craftsmen and artisans, not one of mass production.

Shrinking device size to the nanoscale presents many fascinating opportunities such as manipulating nanoscale objects with nanotools, measuring mass in zeptogram (10^{-21} g) ranges, sensing forces at femtonewton (10^{-15} N) scales, and inducing gigahertz (10^9 Hz) motion, among other new possibilities waiting to be discovered. At this early stage of the various areas of nanotechnologies, scientists are concerned with understanding the basics—how to measure the temperature of nano-objects; how to literally shine a light on them; how to identify, measure, and manipulate clusters of atoms; to understand what makes a material weak or strong and how to exploit this knowledge for developing 'smart' materials; to study the interaction between nanoscale objects and comprehending the specific forces that are occurring at the nanoscale; and ultimately, how to build functioning devices and machines with nanoscale dimensions. This, basically, describes the 'nanotechnology' that is happening in laboratories around the world today.

RSC Nanoscience and Nanotechnology No. 8
Nano-Society: Pushing the Boundaries of Technology
By Michael Berger
© Michael Berger 2009
Published by the Royal Society of Chemistry, www.rsc.org

The way in which the various nanotechnological capabilities are being developed follows in the footsteps of other new scientific fields. Scientists concoct elaborate experiments in their laboratories with expensive instruments and often extremely delicate set-ups that often result in one-of-a-kind results. Theoretical work, models, and simulations are developed to come to terms with strange new phenomena. For instance, as devices evolve from micro- to nanoscale mechanical structures, the adhesive forces (van der Waals forces) become relatively stronger. This effect can sometimes be useful, for instance in binding organic molecules to metallic surfaces, and this could potentially be exploited in semiconductor manufacturing. On the other hand, it could also complicate processes where single atoms and molecules need to be moved.

As we will see below, researchers are even revisiting well-established techniques such as forging and welding to find ways to manipulate nanoscale objects—but with a high-tech twist.

3.1 How to Build a Nanothermometer

One of the problems nanoscientists encounter in their forays into the nano-world is that of accurate temperature measurement. Ever since Galileo invented a rudimentary water thermometer in 1593, accurate temperature measurement has been a challenging research topic and thermosensing technologies have become a field in their own right. Now, that technology has reached the nanoscale, temperature gradients are becoming essential in areas such as thermoelectricity, nanofluidics, design of computer chips, or hyperthermal treatment of cancer. Currently there is no established method for measuring temperature at the nanoscale. Most of today's probes are traditional bulk probes, the kind that is inserted into a sample and measures its temperature. Liquid crystal films that change color depending on temperature also have at least microscale thickness and lateral dimensions.

"The development of a nanoscale thermometer is not only a matter of size; it also requires materials with novel physical properties because all physico-chemical and thermodynamic properties are drastically altered at this tiny scale," says Nicholas Kotov who has reviewed these issues. "The measurement of local temperature in a single cell and in volumes less than 10^{-18} L (an attolitre) is becoming the technological frontier. The development of thermo-meters able to operate within the spatial constraints of the operating environment should be addressed using precise, nanoscale sensing modules, which can potentially be employed in biological applications as a result of their biocompatibility."

Kotov, a professor in Chemical Engineering at the University of Michigan, together with Jaebeom Lee, previously a research fellow in his group, has reviewed current developments of various nanoscale thermometers.

"The accurate temperature regime is critical for medical applications," says Kotov. "All biological systems are temperature sensitive. We are designed to operate at 37° C. If the temperature gets higher or lower, it makes a tremendous change to how we feel. Similarly, large differences occur with biomedical measurements based on protein reactions. For example, a microfluidics diagnostic device that is becoming more common now may show you that you have cancer when you don't, just because the temperature of the process is slightly off."

Carbon nanotubes (CNTs) have proved very useful in fabricating a nanothermometer. In one example, a 7 nm thin film of CNTs forms a fairly simple thermometer with a measurable range of 373–600 K under vacuum. The ambient temperature of the nanotube directly corresponds to the turn-on field and emission current at a given applied electronic field. In another example, the gallium in a gallium-filled carbon/manganese oxide nanotube, analogous in shape to a conventional mercury thermometer but a billion times smaller, serves as a temperature indicator by expanding and contracting inside the nanotube in the range of 303–2478 K.

According to Kotov, nanoparticles and nanocolloids are the natural choice for nanoscale temperature probes because of the size and optical activity that can be used to monitor temperature. Nanothermometers can be made from dynamic nanoscale assemblies. "The dynamic aspect of these assemblies is essential because it allows the thermometer to operate reversibly, *i.e.* it is not a one-shot measurement," says Kotov. "Polymers are excellent choices for making dynamic nanoscale assemblies, and superstructures from particles and polymers are a logical choice for such structures."

Kotov and his group came up with a structure that looks like a planet surrounded by many satellites (a corona assembly). The 'planet' in the center is a gold nanoparticle ∼20 nm in diameter. The 'satellites' are smaller (3–7 nm) cadmium telluride semiconductor particles tethered to the gold particle by a flexible polymeric (PEG) chain. The polymer 'spring' acts like a coiled garden hose that contracts and tightens in the cold and relaxes in the heat. As the polymer responds to heat or cold, the particles attached to the ends move closer together or farther apart and this altered distance between the particles produces a concomitant change in luminescence. Figure 3.1

"Temperature changes make the PEG chains expand or contract and this movement affects the emission of the entire corona assembly," explains Kotov. "The emission becomes weaker when temperature rises and the polymer swells, and stronger when the temperature drops and the PEG contracts." He notes that the optical behavior of the corona assemblies and their response to temperature was explained by using an electromagnetic theory of particle interactions developed by Professor Alexander Govorov from Ohio University.

Kotov points out that these thermometers can be particularly useful in measuring the temperature profiles in fluid media, for instance in lab-on-a-chip devices and in hyperthermal eradication of cancer. In this technique, gold nanoparticles are illuminated with pulsed laser irradiation, which raises the temperature of the attached targeted cancer cells so that they subsequently die. "The temperature change in the tissue is fairly small, but even a difference of 2–3°C is significant in terms of whether the cancer is eliminated completely or not," he says. "Without appropriate temperature profiling in live tissue, in the presence of moving fluids (*e.g.* blood), this is very difficult to predict or

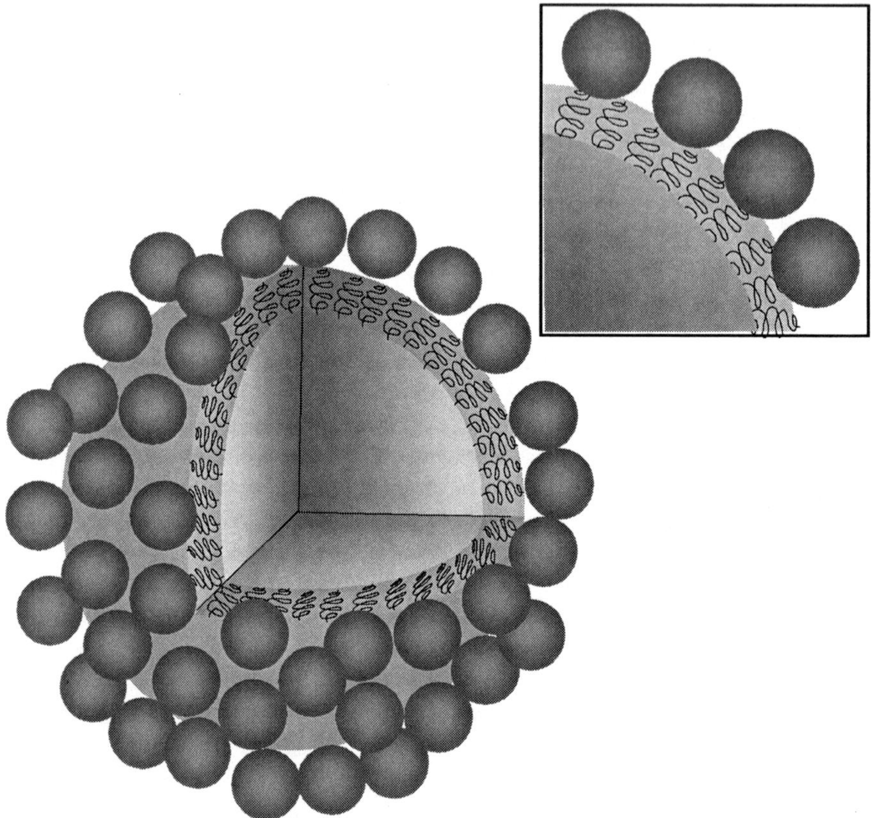

Figure 3.1 Schematic of the nanoparticle superstructure of the nanothermometer. (Image: Dr N. Kotov, University of Michigan)

calculate. Nanothermometers, in principle, allow us to measure it with spatial resolution smaller than a typical cell."

Featured scientist: Nicholas Kotov
Organization: Chemical Engineering, University of Michigan, Ann Arbor, MI (USA)
Relevant publication: Jaebeom Lee, Nicholas A. Kotov: Thermometer design at the nanoscale, *Nano Today*, **2,** 48–51.

3.2 Nanotechnology Reinvents the Wheel

The invention of the wheel has been at the origin of major scientific and technological developments: from the creation of astronomical clocks or calculating machines to motor vehicles. At the molecular level, the smallest at which a wheel can be created, it represents a major challenge for scientists. For years, researchers have been working on the design of molecular machines equipped with wheels. After observing the random rotation of a flat molecular wheel in 1998, designing and synthesizing a monomolecular wheelbarrow in 2003, and then synthesizing a molecular motor in 2005, a European group of researchers managed to operate the first molecular rack with a pinion of 1.2 nm in diameter. They controlled the rotation of a 0.7 nm diameter wheel attached to a 0.6 nm long axle in a molecule. This molecular 'wheel' could revolutionize machinery built at the nanoscale. Nanowheel rotation has been claimed before, but never shown directly.

Initially, manipulation of single molecules with scanning tunneling micro-scopy (STM) led to a hopping motion of the molecules from one adsorption site to the next. Then, a group of French and German scientists demonstrated the first rolling molecular wheel. Figure 3.2

"This is fascinating as it shows that rolling, a motion that is of fundamental importance in the macroscopic world, is also possible at the atomic scale," Leonhard Grill explains. By understanding and controlling how nanosized vehicles move across atomic terrains, nanoscientists may be able to design a new generation of molecular machines equipped with wheels.

Grill, whose research group at the Free University of Berlin focuses on the study of single functionalized molecules on surfaces, together with

Figure 3.2 Illustration of the wheel–axle–wheel molecule. (Image: Dr L. Grill, Free University Berlin)

collaborators from the Nanosciences Group at CEMES-CNRS in Toulouse, shows that triptycene groups, which resemble three-bladed paddlewheels, can act as wheels only 80 nm wide. They fixed one triptycene to each end of a rigid axle, then pushed this primitive molecular vehicle over a copper surface using the tip of an STM. The type of wheel and surface were very carefully chosen. Two notched, 'tireless' wheels were used because of their maximum adherence to the running surface, an ultra-clean copper plate. Its natural roughness presented rows of copper atoms separated by a distance of ~ 0.3 nm, and about one atom high.

The way the molecule moves depends on how it is aligned to the row of copper atoms on the surface. In previous experiments, molecules have hopped across the surface when pushed by the microscope tip. However, when the molecules are aligned properly with respect to the ridges on the copper surface, one of the wheels rotates by 120° when pushed by an STM tip.

While rolling the wheel, the STM operator followed any variations in electrical current passing through it on the control screen, in real time. Depending on the handling conditions of the molecule, the operator could turn the wheels or make the molecule advance without rolling its wheels. "We could also show that hopping and rolling motion can be induced separately in a controlled way from the experimental parameters, in particular from the corrugation of the substrate," says Grill. "It turns out that the surface needs to have some corrugation in the direction of motion in order to enable a rolling motion. On a completely flat surface, the wheel doesn't roll, it hops."

The pathways of the hopping and rolling motions are completely different, which reflects the different mechanisms. Long pathways can be observed in hopping, but only very short ones in rolling, because there the STM tip is already behind the molecular wheel after a 120° rotation, and thus cannot induce any further displacement.

Grill points out that a rolling wheel behaves differently on a surface than a hopping molecule. "Firstly, rolling allows motion only in two directions— forward and backward but not sideways," he says. "Secondly, it might be possible for rolling molecules to overcome nanostructures or step edges on a surface, something that cannot be done by hopping molecules."

This experiment enables an approach to understanding at the monomolecular scale the functionalities that are already known at a macroscopic scale. Without the wheel, some technological advances would not be possible. This analogy may also hold true for molecular vehicles.

Featured scientist: Leonhard Grill
Organization: Experimental Physics Department, Freie Universität
    Berlin (Germany)
Relevant publication: L. Grill, K.-H. Rieder, F. Moresco, G. Rapenne,
    S. Stojkovic, X. Bouju, C. Joachim: Rolling a single molecular wheel
    at the atomic scale. *Nat. Nanotechnol.*, **2**, 95–98.

3.3 Bronze Age Technique Works Just Fine in the 21st Century

Although its historic development can be traced back to the Bronze Age, welding is still used in important areas of modern industry such as construction, manufacturing, and engineering. Spot welding, a type of resistance welding used to weld sheet metal, was originally developed in the early 1900s. The process uses two shaped copper alloy electrodes to concentrate welding current and force between the materials to be welded. The result is a small 'spot' that is quickly heated to the melting point, forming a nugget of welded metal after the current is removed. Perhaps the most common application of spot welding is in the automobile industry, where it is used almost universally to weld the sheet metal forming the car body. Spot welding can be completely automated, and many of the industrial robots found on assembly lines are spot welders. With the continuing development of bottom-up nanotechnology fabrication processes, spot welding may likewise play an important role in interconnecting CNTs, nanowires, nanobelts, nanohelixes, and other nanomaterials and structures for the assembly of nanoelectronics and nanoelectromechanical systems (NEMS). Nanorobotic assembly provides a new way for fabricating such devices by interconnecting as-grown nanomaterials or nanostructures with the capability of 3-D positioning/orientation for site selection.

A team of researchers from ETH Zürich in Switzerland and Zhejiang University in China have demonstrated nanorobotic spot welding using single-crystalline copper-filled CNTs inside a transmission electron microscope (TEM). "We need more effective assembly technologies for strong and sometimes conductive interconnection of nano building blocks for fabricating nanosystems, and nanorobotic assembly is one way," explains Bradley J. Nelson, professor of robotics and intelligent systems at ETH Zürich and the director of IRIS, ETH's Institute of Robotics and Intelligent Systems. He points out that his team's technique is an important step for the interconnection of nano building blocks to assemble NEMS.

Van der Waals forces, electron-beam-induced deposition (EBID), focused-ion-beam chemical vapor deposition (FIB-CVD), high-intensity electron-beam welding, Joule-heating-induced joining, and nanomechanochemical bonding are experimentally demonstrated interconnection strategies, although all have limitations. Van der Waals forces are generally very weak, Joule-heating-induced joining and nanomechanochemical bonding are promising but not yet mature, and the other methods involve high-energy electron or ion beams, which significantly limits their applications. In contrast, Nelson and his collaborators realized the controlled melting and flowing of copper inside nanotube shells by applying very low bias voltages between 1.5 and 2.5 V. The melting is a result of Joule heating, whereas the flowing is caused by electromigration.

"The successful soldering of a copper-filled CNT on to another CNT shows promise for nano spot welding and thermal dip-pen lithography," says Nelson. "Furthermore, nanorobotic manipulation provides the possibility for 3-D site selection, which is essential for complex nanoassembly, and the metallic (copper) connection is significant for nanoelectronics system."

Nelson explains that, compared with the other interconnection processes previously investigated, electrically driven spot welding has several advantages:

(1) A very low current can induce the melting and drive the flow, which is much more efficient than irradiation-based techniques involving high-energy electron beams, FIB, or lasers. Combined with dielectrophoretic assembly, it is possible to solder the tubes on to electrodes for batch fabrication of microelectromechanical systems (MEMS).

(2) The welding site can be readily selected using nanorobotic manipulation, which enables 3-D position and orientation control for continuous mass delivery and will potentially enable 3-D prototyping and assembly.

(3) The melting occurs rapidly, in the millisecond range, several orders of magnitude faster than using a high-intensity electron beam or FIB, which is generally on the order of a minute.

(4) Because both the rate and direction of mass transport depend on the external electrical drive, precise control of the delivery of an ultra-small mass is possible. Time-based control will allow the delivery of attograms of mass.

(5) Copper has good compatibility with the conventional semiconductor industry. The experiments show that it will likewise play an important role in scaled-down systems. Carbon shells provide an effective barrier against oxidation and consequently ensure long-term stability of the copper core.

Nelson points out that the capability to deliver encapsulated materials from carbon shells is of great interest because of potential applications as sources of small numbers of atoms for nanoprototyping, nanoassembly, and injection. "A hybrid approach combining nanorobotic assembly with self-assembly may be an solution for the trade-off between complexity and productivity for nano-systems manufacturing," he says.

> *Featured scientist:* Bradley J. Nelson
> *Organization:* IRIS—Institute of Robotics and Intelligent Systems,
>   ETH Zürich (Switzerland)
> *Relevant publication:* Lixin Dong, Xinyong Tao, Li Zhang, Xiaobin
>   Zhang, Bradley J. Nelson: Nanorobotic spot welding: controlled
>   metal deposition with attogram precision from copper-filled carbon
>   nanotubes. *Nano Lett.*, **7**, 58–63.

3.4 Metal Forging—from Blacksmith to Nanoscientist

Forging is another ancient technology. The earliest known forgings, which date from around 1600 BC, were crudely hammered ornaments made from naturally occurring free metals. The latest state-of-the-art forging techniques use micron-sized hammers to forge nanosized metal shapes to be used as components in nano- and microtechnology systems.

Even today, the smallest free-standing microcomponents from metal, polymer, or ceramics are still in the size range of 10–1000 μm. In recent years, nanoscale fabrication techniques have developed considerably, but the fabrication of free-standing nanosized components is still a great challenge for researchers. The ability to produce nanosized high-strength metallic components by nanoforging opens up new possibilities of producing complex microsystems by assembling free-standing nanoscale components. At this scale they would be of the same dimensions as microorganisms and therefore small enough even to travel through the human body.

"We demonstrated the fabrication of metallic nanocomponents utilizing three basic steps," explains Debashis Mukherji, a scientist at the Institute for Materials at Technical University Braunschweig in Germany. "First, we used metallic alloys to produce a metallic raw stock of nano-objects in large numbers. Using manipulators, these objects are then isolated inside a SEM and placed on a micro-anvil or a die. Finally, the shape of the individual nano-object is changed by nanoforging using a microhammer. In this way, free-standing, high-strength, metallic nano-objects may be shaped into components with dimensions in the 100 nm range."

The miniaturization of components and systems is advancing steadily in many areas of engineering and the trend towards ever smaller components and devices is relentless. Today's integrated circuits and NEMS are fabricated in a top-down approach—they are carved out of a monolithic substrate, the size of which is still on the millimetre scale. In the present state of science and technology, the use of bottom-up approaches to build independent, freestanding, complex devices and machines on the micron and submicron scale is still in the realm of speculative fiction. Since the system/component size ratio in mechanical devices is typically 10:1, fabricating multicomponent machines on a microscale requires individual components on the nanoscale.

Micromanipulation with tools such as micropipettes or lasers is a mature technique for handling microscopic samples in liquids. In contrast, micro- and nano-manipulation in dry environments is still a challenging task because of the absence of fluid to reduce the often strong adhesion forces. In dry environments, scanning probe microscopy (SPM) equipment and mobile microrobots have already been used to push, slide, and roll nanostructures in two dimensions across surfaces, but this is of limited practical application. Apart from standard SPM tips, more advanced tools such as microfabricated grippers could provide better control in manipulation.

Mukherji points out that their new technique has several benefits: "We have now demonstrated that objects of size 400 nm^3 can be shaped at room temperature from a high-strength intermetallic alloy (*e.g.* Ni$_3$Al). This is of significant scientific interest because, despite their very attractive high-temperature properties, several decades of research into the use of Ni$_3$Al type intermetallics for high-temperature gas turbine applications has failed to solve the problem of using these materials in real applications because they are extremely brittle at room temperature, which limits the fabrication of components."

This work demonstrates that raw stock can be prepared, handled, and manipulated to fabricate submicron-size components, which will ultimately make it possible to fabricate complex devices or machines on a micron scale.

Shaped dies are required to form complex shapes such as gear wheels from nanocubes. The required shape can be engraved, for example by electron beam lithography, on to a silicon cantilever anvil. Lithographic techniques have advanced sufficiently that it is now possible to produce detailed nanoscale features on monolithic substrates. A counterpart of the die carved out of a silicon wafer can, for example, be fixed to the flat surface of the tungsten hammer, thereby building a closed die forging set-up.

> *Featured scientist:* Debashis Mukherji
> *Organization:* Institut für Werkstoffe, Technische Universität
>   Braunschweig (Germany)
> *Relevant publication:* J. Rösler, D. Mukherji, K. Schock, S. Kleindiek:
>   Forging of metallic nano-objects for the fabrication of submicron-size components. *Nanotechnology,* **18**, 125303.

3.5   Spooky Technology

Photonic crystal nanocavities—'nanocages' for light—are the focus of much interest in photonics because they can confine photons tightly in a tiny space. Just as semiconductor crystals control the flow of electrons (the basis for all electronics), photonic crystals are a unique material used to construct photonic devices and circuits for manipulating light, *i.e.* photons. A prominent example of a photonic crystal is opal, a naturally occurring gemstone. Photons (behaving as waves) propagate through it or not, depending on their wavelength. Wavelengths of light (streams of photons) that are allowed to travel through the crystal are known as *modes.* Disallowed bands of wavelengths are called *photonic bandgaps.* What's so interesting for researchers is that, once you are able to fully control and manipulate photons, you could not only vastly improve existing applications such as optical data storage, high-precision sensing, and telecommunications, but develop exotic technologies like quantum computing.

Photons move at the speed of light, much faster than electrons. Photonic quantum computers would therefore be able to operate at an unprecedented speeds. However, trapping light (slowing it or stopping it altogether) is a necessary element in replacing electron storage for computer logic, because only when light has been slowed down sufficiently can information be mapped on to it.

In a quantum computer, the fundamental unit of information (called a quantum bit or *qubit*), is not binary as in electronics as we know it. The qubit's properties are a direct consequence of its adherence to the laws of quantum mechanics, which differ radically from the laws of classical physics.

"A qubit can exist not only in a state corresponding to the logical state 0 or 1 as in a classical bit, but also in states corresponding to a blend or superposition of these classical states. In other words, a qubit can exist as a 0, a 1, or simultaneously as both 0 and 1, with a numerical coefficient representing the probability for each state. This may seem counterintuitive because everyday phenomena are governed by classical physics, not quantum mechanics—which takes over at the atomic level."[1]

And that's where it gets decidedly unnerving: The weirdest idea in quantum mechanics is what scientists call *spooky action at a distance*—the notion that certain subatomic particles such as photons can affect one another no matter how far apart they are. Imagine you have two identical marbles in your office in London, and you ship one to Tokyo. Then spin the one in London; the one in Tokyo would start spinning at exactly the same speed but in the reverse direction. (Niels Bohr once remarked that a person who wasn't outraged on hearing about quantum theory didn't understand what had been said.) This phenomenon is called *quantum entanglement* and it led Erwin Schrödinger to the famous thought experiment now known as *Schrödinger's cat*.

To map information onto light beams, scientists would use the *spin* of photons (all elementary particles have a natural orientation, like 'up' or 'down'). *Flipping the spin* has the same effect as switching a transistor on and off. Unfortunately, the entangled state of the photon represents all possible arguments and corresponding values of the function as a linear superposition, so this information is not accessible beforehand. The problem is that, thanks to the laws in the quantum world and what is known as *quantum interference,* the spin of a photon could be up or down simultaneously, until the photon is observed.

Modes, *i.e.* the wavelength of light allowed to travel through a photonic crystal, can be qualified according to their Q factor. This is a measure for the finite photon lifetime and indicates the amount of resistance to resonance in such a system. The average lifetime of a resonant photon in the cavity is proportional to the cavity's Q. To take an analogy from acoustics: when you strike a bell or pluck a guitar string, it vibrates within a small range of frequencies, centering on what is called the *resonant frequency*. The quality factor (Q) refers to how narrow that range is. It is defined as the ratio of the resonant frequency to the range of frequencies over which resonance occurs. A radio receiver with high-Q circuitry, for example, is more effective in separating one station from another.

High-Q nanocavities of optical wavelength size have been difficult to build, as radiation losses increase in inverse proportion to cavity size. Research over the

---

[1] Quoted from Jacob West's excellent introduction to the quantum computer: http://www.cs.caltech.edu/~westside/quantum-intro.html

past few years has shown that record Q values could be realized in nanocavities to confine photons strongly and for a long time. Researchers in Japan have also succeeded in realizing nanocavities with unsurpassed Q factors as high as 2 000 000.

"The next important issue is how to deliberately control storage and release photons from such a high-Q nanocavity," explains Susumu Noda, a professor in the Department of Electronic Science and Engineering and the Vice Director of the Photonics and Electronics Science and Engineering Center at Kyoto University in Japan.

"When the cavity Q increases, the photon lifetime is increased, and the operating speed becomes slow. Thus, the important issue is as follows: when we introduce photons into the nanocavity the Q factor should be small, and once the photons are introduced into the nanocavity the Q factor should be increased rapidly before the photons can leak out. And, if necessary, the Q factor should be decreased again so that the photons can be released quickly from the nanocavity. In other words, dynamic control of the Q factor of the nanocavity on a picosecond timescale is very important."

However, there was no concept of how to achieve such dynamic control of the nanocavity Q until Noda and his collaborators proposed and demonstrated precisely such a concept. The team in Kyoto has successfully demonstrated the dynamic change of the Q factor of a nanocavity from 3000 to 12 000 on a picosecond timescale by using a system composed of a nanocavity, a waveguide with nonlinear response, and a hetero-interface mirror. "We believe that this demonstration is an important step towards realization of the slowing and/or stopping of light, and quantum information processing where nanocavities could be integrated on a chip and the transfer, storage, and exchange of photons would be possible through integrated waveguides," says Noda.

Featured scientist: **Susumu Noda**
Organization: **Quantum Optoelectronics Laboratory, Kyoto University (Japan)**
Relevant publication: **Yoshinori Tanaka, Jeremy Upham, Takushi Nagashima, Tomoaki Sugiya, Takashi Asano, Susumu Noda: Dynamic control of the Q factor in a photonic crystal nanocavity.** *Nat. Mater.,* **6**, 862–865.

3.6   Shedding Light on the Nanoworld

Integrating biochemical analysis MEMS and NEMS has led to the development of a new class of biomedical analytical devices called *lab-on-a-chip*. These combine into one single device a number of biological functions (such as enzymatic reactions, antigen–antibody conjugation, and DNA probing) with proper micro- or even nanofluidic laboratory components (such as sample dilution, pumping, mixing, metering, incubation, or separation) and detection in micro- and nanosized channels and reservoirs. In order to shrink these lab-on-a-chip devices even more, researchers are increasingly finding ways to reduce microsized components to the nanoscale. One problem is the issue of illumination. Today, many lab-on-a-chip devices use external illumination sources such as lasers or light-emitting diodes (LEDs). Being able to fully integrate the excitation and detection mechanisms on lab-on-a-chip devices would allow further size reductions and increase the flexibility of using and handling them. Researchers at Cornell University have electrospun light-emitting nanofibers that, if they can be integrated with micro- and nanofluidic devices, could achieve excitation of light-induced fluorescence and detection within that same device.

"Even though LEDs are relatively inexpensive and can be submillimetre in size, they are still much larger than the typical micro- or nanofluidic device and have to be mounted outside the sensing region," explains José Manuel Moran-Mirabal, a member of the Craighead Research Group at Cornell University. "Some attempts have been made to fabricate on-chip point sources that can be coupled to micro- and nanofluidic devices. However, these point sources are usually much larger than the fabricated devices, can be expensive to fabricate, or do not emit in the visible spectrum. Our work was motivated by the need to have emission sources that can be easily integrated intro micro- and nanofluidic devices. Our goal was to fabricate such a device. Our demonstration of the emission from electrospun light-emitting nanofibers was the first step towards this goal."

The basic impact of this work is in the development of subwavelength emission sources. Although the immediate application would be for lab-on-a-chip devices, such light emitters could also be used in combination with flexible and conventional electronics wherever a tiny light source is needed.

In the past, some subwavelength light sources have been fabricated with dimensions similar to those achieved by the Cornell group, but the processes used to make them were highly involved and more expensive. Using an electrospinning technique for fabricating nanofibers is both easier and cheaper than high-resolution electron beam lithography. Consequently, the nanoscale emission sources that Moran-Mirabal and his colleagues made are easily fabricated and inexpensive.

Electrospinning is a well-developed electrohydrodynamic method used to produce micro- and nanofibers from a variety of dissolved materials without the need for expensive fabrication methods. Within the class of solution processable organic electroluminescent materials, ionic transition metal complexes (iTMCs) have emerged as materials that allow the fabrication of efficient, single-layer light-emitting devices employing air-stable electrodes.

"Because of their characteristics, iTMC light-emitting devices make attractive candidates for on-chip light sources," says Moran-Mirabal. "In this sense, having a light-emitting nanofiber based on an iTMC could provide a point source emission profile, with the axial dimension restricted by the iTMC operational mechanism and the radial dimension given by the diameter of the nanofiber."

The researchers also studied the conductance/emission properties of the fibers and found that the intrinsic properties of the luminescent compound, in particular the bandgap and ion mobility, determine the device's response. Moran-Mirabal describes how the fibers were successfully lit on devices containing gold interdigitated electrodes (IDEs) with interelectrode gaps of 5 μm and 500 nm. "Light emission from the fibers spun on 500 nm IDEs was readily detectable with a CCD camera with voltages as low as 3.2 V and visible to the naked eye at 4 V. Emission from the fibers was found to be highly confined to planar regions 240×325 nm^2 or smaller, with imaged emission areas small enough to be limited by diffraction of the microscope."

The most immediate challenge for the researchers is the stability of the light sources in aqueous solutions. Plans for future research include the development of emitters that have different wavelengths, and multiplexing of light emission into different channels.

Featured scientist: José Manuel Moran-Mirabal
Organization: School of Applied and Engineering Physics, Cornell
   University, Ithaca, NY (USA)
Relevant publication: José M. Moran-Mirabal, Jason D. Slinker, John
   A. DeFranco, Scott S. Verbridge, Rob Ilic, Samuel Flores-Torres,
   Héctor Abruña, George G. Malliaras, H. G. Craighead: Electrospun
   light-emitting nanofibers. *Nano Lett.*, **7**, 458–463.

3.7 Nanoantennas Lighting up Molecules

Instead of using nanofibers, another group of scientists has demonstrated the use of nanoparticles for nano-optical applications. Their inspiration came from radio antennas.

In conventional electronics, the interconnection between locally stored and radiated signals, for example radio broadcasts, is formed by antennas. Antennas play a key role in our modern wireless society. The electromagnetic waves sent and received by antennas are the messages that enable communication between electronics. Antennas with a wide variety of sizes make it possible for us to receive radio broadcasts, watch television, and talk to others using cellphones. For effective communication, the antenna needs to direct signals towards their intended target and, conversely, collect signals from a desired source.

Researchers have shown that the concept of an antenna is equally applicable to directing the visible light sent out by a single molecule. For an antenna to work with visible light, it needs to be millions of times smaller than a conventional radio antenna. In this case, it is only 80 nm long. By placing the antenna near an individual molecule, the light from that molecule is redirected; the molecular message can be steered to a desired target, making efficient communication possible.

These novel nanoantennas have important implications. In (bio)sensing, light can be sent to a molecule and, in return, its response can be aimed at a detector.

Furthermore, the antennas can be part of efficient nanosized light sources. Finally, it is interesting to see that the antenna designs that have revolutionized communication keep finding new applications, this time at the nanoscale.

"We have shown that a metal nanoparticle functions as an optical antenna for a single molecule," says Niek van Hulst. "We demonstrated that the light emitted by a molecule is directed by an optical monopole antenna, just as radio waves are directed by a radio antenna. The direction in which the light is emitted depends on the antenna design, not on the molecule."

Van Hulst, a professor at the Institute of Photonic Sciences (ICFO) in Barcelona, Spain, explains that there are two ways in which these findings advance the field of nano-optics: by increasing the understanding of optical antennas and by experimentally demonstrating them. "First, by studying the most elementary system—an individual molecule coupled to a simple monopole antenna—we can see how an optical antenna works. The resulting direction of emission was not anticipated and will change the interpretation of previous and future experiments. Second, the experimental demonstration serves as a proof of principle and gives a clear guideline of how to proceed and implement the large available library of radio antennas for applications in nano-optics."

Tim Taminiau, a PhD student in van Hulst's group, remarks that the concept of an antenna and directed emission has of course been known for over a century. "Our demonstration on the nanoscale with visible light coming from molecules is a first," he says. "It is the first experiment where a single molecule is brought near a well-defined optical antenna and the effects on the direction of the molecule are studied. There are several differences between working with visible light rather than radio frequencies. For example, metals behave differently at optical frequencies and this has to be taken into account in the antenna design. The experimental demonstration presented here shows that, despite the differences, ideas from traditional antenna theory can be applied in nano-optics."

Taminiau explains that the original motivation of the research team might seem unrelated at first: "We set out to build a better microscope," he says. "By building an optical antenna for a single molecule, one can communicate efficiently with that molecule. The antenna can be moved around; we can select the molecule we communicate with. Since the antenna is so small, we can do this with very high precision. The antenna thus acts like a high-resolution microscope (with a resolution of $\sim 20\,\mathrm{nm}$, the antenna radius). However, to use the antenna in this way, it is first crucial to understand how it works!"

Optical antennas can be used to efficiently send light to and receive light from nanosized systems. Van Hulst points out that one could use optical antennas in a sensor to bring light to a small active area and direct the response obtained towards a detector. "Antennas could be used to create efficient and directed nano-optical light sources," he says. "When the antenna position can be varied they can be used in high resolution (sub-diffraction limited) optical microscopes."

This work opens several avenues of exploration. Since antennas can interact with single molecules, it will be intriguing to see the applications of optical antennas in quantum optics and quantum information. Taminiau points out that the ability of an antenna to enhance the interaction of a molecule—or atom, ion, or quantum dot—with light is a particularly interesting prospect for quantum optics.

"Having demonstrated an elementary antenna, we can now move on to incorporate more complicated and specific antennas into nano-optics," says van Hulst. "One can think of narrow-band cavity antennas or highly directive Yagi–Uda antennas (TV antennas) on the nanoscale. For now, it is interesting to see how old concepts lead to novel tools on the nanoscale."

> *Featured scientists:* Niek van Hulst, Tim Taminiau
> *Organization:* ICFO—Institut de Ciències Fotòniques, Barcelona
>   (Spain)
> *Relevant publication:* T. H. Taminiau, F. D. Stefani, F. B. Segerink,
>   N. F. van Hulst: Optical antennas direct single-molecule emission.
>   *Nat. Photon.*, **2**, 234–237.

3.8   Cut and Paste with Single Molecules

As we saw in Chapter 2, the long-term vision of revolutionary bottom-up nanotechnology is based on two different concepts of molecular assembly technology. One follows Nature's blueprint, which uses molecular recognition for self-assembly of nanoscale materials and structures; the other is human-made and uses instruments to assemble nanoscale building blocks into larger structures and devices. In contrast, the most common nanoscale fabrication techniques used today, for instance in the sub-100 nm semiconductor industry, are top-down approaches where fabrication technologies such as lithography or stamping are used. Although top-down techniques can be highly parallel— semiconductors are a good example—it is not feasible to control single molecules with them.

Researchers in Germany have successfully demonstrated a technique that fills the gap between top-down and bottom-up fabrication. Hermann E. Gaub, head of the Biophysics and Molecular Materials Group in the Physics Department at the Ludwig Maximilians University (LMU) in Munich, together with Elias Puchner and colleagues from the university's Center for Nanoscience and the Center for Integrated Protein Science Munich, combined the precision of AFM with the selectivity of DNA interaction to create freely programmable nanopatterns of DNA oligomers on a surface and in an aqueous environment.

"In the past, great efforts have been put into creating DNA structures like the so-called DNA origami or crystals composed of nanoparticles," Gaub tells us. "However, these approaches exclusively rely on self-assembly and are purely bottom-up. They don't allow control over single molecules and the structures that are formed are predetermined by the design of the experiment."

What the LMU researchers did was to create a DNA scaffold by picking up biotin-bearing DNA oligomers with an AFM tip and depositing them, one by one, in a desired pattern on a surface, basically creating a pattern of attachment points for fluorescent semiconductor nanoparticles conjugated with streptavidin.[2] When the sample with the DNA scaffold is incubated with a solution of fluorescent nanoparticles, a rapid self-assembly process of these particles on the predefined scaffold takes place. The team calls this technique *single-molecule cut-and-paste* (SMCP).

"By using DNA oligomers with biotin as a functional unit, we were able to create freely programmable patterns of binding sites to which, in the next step, streptavidin-conjugated nanoparticles were allowed to bind," explains Gaub. "This new approach establishes independence of the DNA-transporter system and, in principle, allows the assembly of any nano-object that is conjugated with streptavidin in arbitrary structures. Our work therefore expands the toolbox of nanobiotechnology."

Nanoparticle self-assembly guided by specific molecular interactions has been used very successfully in the past to design complex structures with novel functions, promising a rich field of new applications. Gaub and his collaborators have expanded this concept by using a scaffold and demonstrated that molecule-by-molecule assembly of a binding pattern, combined with the self-assembly of semiconductor nanoparticles guided by molecular interactions, is a straightforward and very general way of creating nanoparticle superstructures.

The German scientists point out that the assembly of planar nanoparticle structures of arbitrary design can easily be accomplished in this way. "An expansion into the third dimension appears challenging but achievable," says Gaub. "Covalent cross-linking of the DNA oligomers after hybridization can be employed to stabilize the scaffold, and multifunctionality of the nanoparticle attachment sites may be used to build subsequent layers of structures. This could lead to a new dimension of complexity and novel effects."

Gaub also points out that in this study the team always used the biotin–streptavidin interaction as the coupler, and therefore only single-component structures were assembled. "However, since a multitude of couplers with orthogonal affinities is available, the assembly of multicomponent structures would be straightforward."

This approach is geared not so much towards assembling large numbers of units as towards designing individual assemblies with tailored properties and online control of the arrangement. "Especially appealing is the assembly of active players such as enzymes," says Gaub. "Synthetic biology approaches now become possible, where enzymatic cascades are assembled at will. Coupling between the different enzymes is controlled by vicinity and diffusion of the products, much as it is at the membranes of organelles. With our technique

---

[2]The small bacterial protein streptavidin is commonly used for the detection of various biomolecules and it binds with high affinity to the vitamin biotin. The strong streptavidin–biotin bond can be used to attach various biomolecules to one another or on to a solid support.

the typical distances between the constituents of natural cascades become accessible for the first time."

Research at Gaub's Biophysics and Molecular Materials Group aims towards single-molecule synthetic biology. The observation of single biomolecules, such as enzymes, offers a detailed and very fundamental understanding of biological processes on the nanoscale. In combination with their directed assembly, it might be possible to construct and observe networks of enzymes or signaling cascades.

Notwithstanding its tremendous potential, scientists are only taking the first steps in this direction and there are numerous challenges to be overcome. Further developments of the LMU scientists' SMCP technique might include the improvement of optical resolution by combining SMCP with super-resolution techniques, the confinement of optical excitation to the zeptolitre range by means of nanoantennas, or zero-mode waveguides and biochemical strategies for the modification of enzymes.

Featured scientist: Hermann E. Gaub
Organization: Biophysics and Molecular Materials Group, Physics
   Department, Ludwig Maximilians University, Munich (Germany)
Relevant publication: Elias M. Puchner, Stefan K. Kufer, Mathias
   Strackharn, Stefan W. Stahl, Hermann E. Gaub: Nanoparticle
   self-assembly on a DNA-scaffold written by single-molecule cut-and-
   paste. *Nano Lett.*, **8**, 3692–3695.

3.9   Blowing Bubbles—from Clingfilm to Nanoelectronics

One of the most common methods of film manufacture is *blown film extrusion*. This process, by which most commodity and specialized plastic films are made for the packaging industry, involves extrusion of a plastic through a circular die, followed by bubble-like expansion. The resulting thin tubular film can be used directly or slit to form a flat film.

Nanoscientists have found a way to use this very common and efficient industrial technology to potentially solve the problem of fabricating large-area nanocomposite films. Currently, the problems with making thin film assemblies are either the production cost of using complex techniques such as wet spinning or the unsatisfactory results of unevenly distributed and lumped nanoparticles within the film. The new bubble film technique results in well-aligned and controlled-density nanowire and CNT films over large areas. These findings could finally open the door to affordable and reliable large-scale assembly of nanostructures.

"Our findings demonstrate a method—applicable to all nanotube and nanowire materials—of producing uniform, aligned structures on a size scale several orders of magnitude larger than achieved previously and moreover with much larger sizes possible," Charles Lieber, a professor in the Department of Chemistry and Chemical Biology at Harvard University, explains. "The main motivation of this work was to develop a highly flexible and scalable approach to assemble uniform, aligned arrays of nanowires and nanotubes to open up possible applications in biological sensors and displays, for example."

"The key finding in our work is that homogeneous polymer suspensions of 1-D nanomaterials (CNTs and nanowires) can be expanded into bubbles, much like simple soap bubbles, in which the nanotubes and nanowires are uniformly aligned along the bubble expansion (longitudinal) direction with a density tuned by the solution concentration," adds Anyuan Cao, a scientist in the Department of Civil and Environmental Engineering at the University of Hawaii. "Controlled assembly of these nanomaterials has been a bottleneck for many potential applications, especially for micro- and nanoelectronic systems where regular distribution of aligned nanotubes or nanowires is required."

Previously, researchers have developed methods to produce aligned materials, but only on a centimetre scale or less. Here, the team have demonstrated aligned films of nanowires and nanotubes orders of magnitude larger in scale, and in fact on a scale relevant to many commercial applications. Although blown film extrusion is a well-established technique in industry, this is the first time nanomaterials have been introduced into polymers and blown bubbles with uniform distribution and alignment. The new technique represents a real breakthrough in nanocomposite film fabrication, something researchers have worked on for years but always restricted to a very small scale.

In contrast, the new research demonstrates blown bubble films with well-aligned and controlled-density nanowires and CNTs on flexible plastic sheets up to 225×300 mm, on highly curved surfaces, and also suspended across open frames. "We believe that these films should be scalable to much larger, metre-scale sizes by controlling the steps of bubble formation and expansion, as is currently done with homogeneous polymers, and that our approach could be extended to enable 3-D structures," says Lieber. "Moreover," says Cao, "our blown bubble film approach can assemble aligned structures on virtually any substrate, thus opening up the way to many unique applications on highly flexible substrates."

The approach to creating these bubble films consists of three basic steps: (1) preparation of a homogeneous, stable, and controlled concentration polymer suspension of nanowires or CNTs; (2) expansion of the polymer suspension using a circular die to form a bubble at controlled pressure and expansion rate; (3) transfer of the bubble film to substrates or open frame structures. The films could be transferred to either rigid or flexible substrates during the expansion process.

In the present process, relatively low concentrations of CNTs and nanowires (less than 1%) are dispersed in the polymer suspension. As a result, the distance between individual nanostructures is relatively large (more than $2\,\mu$m). "You need at least 4–5% by weight to get a good mechanical structure, but this is quite achievable, for example using surfactants or alternative polymers. Real advances should now be made quickly by learning from other areas," says Lieber.

Generally, this technique would allow fabrication of large-area thin film products with all types of nanoscale components. The bubble films could be further processed to make arrays of nanoelectronic or optical devices, sensors, and field emitters for flat panel displays. Bubble films coated on plastic substrates could be fabricated into flexible micro- or nanoscale systems. Introducing nanomaterials could also produce reinforced plastic films with better mechanical properties. The regular alignment of CNTs within the film would give it high tensile strength, potentially higher than Kevlar, and would make this an intriguing material for many applications, from airplane wings to protective clothing and armor.

According to Cao, the team has begun exploring several areas including fabrication of nanosystems with distinct electrical or optical properties, and use of different polymers to facilitate subsequent device fabrication. "For example, we have developed bubbles based on the photopolymer PMMA that can be directly integrated into a modern microfabrication technique to pattern electrodes."

He also cautions that their research faces challenges in the near future: "A better understanding of the mechanism of nanostructure alignment during the bubble-blowing process is necessary for further optimizing the process, and materials and instrumentation need to be developed for large-scale production."

> *Featured scientists:* (a) Charles Lieber, (b) Anyuan Cao
> *Organization:* (a) Department of Chemistry & Chemical Biology, Harvard University, Cambridge, MA (USA); (b) Department of Civil and Environmental Engineering, University of Hawaii, Honolulu, HI (USA)
> *Relevant publication:* Guihua Yu, Anyuan Cao, Charles M. Lieber: Large-area blown bubble films of aligned nanowires and carbon nanotubes. *Nat. Nanotechnol.*, **2**, 372–377.

3.10   Sliding Away in the Nanoworld

When two surfaces approach each other in air, they attract. This phenomenon is due to *van der Waals* forces that affect the interaction between atoms and molecules. Without these very weak intermolecular forces, life as we know it would be impossible. They are responsible for a number of properties of molecular compounds, including crystal structures, condensing, melting and boiling points, surface tension, and densities. Intermolecular forces form macromolecules such as enzymes, proteins, and DNA into the shapes required for biological activity.

Van der Waals forces can also be repulsive, for instance when two surfaces approach each other in liquid: the same force that causes attraction in air (and which is responsible for so-called stiction and adhesion) can be made repulsive by choosing the right combination of surface materials and intervening liquid.

This force has the characteristic that it increases very rapidly with very small changes in separation when the surfaces are close to each other. Researchers have shown that if repulsive van der Waals forces exist between two surfaces before they come into contact then friction is essentially precluded and *supersliding* is achieved. This opens the possibility that, in certain material systems, the controlled use of repulsive van der Waals forces could be a way to reduce friction, or even eliminate it.

Friction forces act wherever two solids touch. In simple terms, friction is a force that slows things down—this is one of the reasons why you can't build a perpetual motion machine. In scientific terms, friction between two surfaces is caused by energy loss as atoms from the opposing surfaces smash against each other. In the presence of an attractive/adhesive force the atoms are in intimate contact, and much energy is lost in forcing the atoms to slide past each other. Friction is not just a physical phenomenon, it is also the source of huge economic costs: friction wear causes progressive damage between working machine parts. Friction also causes heat, which represents wasted energy. Lubrication reduces the effects of friction but, of course, lubricants are also a cost factor.

Mark Rutland, a professor of surface chemistry in the School of Chemical Science and Engineering at the Royal Institute of Technology (KTH) in Stockholm, Sweden, explains: "Our work clearly shows that two surfaces experiencing a repulsive surface force, which diverges at small separations, can slide essentially without friction. The number of systems in which repulsive van der Waals forces could occur is limited, but includes metal bearings in a PTFE[3] housing with an organic lubricant, and certain combinations of technically interesting ceramic materials. Recent proposals concerning the control of van der Waals forces, and the fact that they can be enhanced by electromagnetic radiation in the microwave and visible ranges, also open up the possibility of achieving friction-free sliding in a much wider range of systems."

The findings of Rutland and his collaborators are new in two regards: no one has ever used such a force to control friction, or demonstrated such astonishingly low friction on a metal. In fact, says Rutland, only a handful of groups have ever observed repulsive van der Waals forces. Figure 3.3

The advance reported by Rutland and his collaborators—Lennart Bergström, a professor at Stockholm University and head of the Department of Physical, Inorganic and Structural Chemistry, and Dr Adam Feiler, a postdoc with Rutland at KTH's Surface Chemistry Group—is the ability to reduce the friction experienced by a metal to hitherto impossibly small levels (such a reduction is also possible for ceramic materials, and Rutland's team is working on this). The friction measurements presented here are of the same order as the lowest ever recorded friction coefficients in liquid, though achieved by a completely different approach.

---

[3] PTFE is a synthetic fluoropolymer with numerous applications, perhaps best known under the DuPont brand name Teflon™.

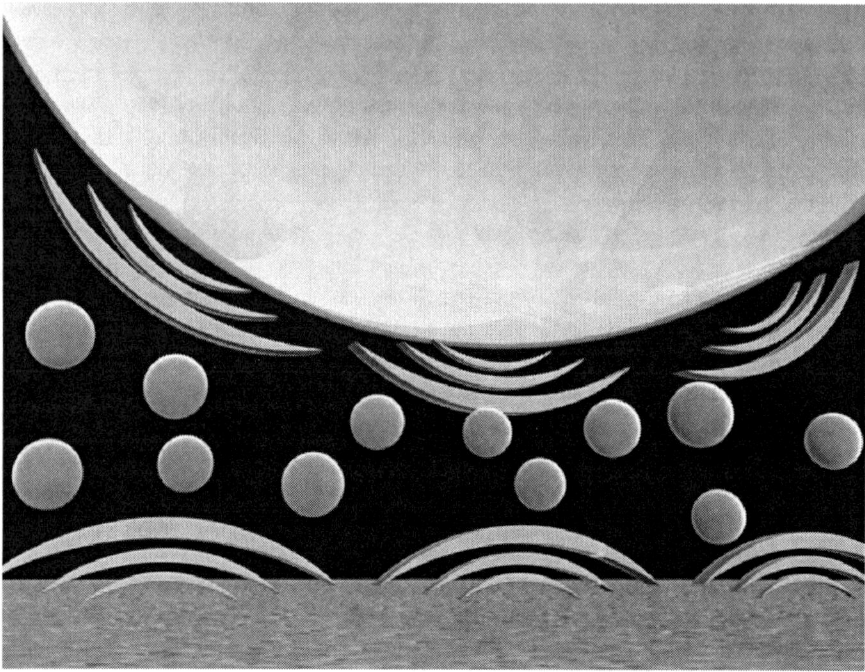

Figure 3.3 An artist's depiction of superlubricity using repulsive van der Waals forces (Image: Fredrik Dahlström, i3D)

"The steeply repulsive van der Waals force acts to prevent the surfaces achieving contact and they tend to slip past one other, rather like two magnets when you try to push them together," says Rutland.

In their experiments, the researchers attached a gold sphere to an AFM cantilever and forced it to interact with a smooth PTFE surface. They found that the normal surface forces are repulsive when cyclohexane is chosen as the intervening liquid between the gold sphere and the PTFE surface. When the refractive index of the liquid is changed, for instance by adding water, the force can be tuned from repulsive to attractive and adhesive.

If these findings could be turned into commercial reality, numerous high-friction systems could benefit: bearings could be made using metal–organic-liquid–PTFE to essentially preclude friction; tools for cutting and drilling could be modified to radically improve performance; in the nanoworld, MEMS and hard disks are obvious areas where one would want to reduce friction. However, the friction reduction demonstrated in these lab experiments can only be achieved with an organic film between the surfaces, which will render its practical application more challenging.

Rutland and his collaborators are working on extending the range of systems where such forces can be induced, and finding new, environmentally friendly, nontoxic organic liquids with suitable dielectric properties.

Featured scientist: Mark Rutland
Organization: School of Chemical Science and Engineering, Royal
  Institute of Technology (KTH), Stockholm (Sweden)
Relevant publication: Adam A. Feiler, Lennart Bergström, Mark W.
  Rutland: Superlubricity using repulsive van der Waals forces.
Langmuir, **24,** 2274–2276.

CHAPTER 4
Learning from Mother Nature

Nature is a successful design lab with millions of years of research experience, so it is quite surprising that scientists haven't tried harder to copy some of Nature's successful and impressive blueprints. The list of commercialized biodesign-inspired products is very short. The most famous is Velcro, the hook-loop fastener that was invented in 1945 by Swiss engineer George de Mestral. The idea came to him after he took a close look at the burrs (seeds) of burdock which kept sticking to his clothes and his dog's hair on their daily summer walks in the Alps. He examined their structure and saw the possibility of binding two materials reversibly in a simple way.

Today, there are quite a number of terms such as bionics, biomimetics, biognosis, biomimicry, or even 'bionical creativity engineering' that refer to more or less the same thing: the application of methods and systems found in Nature to the study and design of engineering systems and modern technology. The use of design concepts adapted from Nature is a promising new route to the development of advanced materials and, increasingly, nanotechnology researchers find nanostructures a useful inspiration for overcoming their design and fabrication challenges. Because biological structures are the result of hundreds of thousands of years of evolution, their designs possess many unique merits that would be difficult to achieve by a completely artificial simulation. However, utilizing them as biotemplates and converting them to inorganic materials could be a highly reproducible and low-cost process for fabricating complex nanostructures with unique functions. Complex functional systems are still out of reach, but the replication of biological structures is making good progress. Good examples are fabricating antireflection nanostructures by replicating fly eyes, or imitating the adhesive capabilities of a gecko's foot. Overall, the ability to replicate biological shapes with nanoscale precision could have profound implications in tissue engineering, cell scaffolding, drug delivery, sensors, imaging, and immunology.

'Reverse engineering' is another term used by researchers in this field—the process of discovering the technological principles of a device or system

RSC Nanoscience and Nanotechnology No. 8
Nano-Society: Pushing the Boundaries of Technology
By Michael Berger
© Michael Berger 2009
Published by the Royal Society of Chemistry, www.rsc.org

through analysis of its structure, function, and operation, often by taking it apart and analyzing its workings in detail. This approach is a common practice among industrial companies who use it to analyze the competition's products, be they cars or MP3 players, to understand where the latest product improvements come from and how individual components are made.

An increasing number of scientists apply a similar approach to Nature's micro- and nanoscale systems. They believe that learning from natural designs is more likely to provide the cues for designing practical nanodevices than simply adopting a 'trial and error' approach. The basic idea is that natural materials and systems can be adopted for human use beyond their original purpose in Nature. Popular examples of this approach are DNA computing and DNA 'nanomachines'.

While you won't be seeing a functional DNA computer sitting on your desktop for quite a while yet, other examples of reverse biophysics have already proved quite useful, for instance the use of individual red blood cells as reliable, ultrasensitive mechanotransducers.

The following section describes several examples where Nature's design principles are inspiring scientists.

4.1 Copying the Greatest Nanotechnologist of All

Current micro- and nanofabrication approaches can reproducibly fabricate precise, regular shapes out of a variety of robust materials, but the morphologies are typically limited to forms such as pillars, lines, pyramids, or other geometric shapes. These shapes do not mimic the structural complexity that can be created by self-assembled organic and biological structures such as viruses.

Joseph M. DeSimone tells us that his research group is able to accurately replicate shapes of naturally occurring nanoscale objects in other materials. DeSimone is a professor of chemistry and chemical engineering at the University of North Carolina at Chapel Hill as well as Director, NSF Science and Technology Center for Environmentally Responsible Solvents and Processes and Director, Institute for Advanced Materials, Nanoscience and Technology.

"Naturally occurring supramolecular objects, such as proteins, micelles, and viruses, exhibit sophisticated morphological shapes or surface motifs that conventional synthetic and fabrication techniques cannot replicate," DeSimone explains. "These structures owe their interesting shapes and shape-related properties largely to noncovalent chemical interactions that can produce unique, 'evolutionarily designed' shapes with nanometer precision."

So far, it has been impossible for scientists to replicate Nature's self-assembly process when trying to control the complex nanoscale shape of organic or inorganic materials. On one hand, the chemical structure of each component has to be carefully designed and precisely synthesized to ensure that the desired morphology is obtained. On the other hand, "Even if the chemical design of the material is sufficient to promote the desired assembly process, factors such as temperature, solution purity, and kinetic limitations can significantly, and often detrimentally, affect the resulting morphology of bottom-up self-assembled materials," says DeSimone. "Alternatively, traditional micro- and nanofabrication approaches can reproducibly fabricate precise, regular shapes out of a variety of robust materials, but the morphologies are typically limited to geometric shapes and therefore cannot easily mimic the structural complexity that can be created by self-assembled organic and biological structures, such as viruses."

DeSimone's group developed a nanofabrication method that is able to reproduce shapes normally associated with self-assembly using robust nanoscale replication methods, thereby combining the morphological sophistication of the natural world with the scalable processing technologies associated with lithography.

"We use an extremely low surface energy, minimally adhesive fluoroelastomer—photocurable perfluoropolyether (PFPE)—to replicate naturally occurring objects," explains DeSimone. "The naturally occurring object, what we call a 'master template', is replicated in the fluoropolymer by pouring the curable fluoropolymer resin over the master template and photopolymerizing the resin into a flexible mold that transfers the details of the master morphology into the mold. The mold precursor resin will spontaneously spread on almost all substrates found in Nature, including organic materials. After spreading on the self-assembled materials to 'mimic' the precise shape of the self-assembled material, the resin is polymerized to capture the shape in the mold."

The mold is then released from the self-assembled master templates and can be used to replicate these shapes into other materials. The fluoropolymer resin is inherently noninteracting with the biological materials, so it does not perturb the fragile self-assembled structure. In this way one can get a very accurate replica of the self-assembled master template, in some cases down to less than 2 nm in height.

Being able to reproduce the sophisticated shapes of the natural world could open the door to biomedical applications where these constructs could potentially can be used in place of, for instance, a live virus. As for next steps, DeSimone's group wants to explore the impact of shape on biological recognition.

"For example, it would be a real breakthrough to show that a replicated bio-interface has some of the same recognition properties as a natural bio-interface," says DeSimone. "Also, the ability to use attached chemical ligands (targeting agents, signaling molecules) and to use molecular imprinting in conjunction with shape-specific nanoreplication could lead to truly biomimetic materials such as artificial viruses."

Featured scientist: Joseph M. DeSimone
Organization: Department of Chemistry, University of North Carolina at Chapel Hill, NC (USA)
Relevant publication: Benjamin W. Maynor, Isaac LaRue, Zhaokang Hu, Jason P. Rolland, Ashish Pandya, Qiang Fu, Jie Liu, Richard J. Spontak, Sergei S. Sheiko, Richard J. Samulski, Edward T. Samulski, Joseph M. DeSimone: Supramolecular nanomimetics: replication of micelles, viruses, and other naturally occurring nanoscale objects, *Small,* **3,** 845–849.

4.2 Biophysics in Reverse

Reliable measurement of miniscule forces is increasingly recognized as a very sensitive way to characterize molecular and cellular interactions in biology and medicine, and in the pharmaceutical industry. Changes in cell mechanical properties have been shown to correlate with metastatic potential and stem-cell differentiation. Forces play a dramatic role in drug targeting and delivery. One important lesson learned in recent years is that biochemistry alone cannot predict how the interaction between two reactants is affected by force—and yet most biomolecular interactions are actually stress-bearing in one way or another.

A technique originally pioneered by Evan Evans at the University of British Columbia and known as the *biomembrane force probe* (BFP) uses the deformation of a red blood cell under tension as a force sensor. This was one of the earliest examples of adopting a biomaterial to a useful task beyond its natural function. BFP spans an arc from theoretical membrane physics and numerical analysis, over innovative instrument development and advanced experimentation, to the eventual adaptation of a well-characterized biological object as a device capable of high-resolution nanomechanical measurements.

"Examining how a red blood cell can be used as a force transducer and could be applied to measure forces between individual molecules and cells or between molecules, we verified the viability of using pipette-aspirated red blood cells as ultrasensitive force-measuring devices," explains Volkmar Heinrich. "In conjunction with our custom-built, cantilever-based 'horizontal atomic force microscope (AFM)', we were able to vary critical parameters such as the aspiration pressure and the position of cell–cantilever contact. Our measurements yielded excellent agreement with our theoretical predictions, proving that among other applications, this red blood cell transducer can be used to reliably calibrate the spring constants of AFM cantilevers."

Heinrich, assistant professor in the Department of Biomedical Engineering at UC Davis, addresses the specific case of the BFP transducer's precision. His group's work has tackled two issues simultaneously: it refines BFP and similar red-cell-based force instruments, and it also presents a new type of cantilever-based force probe that can be viewed as a combination of a horizontal AFM and an automated micropipette-manipulation system.

"It was this combination that allowed us to directly test the red-cell force transducer against the elastic microlevers used in commercial AFMs," says Heinrich. "Our tests have validated that, in conjunction with the correct theory, red blood cells can indeed be used as reliable mechanical nanodevices. At the same time, we were able to turn the tables of conventional nanobiophysics and technology by demonstrating that a biological object could accurately calibrate a human-made device, in this case an AFM cantilever. In fact, a careful look at the data reveals that this method of cantilever calibration may well be more precise than normal calibration techniques."

Heinrich mentions that his team views this particular calibration as an added bonus to their more fundamental innovations in theory and instrument design.

"Of course, it is not the most practical method of calibrating AFM cantilevers, and it is unlikely to become a standard technique. That said, I suspect that whenever we will want to calibrate a cantilever with excruciating accuracy in our lab in the future, we might just fall back on this method."

Heinrich explains that red blood cells exhibit quite impressive mechanical characteristics. They remain in the human blood circulation for 3–4 months. During this time, they are exposed to continuous stress. Each red blood cell has to squeeze numerous times through capillaries that are narrower than its own largest dimension. It is hard to imagine a synthetic material with the same combination of mechanical integrity and flexibility. Therefore, it is perhaps not surprising that researchers have begun to explore alternative uses of this amazing biomaterial.

"When a red blood cell is sucked on to the tip of a micropipette of the right caliber, its mechanical properties dictate that it forms a spherical part outside the pipette and a small cylindrical projection inside the pipette," says Heinrich. "Like a pressurized balloon, the spherical part exhibits a certain amount of springiness. Based on our theoretical work, we are able to precisely predict this springiness—in other words, the relationship between the cell deformation and the force exerted by the cell. Even better, the springiness of the spherical, balloon-like part depends on the suction pressure applied through the pipette. Since we can precisely control this pressure, we can adjust the amount by which the cell deforms when squeezed against another object. Taken together, this means that not only do we know the force that is exerted by the cell on the object, we can even fine-tune it by changing the suction pressure in the pipette."

Skeptics have wondered how well this idea might really work in practice. Are the relevant red-cell physics properly accounted for? Do all red blood cells behave in the same way in this configuration? The UC Davis team's research answers these questions in the affirmative. "The only requirements for a red blood cell to operate reliably as an ultrasensitive force probe are a fluid membrane and cell interior, reasonable lubrication between the membrane and pipette wall, and at least a moderate aspiration pressure. If these conditions are fulfilled, the only other things that one needs to know are the value of the aspiration pressure and the overall cell dimensions. The amazing constitution of red blood cells ensures that the rest is routine."

Heinrich points out that it is important to place this work within a wider perspective. Just over a decade ago, the possibility that one could directly measure the forces acting between molecules appeared remote. The advent of nanotechnology has dramatically changed this perception. Today there are a number of tools that can be used to characterize the nanomechanics of biomolecular and cellular interactions. Examples are optical tweezers, magnetic tractors, cantilever-based instruments like the AFM, and the BFP.

"With new tools came new concepts," he says. "It turns out that the mechanical rules that govern the nanoworld are quite different from our everyday, macroworld experience. New insights have taught us that the accurate characterization of the extremely small forces of 'weak' interactions is more demanding than at first thought. We know now that a large dynamic range of forces is required to properly characterize such weak interactions—ideally this range should be broader than what can be achieved with a single instrument. The future of the field of 'nano-interactions' will see an increased demand for very precise force-probe instruments with complementary dynamic ranges."

> *Featured scientist:* Volkmar Heinrich
> *Organization:* Department of Biomedical Engineering, UC Davis, CA (USA)
> *Relevant publication:* Volkmar Heinrich, Chawin Ounkomol: Biophysics in reverse: using blood cells to accurately calibrate force-microscopy cantilevers, *Appl. Phys. Lett.,* **92,** *153902.*

4.3 For Super-strong Dry Adhesives Look No Further than the Gecko

Animals that cling to walls and walk on ceilings owe this ability to micro- and nanoscale attachment elements. The strongest adhesion forces are encountered in geckos. For centuries, the ability of geckos to climb any vertical surface or hang from a ceiling with one toe has generated considerable interest. A gecko is the heaviest animal that can 'stand' on a ceiling, with its feet above its head. This is why scientists are intensely researching the adhesive system of the tiny hairs on its feet. On the sole of a gecko's toes there are some 1 billion tiny adhesive hairs called setae which are 3–130 im in length, splitting into even smaller spatulae which are about 200 nm in both width and length. It was found that these elastic hairs induce strong van der Waals forces. This finding has prompted many researchers to use synthetic microarrays to mimic gecko feet.

Recent work, mainly from the research groups of Ali Dhinojwala, Pulickel Ajayan, Meyyaa Meyyappan, and Liming Dai, as well as the Max Planck Institute for Metals Research in Germany, has indicated that aligned carbon nanotubes (CNTs) sticking out of substrate surfaces showed strong nanometer-scale adhesion forces. Although CNTs are thousands of times thinner than a human hair, they can be stronger than steel, lighter than plastic, more conductive than copper for electricity and diamond for heat. Applications of bio-inspired artificial dry adhesive systems with aligned CNTs could range from low-tech fridge magnets to holding together electronics or even airplane parts.

Liming Dai and Liangti Qu of the University of Dayton have demonstrated that vertically aligned single-walled CNT (SWCNT) arrays could be used for a successful synthetic approach to mimic gecko foot hairs to develop advanced dry adhesives with additional thermal/electrical management capabilities.

"Although aligned multiwalled CNTs (MWCNTs) have been widely studied for many years, synthesizing vertically-aligned single-walled CNT (SWCNT) arrays has been a recent research effort," Dai explains. "Also, aligned SWCNTs have just recently been used to mimic gecko feet as dry adhesives. Having an extremely high aspect ratio, exceptional mechanical strength, and excellent electronic and thermal properties, the aligned SWCNTs show potential for dry adhesion applications with additional electrical/thermal management capabilities. The smaller nanotube diameter could also allow an aligned SWCNT array to have more contact points per unit surface area than its multiwalled counterpart, leading to an enhanced adhesion force for the aligned SWCNT dry adhesives."

Dai and his group have demonstrated that their vertically aligned SWCNT arrays, measuring 4×4 mm, have a much higher achievable macroscopic adhesive force ($29\,\mathrm{N/cm^2}$) than a natural gecko foot (which achieves about $10\,\mathrm{N/cm^2}$). What this means is that only 150 pieces of these small SWCNT arrays, with a total contact area of about 5×5 cm—much less than the palm of a hand—would collectively be needed to hold a person weighing ~ 70 kg. Dai's group also found that these vertically aligned SWCNT dry adhesives showed fairly reversible semiconducting behaviors under load and excellent thermal resistance, because of the unique thermal and electric properties intrinsically associated with SWCNTs.

Dai says that aligned SWCNT dry adhesives with thermal and electrical management capabilities will open up avenues for many novel applications of CNTs, ranging from optoelectronics through household products (*e.g.* smart dry adhesives that are photosensitive or optoelectronically active) and robotic systems to electronic packaging.

> *Featured scientist:* Liming Dai
> *Organization:* University of Dayton Research Institute, Dayton, OH (USA)
> *Relevant publication:* L. Qu, L. Dai: Gecko-foot-mimetic aligned single-walled carbon nanotube dry adhesives with unique electrical and thermal properties, *Adv. Mater. (Weinheim, Ger.)*, **19**, 3844–3849.

4.4   Nature's Bottom-up Nanofabrication of Armor

Seashells are natural armor plating. They have to be tough because aquatic organisms are subject to fluctuating forces and impacts during motion or through interaction with a moving environment. Nacre (mother-of-pearl), the pearly internal layer of many mollusk shells, is the best example of a natural armor material that exhibits structural robustness, despite the brittle nature of its ceramic constituents. This material is composed of $\sim 95\%$ inorganic aragonite with a small percentage of organic biopolymer. Research at the University of South Carolina has revealed the toughening secrets of nacre: rotation and deformation of aragonite nanograins absorb energy in the deformation of nacre. The aragonite nanograins in nacre are not brittle, but deformable. These findings may lead to the development of ultra-tough nanocomposites, *e.g.* for armor material, by realizing this rotation mechanism.

Super-tough and ultra-high-temperature resistant materials are in critical demand for use in extreme conditions such as jet engines, power turbines, catalytic heat exchangers, military armor, aircraft, and spacecraft. Structural ceramics have largely failed to fulfill their promise of revolutionizing engines by the use of strong materials that withstand very high temperature. The major problem with ceramics as structural materials is their brittleness. Although many attempts have been made to increase their toughness, including the incorporation of fibers, whiskers, or particles, and zirconium oxide phase transformation toughening, currently available ceramics and their composites are still not as tough as metals and polymers. It has proved difficult to solve this problem by conventional approaches.

Nature has evolved complex bottom-up methods for fabricating ordered nanostructured materials that often have extraordinary mechanical strength and toughness. Nacre is one of the best examples. It has evolved through millions of years to a level of optimization not currently achieved in engineered composites. Nacre has a brick-and-mortar-like structure with highly organized polygonal aragonite platelets of a thickness ranging from 200 to 500 nm and an edge length $\sim 5\,\mu m$ sandwiched with a 5–20 nm thick organic biopolymer interlayer. The combination of the soft organic biopolymer and the hard inorganic calcium carbonate produces a lamellar composite with a 2-fold increase in strength and a 1000-fold increase in toughness over its constituent materials. These remarkable properties have motivated many researchers to synthesize biomimetic nanocomposites that attempt to reproduce Nature's achievements and to understand the toughening and deformation mechanisms of natural nanocomposite materials.

Xiaodong Li, who heads the Nanostructures and Reliability Laboratory at the University of South Carolina, and his team examined the role of nanostructures in the amazing properties of nacre. When the group initially reported[1] the discovery of nanosized grains in nacre, their functionality was entirely unknown.

"To reveal the secret recipe of nacre is not an easy job," says Li. "We developed a micromechanical tester that can be used inside an AFM. We performed tensile and bending tests on nacre *in situ* where the nacre surface was imaged simultaneously by the AFM. These discoveries—rotation and deformation of aragonite nanograins—clarify the previous misunderstandings in theoretical models and provide a nanoscale modeling boundary condition. This opens up opportunities to develop nacre-like ultra-tough materials." Li's findings could fundamentally change the way ceramic materials and structural components are prepared and open up new applications of ceramic materials in extreme environments.

---

[1] Xiaodong Li, Wei-Che Chang, Yuh J. Chao, Rizhi Wang, and Ming Chang: Nanoscale Structural and Mechanical Characterization of a Natural Nanocomposite Material: The Shell of Red Abalone, *Nano Lett.*, 2004, **4**(4), pp 613–617

Li points out that Nature has long been using bottom-up nanofabrication methods to form self-assembled nanomaterials that are much stronger and tougher than many synthetic materials. "Mother Nature knows best," says Li. "Nature has evolved highly complex and elegant mechanisms for materials design and synthesis. Living organisms produce materials with physical properties that still surpass those of analogous synthetic materials with similar phase composition. We need to turn our attention to Nature's designs and fabrication of materials. There is still a lot we can learn from it."

Featured scientist: Xiaodong Li
Organization: Nanostructures and Reliability Laboratory, University of South Carolina, Columbia, SC (USA)
Relevant publication: Xiaodong Li, Zhi-Hui Xu, Rizhi Wang: *In situ* observation of nanograin rotation and deformation in nacre, *Nano Lett.*, **6**, 2301–2304.

4.5   Peacock Feathers and Butterfly Wings Inspire Biotemplated Nanomaterials

Photonic crystals—also known as photonic bandgap material—are similar to semiconductors, except that the electrons are replaced by photons (*i.e.* light). By creating periodic structures out of materials that differ in their dielectric constants, it becomes possible to guide the flow of light through the photonic crystals in a way similar to how electrons are directed through doped regions of semiconductors. The photonic bandgap (which forbids propagation of a certain frequency range of light) gives rise to distinct optical phenomena and makes it possible to control light with amazing facility and produce effects that are impossible with conventional optics.

Researchers have been experimenting with several varieties of synthetic photonic crystals (*e.g.* opals made up of polystyrene, silica, and PMMA) as structural matrices to incorporate light-emitting materials such as lasing dyes or quantum dots, in order to achieve hybrid materials with tunable spontaneous emission. These structures are quite promising for enabling directional and tunable emission, which is essential for diverse applications in optoelectronics and optical communications.

Artificial opals are lacking in pattern variety and their fabrication requires very expensive equipment and sophisticated processes. These problems limit their applications. In contrast, natural photonic crystals have various patterns that are quite promising structural matrices for creating novel optical devices. One example is peacock feathers, whose iridescent colors are derived from the 2-D photonic crystal structure inside the cortex. "By varying the lattice constant and the number of periods in the crystal structure, peacock feathers provide several 2-D photonic crystal structures with different colors," says Di Zhang. "In our research, we embedded light-emitting nanoparticles into natural photonic crystals in order to fabricate novel, biomaterial-based optical devices with tunable spontaneous emission."

Zhang, a professor at the State Key Laboratory of Metal Matrix Composites at Shanghai Jiao-Tong University in China, and his group have chosen peacock feathers as the matrix to embed zinc oxide (ZnO) nanoparticles through an *in situ* approach. Zhang explains that both the surface keratin layer and the keratin component connecting melanin rods in the feather cortex can provide reactive sites for the formation of ZnO nanoparticles.

"We assumed that, in an ideal system, the spontaneous emission is tuned both by the embedded nanostructures and by the photonic crystal matrix of the peacock feather, which is famous for its ability to control light in the visible range," says Zhang. "We chose ZnO nanoparticles as the light-emitting entity with defect emission that spans the visible spectrum. Meanwhile, the ordered melanin arrays within peacock feathers are chosen as natural photonic crystals that have the ability to control visible light. In the resulting nanoZnO/peacock feather, the feather not only functions as the support for ZnO nanoparticles, but also should serve as the light controller according to its 2-D photonic crystal structure."

Zhang's group, as well as numerous other research groups around the world, are inspired by the biomineralization processes found in Nature—the process by which living organisms produce minerals. Whereas the fabrication of synthetic crystals often requires elevated temperatures and strong chemical solutions, Nature's organisms have long been able to lay down elaborate mineral structures at ambient temperatures. Being able to duplicate this natural production process would potentially allow for much simpler and greener fabrication technologies than the ones employed today.

Zhang explains that various biomaterials, such as amino acids, dipeptides, DNA, microtubules, and silk fibroins, have been investigated as ideal biomineralization substrates. "These biomaterials mainly serve as the reactive chemical template, the surface modifier, as well as the bottom-up assembly director to control the formation and assembly of nanoparticles and nanoclusters in solution," he says. "However, further treatment, such as spin-casting and Langmuir–Blodgett deposition, is usually needed to obtain solid state products. In our research, we introduce a facile route to fabricate ZnO nanoparticles *in situ* in a solid state biosubstrate."

Natural photonic crystals contain abundant reactive sites according to their chemical components, and thus, the Chinese scientists hypothesized, they could act as the reactive chemical template and surface modifier during the synthesis of ZnO nanoparticles. Furthermore, since the reactive bioresidues involved are dispersed in the solid state structures, this could result in a nanoparticle/biomaterials hybrid nanocomposite that does not require further treatment.

Earlier, the researchers in Zhang's group had already introduced a number of biomaterials and biostructures into their research in order to fabricate functional hybrid nanocomposites with hierarchical nanostructures. They have, for instance, used eggshell membrane, wood, and other plant materials with hierarchical porous structures as templates to synthesize hierarchical porous functional nanomaterials.

"We have observed that the specific hierarchical porous structures influence the gas sensing and photocatalysis properties of biomorphic nanomaterials," says Zhang. "We also investigated silk fibroin fibers with string-like morphology and spherical and rod-shaped bacteria, as well as butterfly wings, peacock feathers, and diatom frustules with ordered photonic crystal structures to integrate and enhance the functionalities of inorganic nanoparticles. The resulting hybrid nanocomposites exhibit outstanding chemical or physical properties and have valuable applications in photocatalysis, gas sensing, ductile ceramics, and semiconductor technology."

Bio-inspired fabrication techniques are multidisciplinary efforts that have developed into an intersection of materials science, soft chemistry techniques, nanotechnology, and biotechnology. Scientists hope that the exploration of these areas will provide new possibilities for the rational design of various kinds of functional nanomaterials with ideal hierarchy and controllable length scales. Zhang expects that bio-inspired strategies integrating biotemplates, biomineralization, and biomimesis will be extensively developed in the coming years to obtain functional nanocomposites with hierarchical architectures and interrelated unique properties.

Featured scientist: Di Zhang
Organization: State Key Laboratory of Metal Matrix Composites,
  Shanghai Jiao-Tong University, Shanghai (PR China)
Relevant publication: Jie Han, Huilan Su, Chunfu Zhang, Qun Dong,
  Wang Zhang, Di Zhang: Embedment of ZnO nanoparticles in the
  natural photonic crystals within peacock feathers, *Nanotechnology,*
  19, 365602.

4.6 Modeling Molecular Force Generation in Cells

In much the same way that our bodies depend on bones for mechanical integrity and strength, each cell within our body is supported mechanically by a skeleton of composite materials called the *cytoskeleton,* which includes protein polymers (such as actin filaments) and motor proteins (myosin). Actin and myosin are key components in muscle contraction and cell motility. The cytoskeleton is an active material that maintains cell shape, enables some cell motion, and plays important roles in both intracellular transport and cellular division. The cytoskeletal system is not at thermodynamic equilibrium, and this nonequilibrium drives motor proteins that are the force generators in cells. Analogously to how our bones are held and moved by muscles, the cytoskeleton is activated by these molecular motors, which are nanosized force-generating enzymes.

In an ongoing effort to design and create a simplified, bottom-up model of the cytoskeleton, researchers have designed and assembled a biomolecular model system capable of mechanical activity similar to that of living cells. This work could serve as a starting point for exploring both model systems and cells in quantitative detail, with the aim of uncovering the physical principles underlying the active regulation of the complex mechanical functions of cells. It could also provide new fundamental insights and offer new design principles for materials science.

"We developed a simplified model of the active cytoskeleton by mixing actin with a cross-linker and myosin," says Fred MacKintosh. "This nonequilibrium, *i.e.* energy-consuming, material exhibits local contractions, similar to how they occur in most living cells." Although there had been many earlier studies of *in vitro* and model networks created from the filamentous proteins or biopolymers that make up the cytoskeleton, these did not include the essential active ingredient of molecular motors that generate internal forces within the network.

MacKintosh, who runs the Theory of Soft Matter and Complex Systems group at Vrije Universiteit Amsterdam in the Netherlands, is motivated by his group's theoretical efforts to develop quantitative models for cytoskeletal networks. In work performed in collaboration with Daisuke Mizuno, Catherine Tardin, and Christoph Schmidt at the University of Göttingen in Germany, the researchers present a quantitative theoretical model connecting the large-scale properties of the cytoskeletal actin–myosin material to molecular force generation.

As well as modeling the molecular force generation, the second major innovation in this work was the ability to directly and simultaneously measure both the mechanical response (by active microrheology) and the fluctuations (by passive microrheology), thereby allowing the researchers to test the fundamentally nonequilibrium nature of the system.

The findings demonstrate that a remarkably simple system, with just three components (myosin, actin, and ATP), can reproduce key phenomena also observed in far more complex living cells. More generally, the results demonstrate an adaptive/active material that can tune its own mechanical properties.

"This really opens the way to constructing more complex bottom-up models of cellular structures and activity, which can then be quantitatively correlated with measurements in cells to elucidate the fundamentals of cell mechanics," says MacKintosh. He points out that one of the surprising findings of this research was the nearly 100-fold stiffening that occurred when the molecular motors became active. This reveals interesting possibilities for the design of 'smart' materials with tunable mechanical properties. This system also represents a sort of micron-scale isotropic muscle that can pull on its environment, much as when a cell crawls.

As for next steps, "One of our main interests, and I think of others following this work, will be to create more complex models of the cytoskeleton and try to mimic and duplicate the activity of real cells," says MacKintosh.

Featured scientist: Fred MacKintosh
Organization: Department of Physics and Astronomy, Vrije
    Universiteit, Amsterdam (The Netherlands)
Relevant publication: Daisuke Mizuno, Catherine Tardin, C. F.
    Schmidt, F. C. MacKintosh: Nonequilibrium mechanics of active
    cytoskeletal networks, *Science*, **315**, 370–373.

4.7 Nanoscale Water Pump Imitating Cell Pores

Nanofluidic channels that confine and transport tiny amounts of fluids are the pipelines that make the cellular activities of organisms possible. Nanoscale channels carry nutrients and water into cells and transport waste and water out. Body temperature, digestion, reproduction, fluid pressure in the eye, and water conservation in the kidney are only a few of the processes in the human body that depend on the proper functioning of cellular water channels. Special proteins called *aquaporins* can transport water through the cell membrane at a high rate while effectively blocking everything else—even individual protons, the nuclei of hydrogen atoms. The aquaporin channels are so narrow that no molecule larger than water can pass through, effectively forcing water molecules through like beads on a chain. A unique distribution of amino acid residues along the pore wall contributes to the channel's ability to move water quickly. To keep out molecules smaller than water there is also a chemical filter, formed by the specific orientation and distribution of the amino acid residues lining the pore. Thus water, and only water, flows freely through the aquaporin nanochannels, the direction of flow depending only on a change of relative pressure inside and outside the cell. This intriguing mechanism has attracted the attention of researchers who see it as a blueprint for the construction of nanoscale water pumps. A molecular dynamics simulation conducted by Chinese researchers proposes a design for such a molecular pump constructed with a CNT.

"Much progress has been made in moving water unidirectionally by designing systems with an imbalance of surface tension or a chemical or thermal gradient, but it is still difficult to make a controllable continuous unidirectional water flow," says Haiping Fang. "Our findings focus on SWCNT with appropriate radii in which water molecules exhibit single-file structure. By arranging discrete charges on the exterior of a SWCNT we can rotate the dipoles of water molecules, which then leads to a smooth concerted water motion through the center of the nanotube."

Since water is charge-neutral, it is not easily driven by an electric field in an environment full of thermal fluctuations. Building a water nanopump that actively transports water through nanochannels is technically difficult or perhaps even impossible using mechanisms comparable to those used in the macro world.

Fang, a professor of physics at the Shanghai Institute of Applied Physics, Chinese Academy of Sciences (CAS), and his collaborators propose a new strategy for building a water nanopump. This pump can push neutral water molecules to move in one direction by precisely placing extra charges on the surface of a nanotube. This strategy is inspired by the structure of aquaporins.

Such a water nanopump would have important applications including the desalination of seawater, chemical separation, water purification, *in vivo* sensing, and drug delivery. Figure 4.1

The method proposed by Fang and his colleagues places three positive charges asymmetrically at the vertical center of the SWCNT, two of $0.5\ e$ and one of $1\ e$ (e is the charge of an electron).

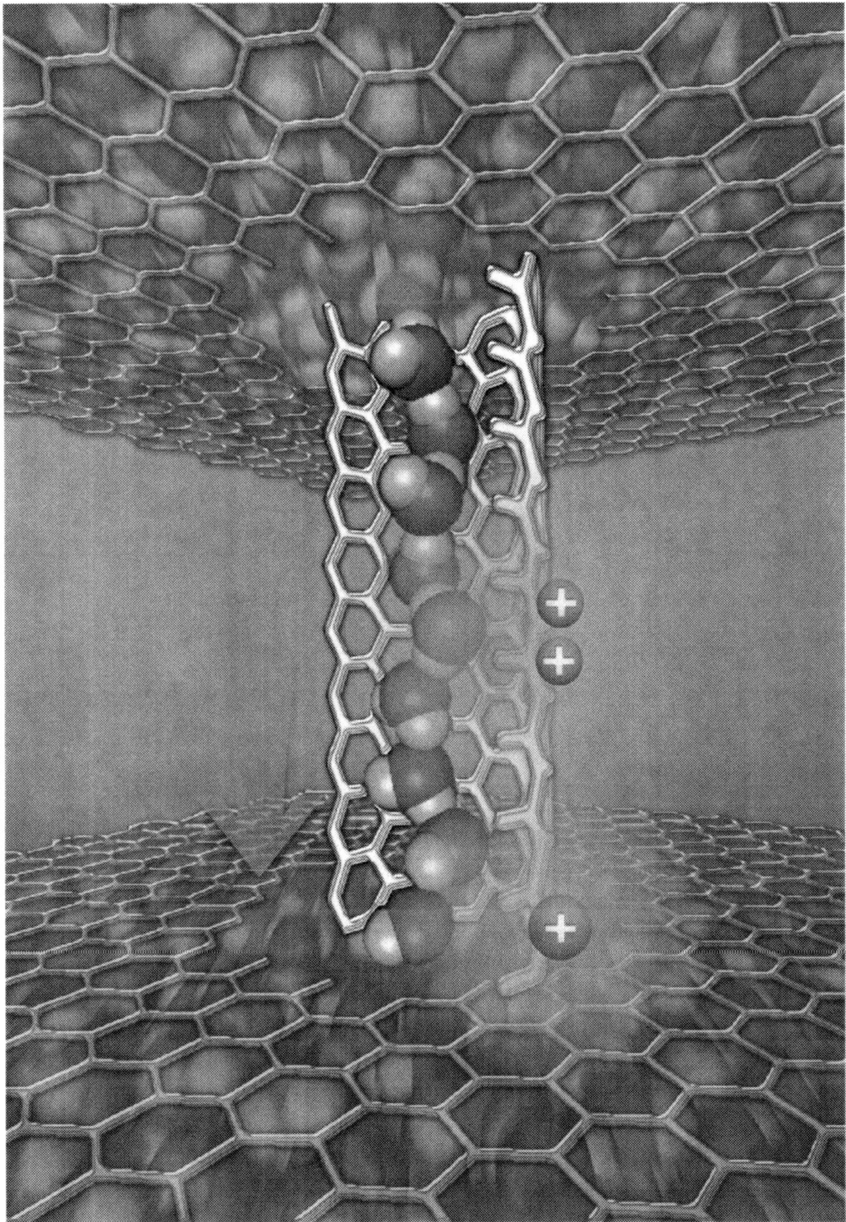

Figure 4.1 Schematic of a charge-driven molecular water pump. Water molecules move in single file through a SWCNT, driven by three electric charges located on the outside of the nanotube. (Image: Dr Haiping Fang, CAS)

"To understand how the particular charge configuration affects the water dynamics in the channel, we compared the behavior of the main system—*i.e.* the three-charge configuration described above—with systems having only some of the charges," Fang explains. "We found that the method of using two $+0.5$ e charges positioned in the middle of the channel is very important. If we merge them into one $+1.0$ e charge the permeation behavior is qualitatively different."

Fang points out that extra energy is required to constrain the charges at their original positions, which is the key to the pumping ability. "We found that the electrostatic forces exerted by water molecules on the three charges are 533 piconewton (pN), 313 pN, and 300 pN. Without the constraint, the charges would be forced away and the net flux would vanish."

According to the Chinese researchers, the pumping ability described here can be directly applied to insulator nanochannels and could also be extended to semiconductor nanopores.

Featured scientist: Haiping Fang
Organization: Shanghai Institute of Applied Physics, Chinese Academy
  of Sciences (CAS), Shanghai (China)
Relevant publication: Xiaojing Gong, Jingyuan Li, Hangjun Lu,
  Rongzheng Wan, Jichen Li, Jun Hu, Haiping Fang: A charge-driven
  molecular water pump, *Nat. Nanotechnol.*, **2**, 709–712.

4.8   Living Transistors with Nanofluidic Diodes

Another example of nanofluidic research based on cell mechanics deals with ion channels. These are proteins with a hole down their middle that are the gate-keepers for cells. In channels with a diameter of more than 100 nm, the interaction between the channel wall and electrolyte solution hardly affects the flow of ions. This changes when the channel diameter becomes smaller, and when it gets down to less than 10 nm, however, things change dramatically. Then, the interaction between the solution and channel wall starts to dominate ionic flow, and ion transport through such narrow, nanoscale channels is dominated by electrostatics. The same is true for biological ion channels where charged amino residues in the selectivity filter determine the ionic flow through the channel, along with the dielectric charge on the channel wall, and the concentrations and potential in the bulk solution. The role electrostatics plays in biological pores has been confirmed by numerous mutation studies where amino acid residues in the selectivity filter were replaced by others. Ion channels have a simple enough structure that they can be analyzed with the usual tools of physical science.

With this analysis in hand, researchers are trying to design practical machines that use ion channels. By exploiting the electrostatics in nanochannels a group of U.S. and Dutch scientists have managed to make a diode. Just as a solid state diode allows current to flow in one direction, the ionic equivalent they designed can be used to direct the flow of ions across a membrane that separates two electrolyte solutions. Now that they know how to manipulate the ion selectivity in these devices, they hope to be able one day to selectively amplify currents carried by individual chemical species—a stunning prospect for molecular nanoelectronics.

Bob Eisenberg at Rush University Medical Center in Chicago has been one of the pioneers in studying the way the potential profile in an ion channel directs the ion flow through the channel. This potential arises not only from the permanent charge present in the selectivity filter of the channel but also from the mobile ion species that, while passing through, reside in the channel. This is one of the key features of the Poisson–Nernst–Planck (PNP) approach developed by Bob Eisenberg, which states that the equations that govern the potential (Poisson) and ionic flux (Nernst–Planck in the ideal case) have to be solved simultaneously using an iterative calculation.

"Our work may challenge us to design and build biodevices with even more sophisticated transport characteristics, for instance one that contains the equivalent of an npn or pnp junction and thus an amplifier, or a pnpn set-up that acts as a thyristor to allow control of huge currents," Eisenberg explains. "Because these devices include ion selectivity, which we know how to manipulate, chances are we should be able to selectively amplify currents carried by individual chemical species. The implications are as staggering as they were when the first electronic diode was converted into an amplifying triode."

"We took the substitution approach, where amino acid residues in the selectivity filter were replaced by others, one step further by creating a second selectivity filter with a net charge of comparable magnitude but opposite sign from that of the original, first selectivity filter," says Henk Miedema. "The two filters, located in the outer membrane protein F (OmpF) pore at an average distance of 2.6 nm apart, create separated regions where either cations or anions accumulate. Our work shows that biological channels can be made into electrostatic devices. Because one or more amino acid residues at any preferred position can be substituted by others, the precision and resolution of the charge distribution (created by chemical modification of a biological pore) are unprecedented compared to those in synthetic nanochannels."

Miedema, a senior scientist at the Biomade Technology Foundation in Groningen, The Netherlands, collaborated with Eisenberg's group at Rush and the whole effort was financed by NanoNed, a nanotechnology program of the Dutch Ministry of Economic Affairs.

The main motivation for this study was to point out the key role electrostatics play in the ion flow through a nanochannel. The resulting work emphasizes the similarity in physics that describes, on the one hand, the flow of electrons and holes in a doped semiconductor and, on the other hand, the flow of ions through a charged ion channel.

"The permanent charge in the selectivity filter of an ion channel plays an analogous role in ion conduction to the role doping plays in the current of quasi-particles in a semiconductor," says Miedema. "If so, we should be able to introduce rectification into an otherwise nonrectifying channel by mimicking the electric field of a semiconductor diode. This was the hypothesis we tested. The observation that the current through this mutant channel not only rectifies but, in addition, shows the voltage dependence we had anticipated, makes us believe that our interpretation in terms of an ionic Nernst–Planck junction is correct."

Miedema says that, due to the nature of coulombic interactions, the role of structure in ion conduction is overstated, whereas the role of electrostatics is understated. "Our study exemplifies this. Even though the structure of the diode mutant was not available, its current profile is as predicted."

"Indeed," says Eisenberg, "our work suggests that the word 'conformation' found on nearly every other page of any work on proteins should be interpreted as the conformation of the electric field rather than the shape of the protein itself. We suspect that changes in the shape of the electric field are the most important conformational changes in ion channel proteins. Certainly, changes in conformation of the electric field have the ability to generate a wide range of the nonlinear properties of biological systems. After all, they are the only conformational changes in semiconductor diodes and transistors which perform an enormous range of nonlinear logical functions."

Ion channels like the ones Eisenberg and Miedema describe may be used in nanofluidic networks composed of membranes in parallel and/or in series, all containing a specific, functionalized ion channel. Such a network would resemble an electronic circuit composed of resistors, diodes, capacitors, and so on. Functionalized ion channels with distinct and addressable features (*e.g.* ion selectivity or rectification) might be a first step to achieving that goal, says Miedema. Although there is still a long way to go, a platform with channels reconstituted in the membranes acting as ion-selective and voltage-dependent valves would be an excellent starting point for such a network.

Eisenberg and Miedema point out that there are two main challenges ahead. "First, to extend the OmpF pore with a third selectivity filter to create the ionic equivalent of a pnp or npn junction. Second, to introduce flexibility and design an np junction that can be turned on and off by an external trigger, for instance light. One way to achieve this is to introduce cysteines (into the originally cysteine-free OmpF) and chemically modify these introduced cysteines with light-sensitive moieties whose charge state depends on the light quality used."

This work is another example that shows how nanotechnology can benefit enormously from scientists' knowledge of the working mechanism of nanoscale biological systems. To avoid reinventing the wheel, nanotechnology can mimic and design devices inspired by nature's nanomachines.

Featured scientists: (a) Bob Eisenberg, (b) Henk Miedema
Organizations: (a) Rush University Medical Center, Chicago, IL (USA); (b) Biomade Technology Foundation, Groningen (The Netherlands)
Relevant publication: Henk Miedema, Maarten Vrouenraets, Jenny Wierenga, Wim Meijberg, George Robillard, Bob Eisenberg: A biological porin engineered into a molecular, nanofluidic diode, *Nano Lett.*, **7**, 2886–2891.

4.9 Moth Eyes Inspire Self-cleaning Antireflection Coatings

Antireflection coatings have become one of the key issues for mass production of silicon solar cells, the most common solar cells on the market today. They are constructed with layers of n-type silicon having many electrons and p-type silicon having many electron holes (portions with missing electrons). When the cell is hit by sunlight, an equal number of electrons and electron holes are generated at the interface between the two silicon layers and, as the electrons migrate from the n-type silicon to the p-type silicon, an electric flow is generated.

One of the problems with silicon solar cells is the high refractive index of silicon, which causes more than 30% of incident light to be reflected back from the surface of the silicon crystals. Solar cell manufacturers have therefore developed various kinds of antireflection coatings (ARCs) to reduce the unwanted reflective losses. Currently, quarter-wavelength silicon nitride thin films deposited by plasma-enhanced chemical vapor deposition are the industrial standard for ARCs on crystalline silicon substrates. Unfortunately, current top-down lithographic techniques for creating subwavelength silicon gratings, such as electron-beam lithography, nanoimprint lithography, and interference lithography, require sophisticated equipment and are expensive to implement, thus contributing to the still high cost of this type of solar cell.

Borrowing from Nature's millions of years old design book, researchers have come up with an antireflection coating inspired by the eyes of moths. "We have developed a simple and scalable spin-coating technique that enables wafer-scale production of colloidal crystals with nonclose-packed (ncp) structures," Wei-Lun Min tells us.

Min is a student in Peng Jiang's group in the Department of Chemical Engineering at the University of Florida. He explains that commercial silicon nitride ARCs are typically designed to suppress reflection efficiently at wavelengths around 600 nm. "The reflective loss is rapidly increased for near-infrared and other visible wavelengths, which contain a large portion of the incident solar energy," he says. "In contrast, our subwavelength-structured moth-eye ARCs directly patterned in the substrates are broadband and intrinsically more stable and durable than multilayer ARCs since no foreign material is involved."

The team in Florida has developed a simple and scalable templating technique for making subwavelength ARCs. Knowing that moths use hexagonal arrays of ncp nipples as antireflection coatings to reduce reflectivity from their compound eyes, they designed a colloidal templating approach for fabricating such moth-eye ARCs on both glass and silicon substrates. Figure 4.2

"The outer surface of the corneal lenses of moths consists of ncp arrays of conical protuberances, termed corneal nipples, typically of sub-300 nm height and spacing," explains Min. "These arrays of subwavelength nipples generate a graded transition of refractive index, leading to minimized reflection over a broad range of wavelengths and angles of incidence." In similar fashion, the

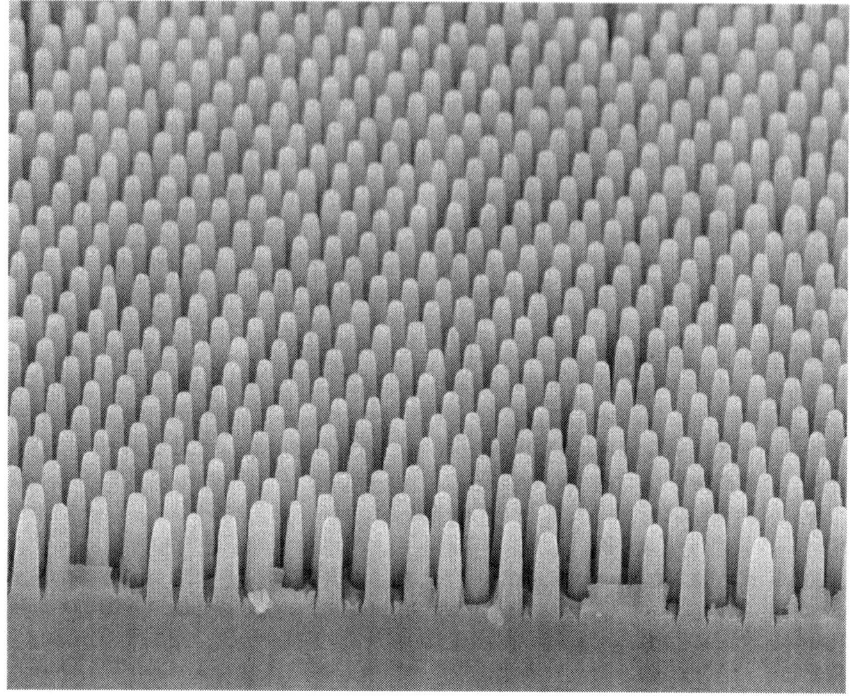

Figure 4.2   Templated silicon pillar arrays. (Image: Wei-Lun Min, University of
Florida)

templated pillar arrays that the researchers designed have a high aspect ratio
and exhibit excellent broadband antireflection and nonwetting properties.

The team points out that transparent polymer moth-eye ARCs with good
antireflective properties have been demonstrated before but, in contrast to their
own ARC, the low aspect ratio ($\sim$0.5) of the previously templated nipples had
made the realization of superhydrophobic coatings difficult.

In their approach, Min and his colleagues used a simple and scalable spin-
coating technique to generate ncp colloidal monolayers of hexagonally ordered
silica particles on silicon wafers. These particles are then used as an etching
mask during a reactive ion etch (RIE). "As the etching rate of silica is much
lower than that of silicon under RIE conditions, silica particles protect the
silicon immediately underneath them from being etched, resulting in the formation of pillar arrays directly on the silicon surface," Min explains the process. Once the silicon pillars are deep enough, the process is stopped by washing
off the templating silica spheres.

Min points out that their technique can be easily utilized to generate wafer-
scale silicon pillar arrays, is compatible with standard industrial manufacturing, and is promising for developing self-cleaning ARCs for a large variety of
technological applications ranging from solar cells and photodiodes to flat
panel displays and optical components.

Featured scientist: Wei-Lun Min
Organization: Department of Chemical Engineering, University of
Florida, Gainesville, FL (USA)
Relevant publication: Wei-Lun Min, Bin Jiang, Peng Jiang: Bioinspired
self-cleaning antireflection coatings, *Adv. Mater. (Weinheim, Ger.)*,
20, 3914–3918.

4.10 Protein Engineering—from the Humble Spider to Next-generation Materials

Much has been written about the fascinating properties of spider silk, a biopolymer that is stronger than steel and more elastic than rubber. The silken threads possess a unique combination of mechanical properties: strength (its tensile strength is about five times as strong as steel of the same density), extensibility (up to 30%), and toughness (its ability to absorb a large amount of energy without breaking). Of course this begs the obvious question: How is it possible that spider silk, produced by little creatures that evolved about 400 million years ago, can be as strong as steel, a modern alloy that plays a critical role in our infrastructure and which still attracts considerable R&D investments in its production technology?

What is perplexing is that the atomic interactions (hydrogen bonds) in spider silk are actually 100–1000 times weaker than those in steel or than those in the superfiber Kevlar, where covalent bonds are used. Hydrogen bonds are the basic chemical bonds that hold proteins together, much like trusses and beams in buildings, and play a key role in controlling the behavior of these structures. In order to attain silk's mechanical properties, most synthetic materials must be much denser, thus much heavier, and consume much more energy during their synthesis and transport.

Analysis performed at MIT's Laboratory for Atomistic and Molecular Mechanics shows that the intriguing strength of spider silk may be made possible by precisely controlling the number and the geometry of hydrogen bonds at a characteristic length scale. The physical concept is that by making many small elements work together cooperatively, the weaknesses of the individual components can be overcome. All this must happen at the nanoscale in order to be effective.

Despite significant advances in scientists' understanding of the nanomechanics of biological materials, several key fundamental questions remain unanswered. What is the strength limit of hydrogen bond assemblies, and how is it possible that protein materials such as spider silk reach strengths that exceed those of steel, despite the weakness of hydrogen bond interactions? The mechanical behavior of protein materials is exceedingly complex, partly due to the hierarchical architecture of these materials that extends from the nano- to the macroscale. Unraveling the mysteries of Nature's protein design blueprints will have a big impact on the future of nanotechnology applications in materials design and engineering.

"Our findings explain how the intrinsic strength limitation of hydrogen bonds in spider silk is overcome by the formation of a nanocomposite structure of hydrogen bond clusters, thereby enabling the formation of larger and much stronger beta-sheet structures," explains Markus Buehler. "We have discovered this by bringing together two independent concepts that were established earlier in the field of proteins. Both are universal properties of proteins, that is, they do not depend specifically on which protein is considered, and both have been clearly confirmed in experiments:

(1) Proteins are highly flexible structures that bend and turn easily, leading to a particular form of elasticity called *entropic elasticity*. Many AFM experiments have shown that this entropic elasticity is indeed the governing mode of deformation.

(2) Protein structures are bound together by hydrogen bonds, one of the weakest types of chemical bond known. These hydrogen bonds contain very little energy. In fact, their energy is so small that if they don't work cooperatively in assemblies they can easily be broken just by the random thermal vibrations at the molecular level."

Earlier studies were unable to combine these two properties of protein structures in elucidating their strength. Neither the entropic elasticity nor the properties of hydrogen bonds in themselves can explain how spider silk can be as strong as steel. Buehler, the Esther and Harold E. Edgerton Professor at MIT's Department of Civil and Environmental Engineering, and his group have achieved a link between the two by returning to a more fundamental level, using thermodynamics as the basic principle.

"Interestingly, the same thermodynamic approach was used more than 80 years ago during the development of the theories of fracture mechanics—but at that time it was applied to ships, and thus far used primarily at much larger scales" says Buehler. "In a different form, but with essentially the same physics, it can also be applied to predict the fracture of proteins." By working out this new model, Buehler's group made a surprising discovery: hydrogen bonds in proteins are special atomic bonds with fascinating properties. The use of only one or two hydrogen bonds in building a protein provides no mechanical resistance, or very little. Using three to four hydrogen bonds, however, leads to a resistance that actually exceeds that of many metals. Using more than four bonds leads to a much reduced resistance. This means that the strength of a protein is maximized at a characteristic number of hydrogen bonds.

"We have not only performed a theoretical and simulation analysis of this phenomenon," says Buehler, "we have also cross-checked our results against a variety of experimental observations from different sources and different materials. For instance, we accessed a large experimental database with thousands of protein structures (from the Protein Data Bank) and analyzed the prevalence of the characteristic 'size' of the protein, focusing on protein structures rich in beta-sheets (these are structures that are often found in structural protein materials). What we found was quite exciting: the strength resistance as a function of size, described above, correlates very closely with the biological prevalence." This suggests that a key evolutionary driving force for the structural features of these proteins may be related to strength. Protein structures of a characteristic dimension are most common in biology. Protein structures smaller or larger than this characteristic dimension are less common.

Buehler explains how these findings are important for several reasons:

(1) They may help us to understand what types of mutations lead to malfunctions of the protein, for instance in genetic diseases that disrupt the hydrogen bonding structures or those that create overly strong structures.

(2) Since hydrogen bonds are an important structural 'glue', in some sense the cement of biology, and since mechanical forces play a critical role in many biological processes, the findings could help us to better understand how elementary biological processes work.

(3) They may enable us to design new materials based on proteins. It is already possible to synthesize novel materials from scratch, using protein elements as building blocks. What is missing are engineering tools that direct this material design process.

(4) The findings may be important to better understand the behavior of other protein materials, for instance those that define hair, protein nanotubes, and amyloids, muscle proteins, or those that make up the cell's cytoskeleton.

Featured scientist: Markus Buehler
Organization: Laboratory for Atomistic and Molecular Mechanics, MIT, Cambridge, MA (USA)
Relevant publication: Sinan Keten, Markus J. Buehler: Geometric confinement governs the rupture strength of hydrogen bond assemblies at a critical length scale, *Nano Lett.*, **8**, 743–748.

CHAPTER 5

Amazing New Materials

Much of nanotechnology today is concerned with producing new or enhanced materials such as better coatings or more effective filter membranes. What is so special about these nanomaterials?

The bulk properties of materials often change dramatically with nanoscale ingredients. There are two main reasons why the properties of materials can be different at the nanoscale:

(1) Nanomaterials have a relatively larger surface area than the same mass of material made up of larger particles. This can make materials more chemically reactive (in some cases, materials that are inert in their larger form are reactive when produced in their nanoscale form, as is the case with aluminium), and affect their strength or electrical properties. For example, a particle of size 30 nm has 5% of its atoms on its surface, at 10 nm the figure is 20%, and at 3 nm 50%. As growth and catalytic chemical reactions occur at surfaces, this means that a given mass of material in nanoparticulate form will be much more reactive than the same mass of material made up of larger particles.

(2) Quantum effects can begin to dominate the behavior of matter at the nanoscale, particularly at the lower end, thus affecting the optical, electrical, and magnetic behavior of materials.

Take carbon, for example. This element comes in many different forms, from the graphite found in pencils to the world's most expensive diamonds. In 1980, we knew of only three basic forms of carbon: diamond, graphite, and amorphous carbon. Then, fullerenes (spherical carbon molecules with at least 60 carbon atoms) and carbon nanotubes (CNTs) were discovered, and all of a sudden this became the research area where many scientists wanted to be.

More recently, there has been quite a buzz about graphene. Discovered only in 2004, graphene is a flat sheet of carbon, one atom thick. Existing forms of

RSC Nanoscience and Nanotechnology No. 8
Nano-Society: Pushing the Boundaries of Technology
By Michael Berger
© Michael Berger 2009
Published by the Royal Society of Chemistry, www.rsc.org

carbon basically consist of sheets of graphene, either bonded on top of each other to form a solid material like graphite, or rolled up into CNTs—think of a single-walled CNT as a graphene cylinder—or folded into fullerenes. Physicists had long considered a free-standing form of planar graphene to be impossible; the conventional wisdom was that such a sheet would always roll up. Despite being isolated only a few years ago, graphene has already appeared in hundreds of research papers. The reason scientists are so excited is that graphene opens up a whole new class of materials with novel electronic, optical, and mechanical properties.

There are many more nanomaterials that scientists create and experiment with—nanowires, nanofibers, nanorods, nanohorns, quantum dots—and in this section we explore some of the issues involved in creating nanomaterials and using them for novel applications.

5.1 Quantum Dots are Ready for Real-world Applications

Quantum dots, also called nanocrystals, are artificial nanostructures that can possess many varied properties, depending on their material and shape. For instance, because of their particular electronic properties, they can be used as active materials in single-electron transistors. Because certain biological molecules are capable of molecular recognition and self-assembly, nanocrystals could also become important building blocks for self-assembled functional nanodevices. The atom-like energy states of quantum dots furthermore contribute to special optical properties such as particle-size-dependent fluorescence wavelength, an effect which is used in fabricating optical probes for biological and medical imaging. So far, the widest range of applications for colloidal quantum dots is in bioanalytics and biolabeling.

The first generation of quantum dots already gave a hint of their potential, but it took a lot of effort to improve basic properties, in particular colloidal stability in salt-containing solution. Initially, quantum dots were used in very artificial environments and these particles would simply have precipitated in 'real' samples such as blood.

"Because of their properties, nanocrystals are particularly interesting with regard to the construction of smaller and faster devices or multifunctional materials on the nanometer scale," explains Wolfgang Parak, a scientist in the Department of Physics at the Ludwig Maximilians University of Munich. "Molecular recognition is a 'lock and key' mechanism realized on a molecular scale: a receptor molecule (the lock) recognizes a certain ligand molecule (the key) with very high selectivity. Thus, only the appropriate ligand will bind to its receptor. Several important classes of receptor–ligand pairs are known: oligonucleotides and their complementary counterpart, antibodies and antigens, and the biotin–avidin system."

"For this purpose, each building block has to be functionalized with ligand molecules," Parak says. "The building block–ligand conjugates will now bind to positions where corresponding receptor molecules are present. In this way the following three types of applications are possible: (1) the assembly of receptor–ligand-mediated groupings of building blocks to form new multi-functional building blocks, (2) the arrangement of ligand-modified building blocks on a surface that is patterned with receptor molecules, and (3) the labeling of specific bioreceptors with ligand-modified building blocks."

It is in this third area that quantum dot-based applications have so far made most progress. When colloidal quantum dots—which are basically fluorescent dye particles—are photo-excited, electron–hole pairs are generated, and when they recombine fluorescent light is emitted. Because of the small size of the interacting particles quantum effects play an important role, resulting in size-dependent fluorescence wavelength. The smaller the particles, the more blue-shifted their fluorescence. In this way all colors in the visible and infrared can be obtained by synthesizing nanoparticles of different sizes.

In contrast to organic fluorophores, colloidal quantum dots are based on inorganic semiconductor nanoparticles and have several advantages. Their emission spectra are quite narrow and symmetric and do not show any 'red tail'. In this way, many different colors can be excited with just one excitation wavelength and can be spectrally well resolved. Their fluorescent lifetime is longer (still measured in nanoseconds, though); and their photobleaching is reduced. Although quantum dots will probably not replace organic fluor-ophores as such, they are expected to become the dominant fluorescent dyes in certain types of applications. In particular, they could have a considerable impact in single-molecule tracing studies, in fluorescence resonance energy transfer (FRET)-based immunoassays, and in tracking the fate of cells in tissues.

Quantum dots can be used in different ways to detect analytes. As active sensor elements, the fluorescence properties of the quantum dots are changed upon reaction with the analyte. Although these applications are fairly simple, they appear to be restricted to sensing just a few reactive small molecules or ions that are able to interact directly with the quantum dots' surface. Moreover, the reactions are usually not highly specific, and they also depend strongly on the surface properties of the quantum dots (including their coating).

"We think that the use of quantum dots as active element in sensors in industrial applications will be limited," predicts Parak. "Much more promising is the use of quantum dots as passive labels in sensor applications, in particular in FRET-based assays".

In passive label probes, selective receptor molecules such as antibodies have been conjugated to the surface of quantum dots. In a first step, antibodies are immobilized on a substrate to which the analyte is added. In a second step, quantum dot-labeled antibodies are used to visualize and quantify the bound analyte. This technique allows for the design of simple multiplexed immunoassays.

Parak thinks that quantum dots have a huge potential for cellular labeling. Here, individual molecules can be fluorescence-labeled with quantum dots to trace the movement of individual membrane proteins. Also, whole structures such as DNA can be quantum dot-labeled and used as fluorescent *in situ* markers.

"Nowadays, the synthesis of quantum dots in organic solvents is so well established that control of their size, shape, and even composition is possible," says Parak. "However, some problems still have to be overcome. One of them is the potential cytotoxicity, especially of cadmium-containing materials".

Coverage of potentially toxic particles with additional shells can seal the cadmium-containing core, but Parak suggests that in the future quantum dots that do not contain cadmium and are therefore more biocompatible will be made available, for example doped zinc selenide particles.

Another problem that is still not completely understood is the process of blinking, *i.e.* when luminescence switches on and off. Blinking limits quantitative single quantum dot based sensor applications.

"There is still plenty of room for further development in all these directions," concludes Parak.

Featured scientist: Wolfgang Parak

Organization: Biophysics and Molecular Materials, Department of Physics, Ludwig-Maximilians-Universität München (Germany)

Relevant publication: Cheng-An J. Lin, Tim Liedl, Ralph A. Sperling, María T. Fernández-Argüelles, Jose M. Costa-Fernández, Rosario Pereiro, Alfredo Sanz-Medel, Walter H. Chang, Wolfgang J. Parak: Bioanalytics and biolabeling with semiconductor nanoparticles (quantum dots), *J. Mater. Chem.*, **17**, 1343–1346.

5.2 Researchers go Ballistic over Graphene

According to Kostya Novoselov, graphene has led to the emergence of a new paradigm of 'relativistic' condensed-matter physics where quantum relativistic phenomena, some of which are unobservable in high-energy physics, can now be mimicked and tested in table-top experiments. Because of its unusual electronic spectrum, graphene represents a conceptually new class of materials that are only one atom thick.

Novoselov, the Royal Society Research Fellow and a member of the Mesoscopic Physics Group at the University of Manchester, is one of the original team, led by Andre Geim, that discovered graphene in 2004.

Experiments with graphene have revealed some fascinating phenomena that excite researchers who are working towards molecular electronics. Geim's team found that graphene remains capable of conducting electricity even at the limit of nominally zero carrier concentration because the electrons do not seem to slow down or localize.

Novoselov says that the electrons moving around carbon atoms interact with the periodic potential of graphene's honeycomb lattice, which gives rise to new quasiparticles that have lost their mass, or 'rest mass' (so-called massless Dirac fermions). That means that graphene never stops conducting. It was also found that these particles travel far faster than electrons in other semiconductors.

In the ultimate nanoscale transistor—dubbed a *ballistic transistor*—the electrons avoid collisions, *i.e.* there is a virtually unimpeded flow of current. Ballistic conduction would enable incredibly fast switching devices. Graphene has the potential to enable ballistic transistors at room temperature.

Efforts to develop graphene electronics are already under way. In addition to Geim's group, researchers at Rensselaer Polytechnic Institute are working on graphene nanoribbons. These ribbons have also been demonstrated by IBM and the Kim group at Columbia University. At Delft's Kavli Institute for Nanoscience they attach graphene to superconductors to get a bipolar transistor for superconducting currents.

"Whichever approach prevails," says Novoselov, "there are two immediate challenges. First, despite the recent progress in epitaxial growth of graphene, high-quality wafers suitable for industrial applications still remain to be demonstrated. Second, individual features in graphene devices need to be controlled accurately enough to provide sufficient reproducibility in their properties.

"The latter is exactly the same challenge that silicon technology has been dealing with successfully. For the time being, to make proof-of-principle nanoscale devices, one can use electrochemical etching of graphene by scanning-probe nanolithography."

Despite all the excitement, don't expect 'graphenium' microprocessors for at least another 20 years. Just look at CNTs. First observed in 1991, they have created tremendous excitement (and hype). But today, almost 20 years later, there are no commercial CNT-based breakthrough products on the market—although they supposedly make your golf balls go straighter (a breakthrough for some golfers at least).

In the meantime, other graphene-based applications might come of age. Besides its electronic properties, graphene is also researched for its mechanical, thermal, and optical properties. Existing graphene samples grown on silicon carbide substrate are already of sufficient quality to be used for gas sensors. At Northwestern University they are working on graphene oxide paper and at Cornell University they are building graphene resonators. Novoselov and Geim themselves have reported that graphene-based sensors could sniff out dangerous molecules. Max Planck scientists in Germany, who work with Geim and Novoselov, are also interested in graphene sheets as membranes for biomolecular sensors. This idea of using graphene as a substrate to study biomolecules originated at the University of Zurich, and there is currently a joint project, Structural Information of Biological Molecules at Atomic Resolution (SIBMAR) there, where graphene could play a role.

"The most immediate application for graphene is probably its use in composite materials," says Geim. "It would allow conductive plastics at less than 1 vol% filling, which in combination with low production costs makes graphene-based composite materials attractive for a variety of uses. However, it seems doubtful that such composites can match the mechanical strength of their nanotube counterparts because of much stronger entanglement in the latter case."

Another enticing possibility mentioned by Geim is the use of graphene powder in electrical batteries, which are already one of the main markets for graphite. "The large surface-to-volume ratio and high conductivity provided by graphene powder could lead to improvements in the efficiency of batteries, taking over from the carbon nanofibers used in modern batteries. CNTs have also been considered for this application, but graphene powder has an important advantage of being cheap to produce."

Novoselov notes that it is just a few years since graphene was first reported, and despite remarkably rapid progress, only the very tip of the iceberg has been uncovered so far. "Because of the short timescale, many experimental groups working now on graphene have not published even a single paper on the subject, which has been a truly frustrating experience for theorists. What has become clear, though, is that graphene is not a fleeting fashion but is here to stay, bringing up both more exciting physics and perhaps even wide-ranging applications."

> *Featured scientist:* Kostya Novoselov, Andre Geim
> *Organization:* Mesoscopic Physics Group, University of Manchester (UK)
> *Relevant publication:* A. K. Geim, K. S. Novoselov: The rise of graphene, *Nat. Mater.*, **6**, 183–191.

5.3 'Cool' Graphene Might be Ideal for Thermal Management in Nanoelectronics

Adding to the buzz about graphene, in a first experimental study of the thermal conductivity of single-layer graphene, researchers found that in addition to its many unique electronic properties, graphene is also an extraordinarily good heat conductor. Measurements revealed that graphene's thermal conductivity near room temperature is in the range 3500–5300 W/mK. It appears that the thermal conductivity of graphene is larger than conventionally accepted experimental values reported for individual suspended CNTs and corresponds to the upper bound of the highest values reported for SWCNT bundles. CNTs are known to have very high thermal conductivity with an experimentally determined room temperature value of around 3000 W/mK for an individual MWCNT and about 3500 W/mK for an individual SWCNT. These values exceed those of the best bulk crystalline thermal conductor, diamond, which has a thermal conductivity in the range of 1000–2200 W/mK. The superb thermal conductivity of graphene is beneficial for its proposed electronic applications and also establishes it as an excellent material for thermal management in optoelectronics, photonics, and bioengineering.

"Our measurement of the thermal conductivity of graphene was the very first one," says Alexander A. Balandin. "It required the development of a completely new experimental technique and data extraction methodology. The measurements were performed for long graphene flakes suspended across a trench in Si/SiO$_2$ substrate and connected to graphitic heat sinks. Before our work, only simple theoretical estimates for the thermal conductivity of graphene were known."

Balandin, a professor in the Department of Electrical Engineering at the University of California, Riverside (UCR) and director of the Nano-Device Laboratory there, explains that the main practical motivation for his group's work was the search for a material with an extremely high thermal conductivity that can be integrated with electronic silicon technology.

"These newly discovered excellent heat-conducting properties, combined with its single-atom thickness and flat geometry, make graphene an excellent candidate for heat removal and thermal management of future nanoelectronic circuits," says Balandin. "It might be easier to integrate graphene layers rather than CNTs with electronic circuits. As the electronics industry moves towards nanoscale designs, one of the most important challenges is the growing power consumption of computer chips. For this reason thermal management in electronic circuits is becoming an integral part of their design, test, and manufacturing".

In spite of the fundamental importance of knowledge about the thermal conductivity of graphene, no experimental data had been reported before Balandin's work. Balandin explains this by the fact that the conventional techniques for measuring thermal conductivity—thermal bridge, 3ω method, or laser flash—are not well suited for single-layer graphene. "The 3ω method is good for measuring the cross-plane thermal conductivity and requires a substantial temperature drop over the thickness of the examined film," he says. "Graphene, with a thickness of one atomic layer and expected high thermal conductivity, cannot satisfy such a requirement. Direct thermal bridge measurements of graphene are possible in principle but very challenging technically."

The UCR team chose an unconventional approach for the non-contact measurement of the thermal conductivity of graphene by using confocal micro-Raman spectroscopy. "Several factors specific to graphene made it possible," says Balandin. "Graphene has clear signatures in Raman spectra. We have also recently discovered that the G peak in the graphene spectrum manifests a strong temperature dependence. The G peak's temperature sensitivity allows one to monitor the local temperature change produced by the variation of the laser excitation power focused on a graphene layer. In a properly designed experiment, the local temperature rise as a function of the laser power can be utilized to extract the value of the thermal conductivity."

However, there are major differences in heat spreading in graphene from that in conventional materials. For this reason, the researchers had to develop a new measurement methodology and derive expressions suitable for the thermal conductivity extraction for graphene.

Balandin points out that the Raman spectroscopy-based measurement of the thermal conductivity does not work very well for bulk crystalline materials with high thermal conductivity because of the rapid escape of heat in 3-D systems.

"This prevents a local temperature rise detectable by Raman spectroscopy for reasonable excitation power levels," he says. "Luckily, graphene is only one atomic layer thick. Thus, if we suspend a graphene layer over a trench and heat it in the middle, the heat is forced to propagate in-plane through the layer (with a thickness of 0.35 nm) toward the heat sink. The extremely small cross-sectional area of the heat conduction channel makes the detection of the local temperature rise possible."

The specific potential application of the UCR team's findings lies with the thermal management of future nanoscale electronic circuits. Graphene layers could be incorporated into the circuits' layered structure to remove heat efficiently from the circuit hotspots to the heat sinks.

The performance and reliability of ultra-large-scale integration (ULSI) are functions of the junction temperature. Even a small temperature increase might result in a significant reduction of device lifetime. One approach to the thermal management problem is finding a material with extremely high thermal conductivity that can be integrated with existing silicon technology. Graphene, or graphene materials a few layers thick, may help solve this problem.

Balandin's group has begun work on a comprehensive theory which would explain their experimental results and predict the value of the thermal conductivity depending on the graphene layer's parameters. More measurements are planned to elucidate the role of boundary scattering and the possibility of graphene integration with silicon CMOS devices and circuits.

> *Featured scientist:* Alexander A. Balandin
> *Organization:* Department of Electrical Engineering, University of California, Riverside, CA (USA)
> *Relevant publication:* Alexander A. Balandin, Suchismita Ghosh, Wenzhong Bao, Irene Calizo, Desalegne Teweldebrhan, Feng Miao, Chun Ning Lau: Superior thermal conductivity of single-layer graphene, *Nano Lett.*, **8**, 902–907.

5.4 Self-Cleaning Stickies

Adhesive tapes are ubiquitous in our lives, whether they take the form of yellow sticky notes, the tape that closes baby diapers, masking connectors on printed circuit boards, or surgical tape in hospitals. Most adhesive tapes will stick to a wide variety of surfaces provided that they are clean and dry. Adhesive tapes are made of two components: a carrier, which is usually paper or plastic, and an adhesive which is either water- or solvent-based. Many modern adhesive tapes use pressure-sensitive adhesives. When pressure is applied to the tape, a strong adhesive bond is formed with the underlying surface. Most tapes have poor ageing properties and deteriorate quickly: with time, after several uses, or as the sticky side becomes dirty, they lose their adhesive ability.

As we have seen before, scientists are very interested in exploring the secret of the gecko's adhesive properties and using this knowledge to create superior synthetic adhesives. But it's not just the stickiness that intrigues researchers: because geckos are able to walk across a dusty or dirty surface and then scale a vertical wall without problems, their feet must also possess some kind of self-cleaning ability. Scientists have managed to mimic the remarkable self-cleaning abilities of the gecko and the lotus leaf, and incorporate this ability into the design of a self-cleaning, CNT-based adhesive material.

"Significant effort in developing synthetic materials inspired by gecko feet show comparable, and in some cases better, shear resistance than natural gecko feet," explains Ali Dhinojwala. "Still, these measurements were done in controlled environments and we only have limited data about the self-cleaning ability of these synthetic materials. For the past few years, my research group has been working on mimicking the remarkable climbing ability of geckos using patterned aligned CNTs. We have achieved shear resistance that is four times higher than that of geckos. Now, for the first time, we have also mimicked the self-cleaning aspect of the gecko foot. By using the patterns of specific dimensions and lengths we have shown that our adhesive tapes can be cleaned by water and mechanical contact. The cleaning by water is very similar to the mechanism that is found on the leaves of plants such as lotus and lady's mantle (*Alchemilla*)."

Dhinojwala, a professor at the University of Akron, together with Pulickel Ajayan's group at Rensselaer, has demonstrated the self-cleaning ability of CNT-based flexible gecko tapes. Previously, the Dhinojwala and Ajayan groups have fabricated hierarchical structures of setae and spatulas found on the gecko foot using aligned multi-walled CNTs. In an earlier paper,[1] they showed that a $1\,cm^2$ area can support nearly 4 kg of weight and that much larger forces can be supported by increasing the area of the tape. The researchers found that the length and diameter of CNTs, size of the pattern, and stiffness of the backing tape are all important parameters that need to be optimized for superior performance.

In subsequent research, Dhinojwala showed that for optimum length and pattern size, these CNT-based synthetic tapes exhibit self-cleaning as well as high shear resistance. "In addition to water, our synthetic tapes can also be cleaned by a contact mechanism similar to that used by the gecko," he says. "After mechanical cleaning, the shear strength recovers to 90% (and 60% for water-cleaned samples) of the values measured before soiling. In comparison, the gecko recovers 50% of the shear stress after eight steps. The ability of these synthetic tapes to self-clean and also retain their shear resistance makes them an excellent example of gecko-inspired adhesives."

For the experiments, MWCNTs with a length of $100\,\mu m$ and an average diameter of 8 nm (2–5 walls) were used. Flexible polymer tape with tacky coating on one side is pressed against the top surface of the CNTs. Upon

---

[1] Liehui Ge, Sunny Sethi, Lijie Ci, Pulickel M. Ajayan, and Ali Dhinojwala: Carbon nanotube-based synthetic gecko tapes, *Proc. Natl. Acad. Sci. U.S.A.*, **104**, 10792–10795.

peeling, the CNTs are transferred from the silicon substrate to the flexible polymer tape.

The CNTs are held by a polymeric glue at the base and this prevents individual structures from collapsing, because of capillary forces. "This process eliminated the use of fluorinated coatings or other nonwetting materials on the CNTs to make these structures superhydrophobic," explains Dhinojwala. "To demonstrate the self-cleaning ability of the synthetic tapes, we soiled the tapes with silica particles—to represent dust—ranging from 1 to 100 μm in size. When rinsed with water, the water droplets roll off very easily, carrying most of the silica particles with them."

The researchers also tested the self-cleaning properties of these synthetic tapes by contact mechanics. After a couple of contacts with mica or glass substrate they observed that the majority of these particles are transferred to the surface of the mica or glass.

"The successful design of a gecko-like adhesive requires the ability to mimic both the stickiness and the self-cleaning," Dhinojwala sums up his findings. "If we want to use these tapes in robotics, we cannot just test them on clean glass. They need to work in real-world dusty environments. In addition, the important element of the gecko design is reversibility and this cannot be achieved without some aspect of self-cleaning."

The scientists envisage a broad spectrum of applications including robotics, space applications, electronics, and sports. Several companies have already shown an interest in this new technique and are in discussions with Dhinojwala on how to commercialize his research.

> *Featured scientist:* Ali Dhinojwala
> *Organization:* Department of Polymer Science, University of Akron, Akron, OH (USA)
> *Relevant publication:* Liehui Ge, Sunny Sethi, Lijie Ci, Pulickel M. Ajayan, Ali Dhinojwala: Gecko-inspired carbon nanotube-based self-cleaning adhesives, *Nano Lett.*, **8**, 822–825.

5.5 Smart Nanofibers at the Flip of a Light Switch

'Smart' is the buzz word used by materials engineers when they describe the future of coatings, textiles, building structures, vehicles, and just any material that you can think of. Materials are made 'smart' when they are engineered to have properties that change in a controlled manner under the influence of external stimuli such as mechanical stress, temperature, humidity, electric charge, magnetic fields, *etc.* Nature of course is full of smart materials that are capable of adapting to new tasks, are self-healing, and can self-assemble autonomously simply out of a solution of building blocks. Duplicating this feat with synthetic materials will one day become a reality, thanks to nanotechnology.

Scientists do not only dream about self-repairing cars or building walls that turn transparent like windows; they are actively working on the first steps towards these goals. Simple smart materials, such as piezoelectric materials and shape memory alloys, are already a reality, but these are not nanotechnology based. Emerging nanotechnologies are now about to give scientists the tools to take smart materials to the next performance level. For instance, the European project Inteltex is developing a new multifunctional textile that could be used as a wallpaper to detect temperature changes or chemical leakage, or that could be used in medical and protective wear to monitor body temperature and mechanical stress. MIT's Institute for Soldier Nanotechnologies works on smart surfaces that switch properties. Nanotechnology-enabled smart materials are still in their very early days, but basic progress is being made. Take for instance a team of Italian researchers, who demonstrated photo-switchable nanofibers based on the reversible transformation between two molecular photochemical states, exhibiting different chemico-physical characteristics.

"Notwithstanding the remarkable advances of nanotechnology, photo-switchable nanomaterials have not been investigated in depth so far," says Dario Pisignano. "Although photochromic nanostructures can be envisaged as nanoscale light sources, switches, detectors, sensors for integrated nanophotonics, and functional elements for nanoactuation and smart surfaces, photo-switchable nanomaterials have basically remained unexplored."

Pisignano, a researcher at the National Nanotechnology Laboratory (NNL) at the Universitá degli Studi di Lecce, finds this surprising given that organic nanofibers can be easily manufactured and controlled with high precision, and photochromic molecules—or generally molecular dopants providing optical functionality—can be straightforwardly incorporated into the nanostructures.

Pisignano and his team realized and characterized one of the first photo-switchable nanomaterials, an electrospun nanofiber. These fibers rely on embedding photochromic molecular systems, which are able to reversibly change their isomeric state upon irradiation at a certain wavelength. The nanofibers are embedded in a polymer matrix and the material can be processed by low-cost, high-throughput methods such as electrospinning.

"We are particularly interested in miniaturized light-emitting sources, especially aiming at the integration of entirely biocompatible, organic-based light sources," says Pisignano. "The good stability and cycling performances we saw with our fibers open up new possibilities for the realization of light-driven nanoscale elements to be incorporated not only in sensors and lab-on-a-chip devices but also optical interconnections. The nanofiber geometry can provide enhanced control of smart surfaces and controlled device function down to the scale of tens of nanometers without the need for nanolithography."

The NNL scientists have just begun exploring the intriguing properties of these light-emitting nanomaterials. Relevant areas of study will include the assembly of complex optical circuits based on active and passive nanofibers and their integration with biological chips for providing optoelectronic sensing functionality.

Featured scientist: Dario Pisignano
Organization: National Nanotechnology Laboratory (NNL), Università
   degli Studi di Lecce (Italy)
Relevant publication: F. Di Benedetto, E. Mele, A. Camposeo, A.
   Athanassiou, R. Cingolani, D. Pisignano: Photoswitchable Organic
   Nanofibers, *Adv. Mater.* (*Weinheim Ger.*), **20**, 314–318.

5.6   Designer Protein-based Smart Materials for Nanomechanical Applications

Elastomeric (*i.e.* elastic) proteins are able to withstand significant deformations without rupturing, before returning to their original state when the stress is removed. Consequently, these proteins confer excellent mechanical properties on many biological tissues and biomaterials. Depending on the role performed by a specific tissue or biomaterial, elastomeric proteins can behave either as springs or as shock absorbers. Scientists in Canada have engineered the first artificial 'chameleon' elastomeric proteins that mimic these two different behaviors and combine them into one protein. Under the control of a molecular regulator, these designer proteins exhibit one of the two distinct mechanical behaviors—spring or shock absorber—which closely mimic the two extreme behaviors observed in naturally occurring elastomeric proteins.

"Smart materials that can change their behaviors in response to environmental changes have always been of interest to us," says Hongbin Li. "Our discovery that the mechanical stability of the small protein GB1 can be regulated by the binding of an antibody fragment provided the opportunity to use protein engineering techniques to turn GB1 into a chameleon protein that can display two different elastic behaviors in response to environmental stimuli."

Li, an assistant professor and Canada Research Chair in the Department of Chemistry at the University of British Columbia (UBC) in Vancouver, explains that these results provide a new concept to tune the mechanical performance of proteins at the single molecule level.

Elastic proteins are important structural and functional components in living cells. They serve as molecular springs in tissues to establish elastic connections and provide mechanical strength, elasticity, and extensibility. They are important not only for their functions in various biological processes but also as building blocks for bottom-up construction of smart materials and mechanical devices on the nanoscale.

"Our recent work represents a step forward for the bottom-up construction of smart mechanical materials," says Li. "Until now, tuning the mechanical properties of proteins in a reversible fashion has been a challenging task for scientists." He notes that his group had previously demonstrated that protein–protein interactions can be used to tune the mechanical stability of proteins.[2]

In their work, using single-molecule atomic force microscopy (AFM), the team showed that the binding of fragments of IgG antibody to GB1 can significantly enhance the mechanical stability of this protein. The regulation of the mechanical stability of GB1 by IgG fragments does not occur through direct modification of the interactions in the mechanical key region of GB1; instead, it is accomplished via long-range coupling between the IgG binding site and this region.

"However," explains Li, "both the protein–protein interactions and the mechanical stability of proteins are sensitive to the structural alternation of proteins. It was unknown whether these two properties can be tuned independently in the same protein. We have successfully demonstrated that re-engineering the region of a protein that is responsible for mechanical stability has little effect on its interaction with another protein. Therefore, the resulting proteins can display dual mechanical properties under the regulation of protein–protein interactions."

In their experiments, the scientists used the technique of proline mutagenesis to construct a mutant protein (GV54P) where the mechano-active site is disrupted but the binding affinity to the Fc receptor is retained.

"Our results demonstrate that GV54P exhibits two distinct mechanical compliances depending on the presence of hFc," says Li. "In the absence of hFc, GV54P is mechanically compliant and functions as an entropic spring; however, upon binding of hFc, GV54P exhibits significant mechanical stability and unfolds sequentially upon stretching."

While the UBC team's study is fundamental in nature, the concept it illustrates may help engineer protein-based smart materials for nanomechanical and biomedical applications. One example that Li mentions is that it would be possible to use elastomeric proteins to engineer smart hydrogels that can change their mechanical and physical properties upon binding of regulator molecules.

Single-molecule force spectroscopy and protein engineering have made it possible to engineer proteins with well-defined mechanical properties. Designing proteins that are sensitive to environmental stimuli, like the chameleon proteins engineered by Li and his group, will be of tremendous importance for future applications of elastomeric proteins.

"The technical challenge we are facing now is the high molecular weight of the molecular regulator," says Li. "The large size of the IgG antibody fragments makes them difficult to work with. We have begun experimenting with different techniques to design chameleon elastomeric proteins that can be regulated by small molecules such as metal ions."

---

[2] Yi Caoa, Teri Yooa, Shulin Zhuanga and Hongbin Li: Protein–protein interaction regulates proteins' mechanical stability, *J. Mol. Biol.*, **378**, 1132–1141.

> *Featured scientist:* Hongbin Li
> *Organization:* Department of Chemistry, University of British
> Columbia, Vancouver, BC (Canada)
> *Relevant publication:* Yi Cao, Hongbin Li: Engineered elastomeric
> proteins with dual elasticity can be controlled by a molecular
> regulator, *Nature Nanotechnology*, 3, 512–516

5.7 Nanotechnology that will Rock You

Forget boxy loudspeakers. Researchers have found that just a sheet of CNT thin film could become a practical magnet-free loudspeaker simply by applying an audio-frequency current through it. These loudspeakers—which are only tens of nanometers thick, transparent, flexible, and stretchable—can be tailored into many shapes and mounted on a variety of insulating surfaces such walls, ceilings, pillars, windows, flags, and clothes, without area limitations. These CNT loudspeakers can generate sound with wide frequency range, high sound pressure level, and low total harmonic distortion.

Back in 2002, a group of Chinese researchers developed a technique for the creation of continuous CNT yarns as much as 30 cm long. They did this simply by drawing the yarns out from super-aligned arrays of CNTs. The CNTs in super-aligned arrays are much better aligned than those in ordinary vertically aligned CNT arrays. The key feature of a super-aligned CNT array is that continuous thin films or ribbons, which are composed of a thin layer of parallel pure CNTs, can be directly drawn from it in solid state. These thin films are transparent and conductive, with CNTs aligned parallel to the drawing direction.

Developing this work further, in 2005 the same group of scientists successfully synthesized super-aligned CNT arrays on 4-inch (10 cm) silicon wafers. One wafer of super-aligned CNTs is capable of being transformed into a continuous thin film up to 10 cm wide and 60 m long.[3]

"Near the end of 2007, we found that a piece of CNT thin film could emit loud sound simply by applying an audio-frequency current through it," says KaiLi Jiang. "But the sound frequency is double that of the input. We attributed this to the thermoacoustic effect. The alternating current periodically heats the CNT thin films, resulting in a temperature oscillation. The temperature oscillation of thin film excites the pressure oscillation in the surrounding air, resulting in the sound generation."

Jiang, an associate professor in the Department of Physics and Tsinghua–Foxconn Nanotechnology Research Center at Tsinghua University in Beijing, China, explains that the thermoacoustic effect, which has been studied for more than 200 years, has led to the invention of thermoacoustic engines and even

---

[3] X. Zhang, K. Jiang, C. Feng, P. Liu, L. Zhang, J. Kong, T. Zhang, Q. Li, S. Fan: Spinning and processing continuous yarns from 4-inch wafer scale super-aligned carbon nanotube arrays, *Adv. Mater. (Weinheim, Ger.)*, **18**, 1505–1510.

loudspeaker-driven refrigerators. "But we had not heard of the thermoacoustic effect induced by alternating current (AC)," says Jiang. "We thought we were the first to discover such a thermoacoustic effect induced by AC, until in April 2008 one of us found papers published more than 100 years ago reporting that ultra-thin metal wires or foils made of platinum and fed by AC could emit sound. The inventors called their device the 'thermophone'. It appears that these findings have been ignored for more than 80 years because of the very weak effect that was initially demonstrated."

Jiang explains that the reason why platinum foil sounds weak but CNT thin film sounds loud has to do with the heat capacity per unit area (HCPUA)—a key parameter that determines the sound generation efficiency. According to the theory proposed by Arnold and Crandall,[4] the sound efficiency is inversely proportional to the HCPUA value. For platinum foil 700 nm thick the HCPUA is 260 times that of the Chinese team's CNT film. The sound emitted by platinum foil is therefore roughly 260 times weaker than that emitted by CNT thin film with the same power input.

To fabricate a simple loudspeaker, Jiang and his collaborators placed the as-drawn CNT thin film directly on to two electrodes. By layering several thin films together the loudspeaker area can be increased with no limitation on its size. The films can also be tailored into arbitrary shapes or placed on arbitrarily curved surfaces to make special-purpose loudspeakers.

The CNT loudspeakers have excellent acoustic performance and, when hooked up to a simple amplifier, they exhibit all the functions of a voice-coil loudspeaker as well as the merits of being magnet-free and without moving components. In contrast to conventional loudspeakers, they possess several remarkable features—they are stretchable, transparent, and flexible—that will allow the fabrication of novel and unique loudspeaker applications. An example might be laptop computers where the current audio system could be replaced by a transparent loudspeaker film simply placed over the screen.

An additional advantage compared to conventional loudspeakers is that the CNT loudspeakers do not vibrate and are damage tolerant. They will work even if part of the thin film is torn or damaged.

Featured scientist: Kaili Jiang
Organization: Department of Physics and Tsinghua–Foxconn
    Nanotechnology Research Center, Tsinghua University, Beijing (PR
    China)
Relevant publication: Lin Xiao, Zhuo Chen, Chen Feng, Liang Liu,
    Zai-Qiao Bai, Yang Wang, Li Qian, Yuying Zhang, Qunqing Li,
    Kaili Jiang, Shoushan Fan: Flexible, stretchable, transparent carbon
    nanotube thin film loudspeakers, *Nano Lett.*, 2008, **8** (12), 4539–
    4545.

---

[4] H. D. Arnold and I. B. Crandall: The thermophone as a precision source of sound, *Phys. Rev.*, **10**, 22–38.

5.8 Drill, Baby, Drill

As we will see later, nanotechnology applications could provide important technological breakthroughs in the energy sector and have a considerable impact on creating the sustainable energy supply that is required to make the transition away from fossil fuels. Gratifying though it would be to write about all the clean and green applications that will be enabled by nanotechnology, the harsh reality is that dirty energy is still fuelling our way of life.

No matter whether you are a member of the 'drill, baby, drill' crowd or if you are actively involved in saving energy and think that the development of renewable energies cannot come fast enough, we have to live with the fact that the world's energy production will continue to depend on oil, gas, and coal for quite a few more years. But even here, nanotechnology applications might offer some improvements.

Nanomaterials in the form of CNT rubber composites could help to enhance oil production efficiency by allowing us to probe and drill deeper wells. This in turn might allow the better exploitation of existing oil fields and perhaps weaken the argument for new drilling in environmentally sensitive areas.

While there is a hot debate going on as to whether the world is close to, or already has reached, 'peak oil' (the time when global oil production begins a terminal decline), oil companies today are faced with increased production difficulties. The problem is that the oil and gas industry has already picked much of the low-hanging fruit when it comes to exploring oil reservoirs. Much of the remaining oil resources will have to be produced from residuals that are increasingly difficult to recover—primary recovery can typically extract only 10–30% of the oil in places—and in deeper and less accessible reservoirs.

Some of the technical challenges in recovering untapped oil resources have to do with the extreme heat and pressure that oil drilling equipment is exposed to at increasing depths. One of the materials that is being stretched to its limits in these extreme conditions is rubber. Rubbers are almost exclusively used as sealing materials in oil probing and excavation, typically as O-rings and sealants between the various joining modules of a drill or probe.

The performance limits of modern rubbers are typically reached when temperatures exceed 200 °C or pressures go above 200 MPa. Commonly used O-rings in oil exploration typically operate under temperatures and pressures such as 175 °C and 135 MPa. Formulating rubber that has the ability to withstand higher temperatures and pressures has been a serious technological challenge in oil and gas exploration.

Morinobu Endo, a professor of electrical and electronic engineering at Shinshu University in Japan, explains that fluorine rubber filled with carbon black has traditionally been used as a sealant in oil exploration. "Although several publications regarding CNT-filled rubber composites have appeared, no one, to the best of our knowledge, has found a way to significantly improve the heat resistance and durability of CNT–rubber composites. This has been due to the great difficulty of dispersing CNTs homogeneously, and the lack of strong binding interactions between the filler and the rubber matrix."

In order to accomplish the homogeneous dispersion of CNTs within the rubber, Endo and an international team of collaborators developed a low-temperature milling process to attain enhanced elasticity and shear force. The result is an extreme-performance rubber nanocomposite material that is able to withstand temperatures of up to 260 °C and pressures as high as 239 MPa.

"Based on our team's estimates of the depths and temperatures of oil resources, the development of an extreme rubber sealant having the enhanced performance of 100 °C higher temperature and 70 MPa higher pressure durability, as compared to those of the currently used O-rings, will contribute to doubling the current average oil recovery efficiency by incorporation with other related technological innovations," says Endo.

Endo collaborated with professors Kenji Takeuchi, Takuya Hayashi, and Yoong Ahm Kim from his university as well as Mauricio Terrones from IPI-CYT in Mexico, Mildred Dresselhaus at MIT, and scientists from Nissin Kogyo, Schlumberger, and Fukoko. The scientists point out that the use of their novel CNT–rubber nanocomposite material is not limited to oil exploration but could be suitable for a wide range of innovative applications ranging from factory tools to environmental applications to aerospace and space technologies.

To develop their rubber nanocomposite, the researchers used a low-temperature roll mill process to disperse MWCNTs in a synthetic rubber matrix (fluoroelastomers/FKM). Endo explains that the three key issues in making the nanocomposite are the use of MWCNTs embedded in fluorinated rubber, the surface modification of these nanotubes, and the formation of a cellulation structure in the composite.

The temperature of the roll mill was maintained at less than 20 °C, to obtain enhanced elasticity and shear force. Endo explains that this process allows three important things: "(1) The elastomer molecules of the matrix fill the voids created by the physically intermingled nanotubes, thus effectively breaking their intrinsic agglomeration during the mixing process; (2) the rubber matrix exhibits good wettability with CNTs; and (3) the rubber matrix displays extreme elasticity."

To demonstrate an innovative end use of their novel nanocomposite, Endo and his collaborators developed a rubber sealant for oil exploration and probing purposes, which can be used under extremely harsh conditions. The pressure resistance of their nanocomposite O-rings was ~80–100 MPa higher than that of conventional sealants, corresponding to a water depth of 8000 m.

"Our experimental system was limited in its ability to simulate further extreme conditions, but we believe that our new rubber composite may actually exceed our claims," says Endo. He cautions, though, that the team has not yet developed an effective mass production technique for this CNT rubber nano composite, so practical applications may have to wait a while.

Featured scientist: Morinobu Endo
Organization: Department of Electrical and Electronic Engineering,
Shinshu University, Wakasato (Japan)
Relevant publication: Morinobu Endo, Toru Noguchi, Masaei Ito,
Kenji Takeuchi, Takuya Hayashi, Yoong Ahm Kim, Takashi
Wanibuchi, Hiroshi Jinnai, Mauricio Terrones, Mildred S.
Dresselhaus: Extreme-performance rubber nanocomposites for
probing and excavating deep oil resources using multi-walled carbon
nanotubes, *Adv. Funct. Mater.*, **18**, 3403–3409.

5.9   'Nailing' Superlyophobic Surfaces

When raindrops splash against your window you probably just think the weather has turned bad again. Physicists and material engineers, on the other hand, are quite fascinated by this process of 'wetting.' What happens when a fluid is brought in contact with a solid surface is much more complex than you might guess from just looking at your wet window. In physical terms, the process of wetting is driven by the *minimum free energy* principle—the liquid tends to wet the solid because this decreases the free energy of the system (in this case the system consists of a liquid plus solid). For low-surface-tension liquids the minimum free energy is achieved only when the liquid completely wets the solid. Understanding these mechanics, and using nanotechnology to structure surfaces to control wetting, has a far-reaching impact for many objects and products in our daily lives—by preventing wear on engine parts or fabricating more comfortable contact lenses, better prosthetics, and self-cleaning materials.

The primary measurement to determine wettability is the angle between the solid surface and the surface of a liquid droplet on the solid's surface. A droplet of water on a hydrophobic surface has a high contact angle, but a liquid spread out on a hydrophilic surface has a small one. Surfaces where the contact angle is approaching 180° are called *superhydrophobic;* surfaces where the contact angle is approaching 0° are called *superhydrophilic.* Advanced material engineering techniques can structure surfaces that allow dynamic tuning of their wettability all the way from superhydrophobicity to almost complete wetting—but these surfaces only work with high-surface-tension liquids. Unfortunately, almost all organic liquids that are ubiquitous in human environment, such as oils, solvents, and detergents, have fairly low surface tensions and thus readily wet even superhydrophobic surfaces. Researchers are about to create surfaces that will extend superhydrophobic behavior to all liquids, no matter what their surface tension is.

Contamination of superhydrophobic surfaces with low-surface-tension organic liquids is one of the leading reasons why superhydrophobic surfaces are not widely used in practical applications. If engineers were to succeed in creating a surface that repels any liquid, the practical implications would obviously be substantial. Such materials would have the ultimate self-cleaning and self-decontaminating surface. Other applications include control of low-surface tension liquids in microfluidic and lab-on-a-chip devices, which often require organic solvents to operate.

"Solid–liquid surface systems can't always attain their minimum free energy state; sometimes the system is 'captured' in a state which is not a minimum free energy state," explains Tom N. Krupenkin. "These states are called *metastable*. Unlike the minimum-free-energy state, which is usually unique, a broad variety of metastable states with various properties can exist. In particular, for the case of low-surface tension liquids, a minimum-free-energy state is always a wetted state but a metastable state can be a nonwetting state.

"My idea was to create a surface with a special type of nanoscale topography that would 'lock' the liquid in a metastable nonwetting state, thus preventing the surface from being wetted. Nanonails represent an example of such nanotopography. Moreover, we were able to electrically switch this metastable non-wetting state on and off, thus producing a surface that can be dynamically switched from superlyophobic (*i.e.* nonwettable) to a highly wettable hydrophilic state."

Krupenkin, an associate professor in the Department of Mechanical Engineering at the University of Wisconsin-Madison, believes that his team, in collaboration with researchers from Bell Labs at Lucent Technologies, was the first to use metastable states to control wetting properties of solids.

Since there is a much wider range of potentially available metastable states than of thermodynamically stable states, the approach proposed by Krupenkin and collaborators can greatly broaden the range of control mechanisms over the wetting behavior of solid surfaces.

Researchers call surfaces that extend superhydrophobic behavior to all liquids—no matter what their surface tension—*superlyophobic* (the word lyophobic means 'solvent-fearing'). Krupenkin and his team have managed to experimentally demonstrate tunable nanostructured surfaces that are capable of undergoing a transition from profound superlyophobic behavior to strong wetting.

Krupenkin notes that in the initial state, with no voltage applied, these surfaces exhibit contact angles as high as 150° for a wide variety of liquids with surface tensions ranging from 21.8 mN/m for ethanol to 72.0 mN/m for water. Upon application of an electrical voltage, contact angles of the liquids on these surfaces can be reduced to less than 30°.

"Our proposed approach opens a pathway to a simple, material-independent method for creating electrically tunable superlyophobic surfaces and thereby extends many of the potentially attractive properties usually associated with tunable superhydrophobic surfaces—high liquid droplet mobility, controllable chemical reaction initiation, tunable drag reduction, *etc.*—to a much broader range of materials and applications," says Krupenkin.

He points out that, in order to extend dynamically tunable superhydrophobic nanostructured surfaces to the superlyophobic domain, one needs to understand the reasons that prevent traditional superhydrophobic surfaces from being able to 'repel' low-surface tension liquids.

"When a droplet of a liquid is placed on a very rough surface, such as a fractal-like surface or a surface covered with a carpet of high-aspect-ratio micro- or nanoscale spikes, bumps, or posts, it assumes a state that minimizes its free energy," explains Krupenkin. "This can be achieved by the droplet either wetting the rough surface and thus substantially increasing the total area of the liquid–solid interface or by minimizing the liquid–solid contact area by dewetting most of the surface."

To create their superlyophobic surface, the researchers designed a special 3-D surface topography that resembles arrays of nails. Each nanonail consists of a conductive silicon stem and a dielectric silicon oxide nail head. The resulting structure is covered with a thin conformal layer of low-surface-energy fluoropolymer.

During electrically induced transitions from the nonwetting to the wetting state, the presence of the conductive nail stem allows the researchers to generate high electrical field values in the immediate vicinity of the liquid–air interface. Unless high pressure is applied, the liquid has to stay on top of the nanonails and is unable to penetrate inside and wet the nanostructured layer.

"Obviously, the nanonail geometry is not the only structure that allows one to lock the liquid in the nonwetting state," says Krupenkin. "Essentially, any geometry that features 'overhang' analogues to a nail head would produce similar results."

Expanding their work, the researchers want to better understand the stability of these surfaces with respect to environmental factors such as temperature, mechanical damage, vibrations, *etc.* Krupenkin says that he would also like to extend this technique to a wider range of materials, most importantly plastics, with the goal of eventually producing practically usable superlyophobic films or coatings.

Featured scientist: Tom N. Krupenkin
Organization: Department of Mechanical Engineering, University of Wisconsin-Madison, Madison, WI (USA)
Relevant publication: A. Ahuja, J. A. Taylor, V. Lifton, A. A. Sidorenko, T. R. Salamon, E. J. Lobaton, P. Kolodner, T. N. Krupenkin: Nanonails: a simple geometrical approach to electrically tunable superlyophobic surfaces, *Langmuir*, **24**, 9–14.

CHAPTER 6
Tiny Steps Towards Tiny Machines

Science fiction style robots like *Star Wars'* R2-D2 or the NS-5 model in *I, Robot* belong firmly to the realm of Hollywood—as do scary 'nanobots' *à la* Michael Crichton's *Prey*. Staying with both feet firmly on scientific ground, robotics can be defined as the theory and application of robots—completely self-contained electronic, electric, or mechanical devices—to an activity like manufacturing. Scale that robot down to a few billionth of a meter and you are talking *nanorobotics*. The field of nanorobotics brings together several disciplines, including nanofabrication processes used for producing nanoscale robots, nanoactuators, nanosensors, and physical modeling at nanoscales. Nanorobotic manipulation technologies, including the assembly of nanometer-sized parts, the manipulation of biological cells or molecules, and the types of robots used to perform these tasks also form a component of nanorobotics. Nanorobotics might one day even lead to the holy grail of nanotechnology where automated and self-contained molecular assemblers not only are capable of building complex molecules but build copies of themselves—*self-replication*—or even complete everyday products.

Whether this will ever happen is hotly debated. To understand where both sides stand, read the famous 2003 debate[1] where Drexler and Smalley make the case for and against molecular assemblers. In contrast, today's nanorobotics research deals with more mundane issues such as how to build nanoscale motors and simple nanomanipulators.

Nanorobots are quintessential nanoelectromechanical systems (NEMS) and raise all the important issues that must be addressed in the design of

---

[1] Read the entire debate here: http://pubs.acs.org/cen/coverstory/8148/8148counterpoint.html

---

RSC Nanoscience and Nanotechnology No. 8
Nano-Society: Pushing the Boundaries of Technology
By Michael Berger
© Michael Berger 2009
Published by the Royal Society of Chemistry, www.rsc.org

NEMS: sensing, actuation, control, communications, power, and interfacing across spatial scales and between organic and inorganic materials. Because of their size, which is comparable to that of biological cells, nanorobots have a vast array of potential applications in fields such as environmental monitoring or medicine.

6.1 Nanopiezotronics—a Pathway to Self-powering Nanodevices

Electric polarization resulting from mechanical deformation is perceived as *piezoelectricity*. The word is derived from the Greek *piezein*, which means to squeeze or press, and the phenomenon was discovered by the brothers Pierre and Jacques Curie in 1880. Piezoelectricity is a coupling between a material's mechanical and electrical behavior. When a piezoelectric material is squeezed, twisted, or bent, electric charges collect on its surfaces. Conversely, when a piezoelectric material is subjected to a voltage drop, it mechanically deforms. Many crystalline materials exhibit piezoelectric behavior and when such a crystal is mechanically deformed, the positive- and negative-charge centers are displaced with respect to each other.

The piezoelectric effect finds useful applications in the production and detection of sound, generation of high voltages, electronic frequency generation, microbalance, and ultrafine focusing of optical assemblies. The traveling-wave motor used for the autofocus function in cameras is a type of piezoelectric motor.

The term *nanopiezotronics* was coined by Zhong Lin Wang at Georgia Tech to describe the coupled piezoelectric and semiconducting property of nanowires and nanobelts for designing and fabricating novel electronic devices such as nanotransistors and nanodiodes. These devices could provide the fundamental building blocks that would allow the creation of a new area of nanoelectronics. As a new research field, nanopiezotronics refers to the generation of electrical energy at the nanoscale via mechanical stress to the nanopiezotronic device. For example, the bending of a zinc oxide (ZnO) nanowire transforms that mechanical energy into electrical energy that could be harvested. This new approach has the potential of converting biological mechanical energy, acoustic/ultrasonic vibration energy, or biofluid hydraulic energy into electricity, demonstrating a new pathway for self-powering wireless nanodevices and nanosystems.

The nanopiezotronic mechanism takes advantage of the fundamental property of nanowires or nanobelts made from piezoelectric materials: bending the structures creates a charge separation, positive on one side and negative on the other. The connection between bending and charge creation has also been used to create nanogenerators that produce measurable electrical currents when an array of ZnO nanowires is bent and then released.

Wang explains how nanopiezotronics, based on nanowires and nanobelts as the fundamental building blocks, has unique advantages that will help make integrated nanosystems possible:

- "Nanowire-based nanogenerators can be subjected to extremely large deformation, so they can be used for flexible electronics as a flexible/foldable power source.
- The large degree of deformation that can be withstood by the nanowires is likely to result in a larger volume density of power output.
- The material we used—wurtzite ZnO—is a biocompatible and biosafe material; it has great potential as an implantable power source within the human body.
- The flexibility of the polymer substrate used for growing ZnO nanowires and nanobelts makes it feasible to accommodate the flexibility of human muscles so that the mechanical energy (body movement, muscle stretching) in the human body can be used to generate electricity.
- ZnO nanowires and nanobelts nanogenerators can directly produce current as a result of their enhanced conductivity in the presence of oxygen vacancies.
- ZnO is an environmentally 'green' material. The phenomenon that was demonstrated for ZnO can also be applied to other wurtzite-structured materials, such as gallium nitride (GaN) and zinc zulfide (ZnS)."

In comparison to conventional electronic components, nanopiezotronic devices operate much differently and exhibit unique characteristics. In conventional field-effect transistors (FETs), an electrical potential—called the *gate voltage*—is applied to create an electrical field that controls the flow of current between the device's source and its drain. In the nanopiezotronic transistors developed by Wang and his research team, the current flow is controlled by changing the conductance of the nanostructure by bending it between the source and drain electrodes. The bending produces a gate potential across the nanowire and the resulting conductance is directly related to the degree of bending applied.

Wang and his group anticipate that nanopiezotronics will have a wide range of applications in electromechanical coupled sensors and devices, nanoscale energy conversion for self-powered nanosystems, and harvesting/recycling of energy from the environment. So far, they have been able to demonstrate FETs, diodes, sensors, and current-producing nanogenerators that operate by bending ZnO nanowires and nanobelts.

According to Wang, the physical principle for creating and separating the piezoelectric charges in the nanobelt is the first half of a nanogenerator for converting mechanical energy into electricity. "Then, the preserving and discharging is the second half of the nanogenerator, which is a coupling process of piezoelectric and semiconducting properties."

It is important to find various approaches that are feasible for harvesting energy and recycling energy from the environment to self-power a nanosystem so that it can operate wirelessly, remotely, and independently with a sustainable energy supply.

Wang says that the principle he and his group demonstrated for the piezoelectric nanogenerator could be the foundation for self-powered nanosystems. "It also has the potential to harvest or recycle energy from the environment or recycle energy that is wasted, such as the energy when walking." He gives two examples:

(1) Piezotronic nanosensors can measure nanonewton forces by examining the shape of the structure under pressure. Implantable sensors based on this principle could continuously measure blood pressure inside the body and relay the information wirelessly to an external device similar to a watch. The device could be powered by a nanogenerator harvesting energy from blood flow.
(2) Other nanosensors can detect very low levels of specific compounds by measuring the current change created when molecules of the target are adsorbed to the nanostructure's surface.

"Utilizing this kind of device, you could potentially sense a single molecule because the surface area-to-volume ratio is so high," says Wang.

> *Featured scientist:* Zhong Lin Wang
> *Organization:* School of Materials Science and Engineering, Georgia Institute of Technology, Atlanta, GA (USA)
> *Relevant publication:* Z. L. Wang: Nanopiezotronics, *Adv. Mater.* (*Weinheim, Ger.*), **19**, 889–892.

6.2 Nanoelectromechanical Systems Start to Take Shape

In his famous 1959 speech 'Plenty of Room at the Bottom', Richard Feynman offered a prize of $1000 "to the first guy who makes an operating electric motor—a rotating electric motor which can be controlled from the outside and, not counting the lead-in wires, is only 1/64 inch cube." Feynman had hoped his reward would stimulate some new fabrication technology but he was quite taken aback when one year later, Bill McLellan, using amateur radio skills, built the motor with his hands using tweezers and a microscope (and many, many hours of fiddling around). McLellan's 2000 rpm motor weighed 250 µg and consisted of 13 parts. In the almost 50 years since, not only has the field of microelectromechanical systems (MEMS) caught up with Feynman's bet and achieved commercial production capabilities of motors many times smaller than McLellan's, but researchers have begun exploring another level of miniaturization—NEMS. Efficient actuation, the creation of mechanical motion by converting various forms of energy to rotating or linear mechanical energy, is an important—and today still frustrating—issue in designing NEMS. Research on building functional nanoscale electromechanical systems is well under way, as just demonstrated by another achievement by scientists at Caltech—the place where Feynman gave his speech and where McLellan's motor still is on display.

The two principal components common to most electromechanical systems, irrespective of scale, are a mechanical element and transducers. The mechanical element either deflects or vibrates in response to an applied force. Depending on its type, the mechanical element can be used to sense static or time-varying forces. The transducers in MEMS and NEMS convert mechanical energy into electrical or optical signals and vice versa.

"NEMS have shown great promise as highly sensitive detectors of mass, displacement, charge, and energy," says Michael L. Roukes. "However, as electromechanical devices are scaled downward, transduction becomes increasingly difficult, hampering efforts to create finely controlled integrated systems. In spite of substantial progress in the field, an efficient, integrated, and customizable technique for actively driving and tuning NEMS resonators has remained elusive."

What NEMS researchers have found is that conventional approaches such as magnetomotive, electrostatic, and electrothermal techniques suffer from either low power efficiency, limited potential for integration, or poor nanoscale control over electromechanical coupling.

Roukes, a professor of physics, applied physics and bioengineering at California Institute of Technology's Kavli Nanoscience Institute, together with his colleague Sotiris Masmanidis, describes how the piezoelectric effect can give rise to interesting electromechanical actuation phenomena that are only strongly manifested in NEMS. Their work is a collaborative effort with Iwijn De Vlaminck from IMEC in Belgium and researchers from the National Institute for Nanotechnology in Edmonton, Canada.

"The traditional view calls for well-defined, alternating layers of electrodes and piezoelectric materials," says Masmanidis. "But this view breaks down in nanoscale devices made from semiconductors where charge depletion smears out the boundary between piezoelectrically inactive (electrically conducting) and active (electrically insulating) regions. This effect is only significant when the total device thickness approaches the semiconductor's charge depletion width."

To demonstrate the concept of depletion-mediated NEMS actuation, the team fabricated cantilevers and doubly clamped beams out of a 200 nm thick diode consisting of gallium arsenide (GaAs). The piezoelectric effect inherent to this material afforded a highly efficient means of resonantly exciting these devices with an alternating voltage.

"Our experiments revealed that the efficiency of driving the devices can be tuned with DC voltage, as well as by tailoring the doping profile of the diode" says Masmanidis. "This novel form of control offers a new paradigm in designing NEMS."

Besides depletion-mediated actuation, the team demonstrated some additional functionality with these devices. One remarkable feature of NEMS fabricated from piezoelectric materials is voltage-induced resonance-frequency control; another is a nanomechanical logic element based on the crystallographic anisotropy of piezoelectricity. "The integration of a reliable and customizable frequency-tuning method adds a useful layer of functionality that has so far been absent in NEMS," says Roukes.

Masmanidis points out that the electromechanical coupling property of their device relies entirely on intrinsic material properties. This should facilitate the creation of compact, tunable NEMS arrays for applications ranging from mass and charge detection for chemical species sensing, through high speed and low power switches, to nanomechanical logic gates for new computational paradigms. In particular, the ability to regulate actuation efficiency through depletion-mediated strain in the semiconductor heterostructure's low–operating-power regime raises the prospect of developing efficient, high-speed electromechanical switches.

"In principle, the reversibility of piezoelectric phenomena also offers the potential for ultrasensitive electrical measurement of nanomechanical motion," says Roukes. "Finally, all the concepts we presented are transferable to a wide variety of other materials beyond GaAs—such as aluminium nitride (AlN), silicon carbide (SiC), or ZnO—which may provide enhanced electrical and mechanical properties."

> *Featured scientists:* Michael L. Roukes, Sotiris Masmanidis
> *Organization:* Kavli Nanoscience Institute, California Institute of Technology, Pasadena, CA (USA)
> *Relevant publication:* Sotiris C. Masmanidis, Rassul B. Karabalin, Iwijn De Vlaminck, Gustaaf Borghs, Mark R. Freeman, Michael L. Roukes: Multifunctional Nanomechanical Systems via Tunably Coupled Piezoelectric Actuation, *Science*, **317**, 780–783.

6.3 The Promise of Industrial-scale Nanofabrication Techniques

Fundamental nanotechnology research in labs around the world advances rapidly, as witnessed by the hundreds of new nanoscience-related research papers that get published every month. The big bottleneck in getting these new technologies from the lab translated into commercial products is the lack of suitable large-scale fabrication techniques. Almost all laboratory experiments involve elaborate set-ups and are quite tricky processes that require a lot of skill and expertise on the part of the researchers. Nanotechnologies today are more an art than a basis for industrial technologies. Think of a 15th century monk spending 10 years painstakingly writing and painting every letter, dot, and adornment of a single bible—that's where nanotechnology is today. Where we need to get to is something that resembles modern high-speed printing machines that print thousands of books an hour.

Take nanowires, for instance. Researchers have used nanowires to create transistors like those used in memory devices and prototype sensors for gases or biomolecules. A common approach in the lab is to grow nanowires like blades of grass on a suitable substrate, mow them off and mix them with a fluid to transfer them to a test surface. When the carrier fluid dries, the nanowires are left behind like tumbled jackstraws. Using scanning probe microscopy (SPM) or similar tools, researchers hunt around for a convenient, isolated nanowire to work on, or place electrical contacts without knowing the exact positions of the nanowires. This is not a technique suitable for mass production.

A group of researchers have developed a novel technique that allows them to grow nanowires selectively on sapphire wafers in specific positions and orientations accurately enough to attach contacts and layer other circuit elements, all with conventional lithography techniques. This fabrication method requires a minimum number of steps and is compatible with the processes in today's microelectronics industry.

"One of the promises of nanotechnology is nanodevice fabrication from very small building blocks such as nanocrystals," explains Babak Nikoobakht. "If you look at the scientific literature there are many, many nanowire growth methods—but methods that are suitable to make large numbers of nanodevices are as good as nonexistent. At least you can't find any actual examples of mass production anywhere. My motivation was to fill the current gap and convince people that there are other ways to solve the current issues of nanodevice fabrication."

In his work, Nikoobakht, a researcher at the U.S. National Institute of Standards and Technology's (NIST) Surface and Interface Research Group in Maryland, explores the ability to electrically address the coordinates of millions of horizontally grown nanowires on a given surface.

"Our method allows us to control the nanowire's location and its orientation," says Nikoobakht. "Currently, for large-scale nanowire device fabrication, it is necessary to grow nanowires in standing positions. These nanowires have to be transferred to a surface and aligned. This requires multistep treatments and brings complexities and inefficiency issues, especially when industrial scaling is the goal. Our method has a minimum number of steps. It combines a chemical growth method with well-known optical lithography techniques. Nanowires are grown where they need to be."

The NIST researchers used conventional semiconductor manufacturing techniques to deposit small amounts of gold in precise locations on a sapphire wafer. In a high-temperature process, the gold deposits bead up into nanodroplets that act as nucleation points for crystals of ZnO, which is a technologically important material—as sensor, piezoelectric component, UV light emitter, and transparent semiconductor in the visible spectrum. A slight mismatch in the crystal structures of ZnO and sapphire induces the semiconductor to grow as a narrow nanowire in one particular direction across the wafer.

Because the starting points and the growth direction are both well known, it is relatively straightforward to add electrical contacts and other features with additional lithography steps.

This appears to be the first technique that allows nanodevice fabrication on the original position of nanowires and this is done using a combination of a bottom-up chemical approach and conventional optical lithography. It is scalable, *i.e.* this could be done on 10–15 cm wafers. It is simple because nanowires are grown horizontally and there is no need to remove, align, or treat them.

"We have shown that these nanodevices could work as FETs, and that one can make them in large numbers; but more time is needed to answer all the questions about their performance, stability, and so on," says Nikoobakht. "ZnO and sapphire are transparent materials in the visible region of the electromagnetic spectrum and people see a strong potential for such materials in transparent circuitry and displays. Sensing is another application, with powerful sensors that are better than the current thin film technology. Light emitters and flash memory devices would be other applications."

Nikoobakht points out that, in order to demonstrate the potentials of this method, the NIST team fabricated nanowire FET devices with one, two, or several nanowires. He also notes that, using this platform, an interesting distinction between electrical properties of individual and ensembles of nanowires was found.

Compared to vertical, standing nanowire fabrication techniques, horizontal methods are a fairly new approach and it will take time for other researchers to realize their potential. Nikoobakht is convinced that this method for growing nanocrystals could have quite a significant impact on a wide range of applications such as single photon sources, LEDs, photodetectors, transparent FETs, and displays.

"The important part is that the nanowires are grown where the devices will be fabricated and there is no need to transfer them to a different substrate," concludes Nikoobakht. "This approach in its current state seems to be a promising methodology for parallel nanodevice fabrication at technologically relevant scales."

Featured scientist: Babak Nikoobakht
Organization: Surface and Interface Research Group, National Institute of Standards and Technology's (NIST), Gaithersburg, MD (USA)
Relevant publication: Babak Nikoobakht: Toward industrial-scale fabrication of nanowire-based devices, *Chem. Mater.*, **19**, 5279–5284.

6.4 Swimming Microrobots Propelled by Bacteria

It's not quite like Isaac Asimov's science fiction classic *Fantastic Voyage*, where five people travel in a microscopic submarine inside a person's bloodstream, but scientists have talked for quite some time about micro- and nanorobotic devices that can travel inside the human body and carry out a host of complex medical procedures such as monitoring, drug delivery, and cell repair. Recent developments in micro- and nanoscale engineering have led to the realization of various miniature mobile robots. It turns out that the most significant bottleneck for further miniaturization of these devices down to micron scale is the miniaturization of the on-board actuators and power sources required for mobility.

Nature has already provided remarkable solutions to this problem by evolving chemically powered molecular motors. Such biomotors seem to be one of the most promising choices for on-board actuation. They are advantageous over human-made actuators because they are much smaller and are capable of producing more complicated motions. More importantly, they convert chemical energy to mechanical energy very efficiently. However, the major drawback of biomotors is that isolating and reconstituting them are complicated tasks with low yield and it is difficult to interface them with electronic circuitry. This has led scientists to experiment successfully with a new approach by using flagellar motors (the propulsion system of bacteria), still inside the intact cells, as actuators.

"We demonstrated the feasibility of using bacterial flagella as bioactuators by propelling 10 μm polystyrene beads with several *S. marcescens* bacteria attached to them," explains Metin Sitti. "The biomotors are still inside the intact cell which has several advantages: there is no need for purification and reconstitution; a simple nutrient such as glucose is provided, and ATP[2] or ion gradients are generated by the cell. Most importantly, sensors are already present in the cell and integrated with the motors. Lastly, more complex organelles can be used, hence more sophisticated motion can be generated."

Sitti is associate professor at the Department of Mechanical Engineering and Robotics Institute at Carnegie Mellon University in Pittsburgh. He points out that, although there have been examples of using bacteria or algae as nano-actuators by the research groups of Howard Berg and George Whitesides at Harvard, there has been no work using these actuators for microrobot propulsion and controlling their motion on/off using chemical stimuli. Sitti and PhD student Bahareh Behkam have show in a series of experiments that several bacteria, randomly attached to a 10 μm polystyrene bead, produce a large enough propulsive force to propel the bead forward. On/off motion control of the bead is achieved by introducing copper ions and subsequently EDTA,[3] prompting the bacterial flagellar motor to halt and resume motion without causing any damage.

This method of motion control is reversible and can be used repeatedly. Bacteria demonstrate an immediate response to the metal ion and chelating agent. However, when the sample volume is large, the response time is affected

---

[2]Adenosine triphosphate (ATP) is the main energy storage and transfer molecule in cell metabolism.

[3]Ethylenediaminetetraacetic acid, normally referred to as EDTA, is a widely used chelating agent— a substance whose molecules can form several bonds to a single metal ion and thus effectively 'remove' the ion from solution.

by the diffusion rate of these chemicals. This problem can be overcome by local introduction of the chemicals. Figure 6.1

Sitti notes that the mechanism of adhesion of bacteria to surfaces is not well understood, though it is speculated that it occurs in two steps: (1) Reversible adhesion first occurs within a few seconds. This is due to van der Waals forces, electrostatic forces, or acid–base interactions. (2) Irreversible attachment happens after the reversible attachment is made.

In their experiments, Sitti and his team observed that bacteria in a motility medium randomly interact with the microbeads and sometimes attach to them. "Different beads demonstrated different behaviors," says Sitti. "Some of the beads did not have any bacteria attached to them and were not mobile. For the 35 experiments we carried out, on average 60% of the beads were mobile."

Although significant displacement of the polystyrene bead was consistently observed, the net displacement and speed of the beads were not identical across multiple experiments. This is largely due to the fact that wild-type bacteria were used in these experiments, and their flagellar motors demonstrate random "run" and "tumble" behavior. In addition, for any given bead, different numbers of bacteria are attached to the bead and flagellar motors, with each bacterium demonstrating different behaviors at any instant in time, hence it would be nearly impossible to have two beads move on similar paths. Further, the orientation and spacing of the adhered bacteria were not controlled.

After demonstrating the propulsion of the bead by bacteria, devising a speed control technique was the next significant step towards developing a

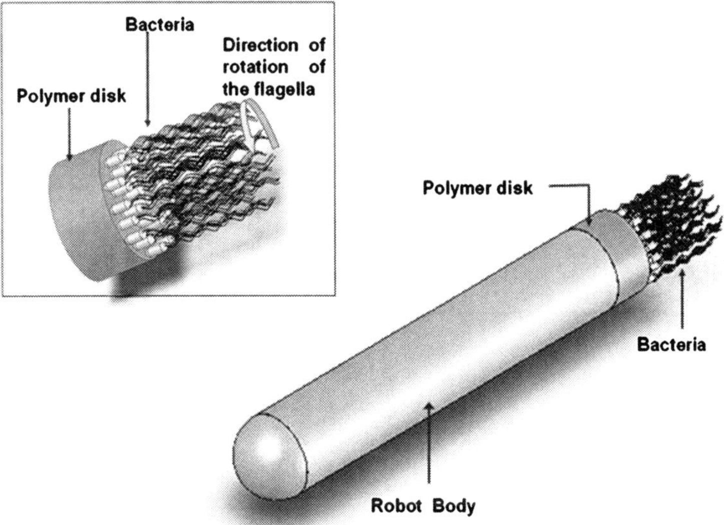

Figure 6.1 Conceptual drawing of the hybrid swimming robot; robot body bonded to the propulsive element. (Illustration: Dr M. Sitti, Carnegie Mellon University)

microrobotic system. Chemical and optical stimuli can be used to modulate the speed of the bacteria, but neither of these two stimuli can cause the flagellar motors to halt.

Sitti's team used the fact that the absence of a chelating agent in bacterial suspensions leads to paralysis of the bacteria. The reason for this is that in the absence of the chelating agent the heavy metal ions that are naturally present in water bond to the rotor of the flagellar motors of the bacteria and prevent their motion.

"By taking advantage of this fact we can purposely paralyze the bacteria, but only temporarily and in a reversible fashion," says Sitti. "It is important to note that unlike the chemical stimuli which interact with chemoreceptors causing them to signal to the motor to modulate its speed, heavy metal ions bond directly to the rotor of the flagellar motor, impairing its motion instantaneously." When an EDTA solution was added, the polystyrene bead resumed its motion immediately.

Sitti and his team are aiming to fabricate microscale swimming robots to swim within low-velocity and stagnant liquids of the human body. "This propulsion method will be used to develop hybrid biotic/abiotic swimming microrobots for potential biomedical applications such as noninvasive screening for diseases and targeted drug delivery in areas of the body with stagnant or low-velocity fluid flow. In addition, we could use them to detect leakages inside tiny pipes such as in nuclear plants and space shuttles."

Although these results are very promising first steps toward a swimming microrobot, there is a lot of work to be done. "We are at the very beginning stage of 1-D robot propulsion," Sitti says. "We need to steer the beads in 3-D, and fabricate and integrate an onboard on/off chemical stimulus control system (using microfluidic channels and valves), CMOS electronics-based simple processor and sensors, and wireless communication type microcomponents on the robot body to build a complete hybrid robot and conduct biomedical applications."

> *Featured scientists:* Metin Sitti, Bahareh Behkam
> *Organization:* Department of Mechanical Engineering and Robotics
>   Institute, Carnegie Mellon University, Pittsburgh, PA (USA)
> *Relevant publication:* Bahareh Behkam, Metin Sitti: Bacterial
>   flagella-based propulsion and on/off motion control of microscale
>   objects, *Appl. Phys. Lett.*, **90**, 023902.

6.5   On the Way to Building Nanomachinery

Machines usually require various components such as bearings, gears, couplings, or pistons. As machines shrink to the micro- and ultimately the nanoscale, their components of course need to shrink with them. One of the major obstacles to the realization of sophisticated nanomachines is the lack of effective processes for building free-standing nanocomponents of specific shapes and sizes. Self-assembly methods produce both organic and inorganic nano-objects with high yields through bottom-up approaches. In most cases, however, the shapes are confined to rather simple forms such as spheres, rods, triangles, or cubes and are not suitable for the elementary components of intricate nanomachines. Meanwhile, top-down approaches including electron beam lithography and micro-contact printing focus on surface patterning or fabrication of suspended objects, although they can fabricate sophisticated nanostructures. So far, the fabrication and assembly of nano-objects with specific shapes and sizes that can act as elementary components for movable NEMS is only at the conceptual stage. Techniques being developed in South Korea offer a step toward the realization of complex nanomachines, designed to perform specific tasks, with overall dimensions comparable to those of biological cells.

"We have fabricated freestanding gold nanogears less than 500 nm in overall size and 60–70 nm in thickness, and assembled them tooth to tooth without damage," says Dong Han Ha. "This is the first case that we know of in which free-standing nanocomponents with precisely designed shape and size have been fabricated and assembled. The freestanding nanogears with one center hole surrounded by six teeth were reproducibly obtained by selectively etching chemically synthesized single-crystalline gold nanoplates using gallium focused ion beam (FIB). The fabrication process is very simple and it takes only 10–20 s to fabricate a nanogear under real time monitoring."

Ha is a principal researcher in the Division of Advanced Technology at the Korea Research Institute of Standards and Science (KRISS). Together with his colleagues from KRISS he has demonstrate that blunt AFM tips on stiff cantilevers are effective for the nondestructive manipulation of ductile and flat nano-objects having large contact areas with the substrate.

The KRISS team first synthesized gold nanoplates that were used as the raw material for the fabrication of freestanding nanogears. "The in-plane size and thickness of the nanoplates can be adjusted by controlling the synthesizing condition," says Yong Ju Yun, a postdoctoral researcher in Ha's team. "A lot of high-quality nanoplates motivated us to face the challenge of overcoming the limitations of previous methods, that is, the motivation for our work was to fabricate free-standing nanocomponents with precisely designed shape and size using a lot of high-quality nanoplates by combining bottom-up and top-down approaches."

The nanogears are fabricated by selectively etching the nanoplates sprinkled on a silicon dioxide (SiO_2) substrate using FIB. "We first checked whether or not the nanogears could be separated from the fabrication sites using a micromanipulator equipped with a tungsten probe," says Ha. "In most cases, the nanogears were severely bent or torn accidentally by the tungsten probe before they started to move. It seems that the nanogears bonded to the substrate surface during the FIB process because of the instantaneous melting and re-solidification of gold nanoplates along the brims of etched regions."

To remove the nanogears, and in order to weaken the bond between them and the substrate, the scientists slightly etched the underlying SiO_2 layer by dipping the sample into a buffered oxide etch solution for 20 s and rinsing it with deionized water at room temperature. Afterwards, oxygen plasma ashing was carried out for 1 min to remove residual organic materials on the substrate surface.

"After etching the SiO_2 layer, we could easily flip, pick up, translate, and rotate the nanogears without damaging them at all, which also confirms that both sides of the nanogears are clean and flat," says Ha. "However, the micromanipulator was not suitable for precisely controlling the position and angle of the nanogears in order to assemble them tooth to tooth. For a precise manipulation of the nanogears we had to use an AFM in contact mode." Figure 6.2

Ha expects that the precise manipulation processes combined with the fabrication techniques for freestanding nanocomponents with arbitrarily designed shapes and sizes will offer the first step to the realization of the development and post-repair of sophisticated nanoelectromechanical systems like nanobots. "Moreover," he says, "our work will also promote the study on the effects of size and shape on the various properties of nanostructures." Ha and his team are already working on the next step: trying to rotate the nanogears after setting them on fixed shafts.

The main challenges for researchers in this field today lie in the study and the development of efficient 3-D assembling and driving techniques. An assembly process using micromanipulators or AFMs is very tedious and time consuming and does not lend itself to industrial-scale assembly processes.

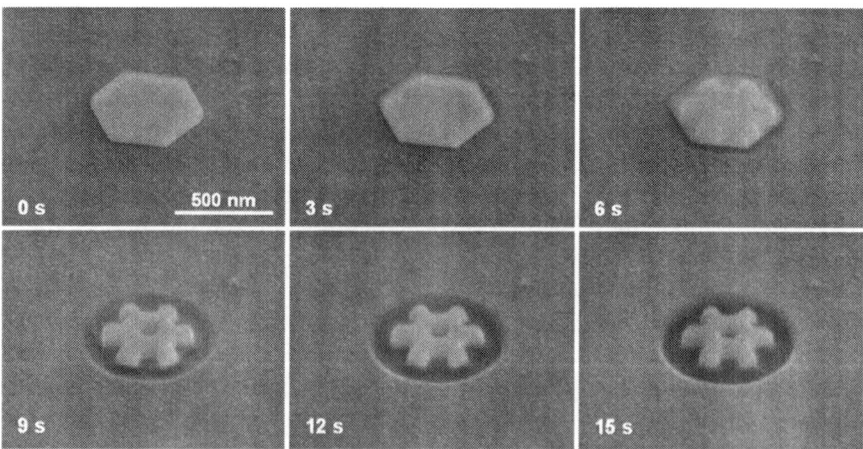

Figure 6.2 Successive images of the fabrication of a nanogear by selective etching of a gold nanoplate using a Ga+ focused ion beam. The time on each frame denotes the elapsed etching time. (Image: Dr D. H. Ha, KRISS)

Featured scientist: Dong Han Ha
Organization: Division of Advanced Technology, Korea Research
    Institute of Standards and Science (KRISS), Daejeon (South Korea)
Relevant publication: Yong Ju Yun, Chil Seong Ah, Sanghun Kim,
    Wan Soo Yun, Byong Chon Park, Dong Han Ha: Manipulation of
    freestanding Au nanogears using an atomic force microscope,
    Nanotechnology, **18**, 505304.

6.6 A Gripping Tale of Nanomanufacturing

Sporadic automation of certain nanofabrication tasks has already begun (for instance, automated microrobotic injection of foreign materials into biological cells) but we are a long way from industrial-scale, automated mass production processes for nanoscale devices. It has long been a dream for nanotechnologists that robots could one day be used in an assembly-line type of process to manufacture nanodevices. Researchers are beginning to develop the first rudimentary nanomanipulation devices that could lead to future automated manufacturing systems. A team of scientists in Canada have reported the first demonstration of closed-loop force-controlled grasping at the nanonewton level.

"We have demonstrated force-controlled micrograsping of highly deformable cells at a 20 nN force level," says Yu Sun, an assistant professor in mechanical and industrial engineering at the University of Toronto. "The contact force feedback of the MEMS-based microgripper enables the micromanipulation system to conduct rapid contact detection at a nanonewton force level and protects the microgripper from breakage. The work clearly explains the importance of the availability of force feedback along multiple directions. The implication is that these microgrippers get us closer to reliable, autonomous micromanipulation."

Sun and his team are applying these force-feedback microgrippers to characterize elastic and viscoelastic properties of polymeric microcapsules ($<20\,\mu m$ in diameter) used for drug delivery and cell encapsulation. He explains that they are also further improving the force-sensing resolution of these microgrippers for sub-nanonewton force measurements. "This will enable a new, easy-to-operate technique for our fundamental cell mechanics studies, for instance for distinguishing malignant cells from benign cells and for correlating mechanical properties of cells to disease states," he says.

In order to facilitate the maneuvering of nanoscale materials (*e.g.* pick–transport–place), Sun's group is further miniaturizing these microgrippers, concentrating not so much on the overall device size, but more on reducing the gripping tip thickness and integrating novel mechanisms to facilitate nanoobject release by counteracting undesired adhesion forces.

"These nanogripping devices promise highly reproducible pick–place of nanoscale objects, a capability the nanorobotics community has been striving for," says Sun.

For their microgripper design, the researchers used a V-beam electrothermal actuator to control the opening of the active gripping arm for object grasping. With an applied voltage, the V-beams are heated and expand to produce motion. The microgripper used is a commonly closed type with an initial opening of 5 µm. When actuated, the active gripping arm is pulled open. In order to prevent a high temperature at the gripping arm tips, electrical and thermal isolation on the device silicon layer is implemented, and many heat-sink beams are used. The temperature rise at the gripping arm tips caused by the integrated electrothermal microactuator was determined to be tolerable by biological cells.

To verify the effectiveness of their force-controlled manipulation system, the team selected biological cells for manipulation because of their high delicacy, high deformability, variations in size, and mechanical properties. The two key steps in the process are contact detection as the gripper approaches a cell, and the grasping of the cell.

"Contact detection is important to protect both the microgripper and the cell from damage," says Sun. "Without the integrated contact force sensor, this process would be extremely time consuming and dependent on operator skill. When the monitored contact force level reaches a preset threshold value, it indicates that contact between the gripping arm tips and the substrate is established. Subsequently, the microrobot stops lowering the microgripper further and moves the microgripper upward until the contact force returns to zero. After the initial contact position is detected, the microgripper is positioned a few microns above that."

Sun explains that, in order to achieve reliable micrograsping, a control system was implemented by using gripping force signals as feedback to form a closed loop. "Enabled by the monolithic microgripper with two-axis force feedback, our system demonstrates the capability of rapidly detecting contact, accurately tracking nanonewton gripping forces, and performing reliable force-controlled micrograsping to accommodate size and mechanical property variations of objects."

Besides force-controlled manipulation of biomaterials in liquid, these grippers can also find important applications in mechanical characterization of biomaterials and in electronic component handling as well as the assembly of micro objects.

Featured scientist: Yu Sun
Organization: Institute of Biomaterials and Biomedical Engineering, University of Toronto (Canada)
Relevant publication: Keekyoung Kim, Xinyu Liu, Yong Zhang, Yu Sun: Nanonewton force-controlled manipulation of biological cells using a monolithic MEMS microgripper with two-axis force feedback, *J. Micromech. Microeng.*, **18**, 055013.

6.7   Playing Nano-pinball in the Atomic Café

The concept of a 'machine'—a mechanical or electrical device that transmits or modifies energy to perform a certain task—can be extended to the nanoworld. On the nanoscale, the nanomachine components would be atomic or molecular structures each designed to perform a specific task which, taken all together, would result in a complex function. However, these nanomachines cannot be built just by further miniaturizing machine blueprints from the macroworld. Computer chips are an example where quantum size and coherence effects, high electric fields creating avalanche dielectric breakdowns, heat dissipation problems in closely packed structures, and the relevance of single-atom defects all present roadblocks along the current road of miniaturization.

Functional nanomachinery will need to take into account the quantum effects that dominate the behavior of matter at the nanoscale, affecting optical, electrical, and magnetic behavior.

Scientists in The Netherlands have shown how an atomic-scale mechanical device consisting of two moving parts, each composed of only two atoms, can be controlled by an external electrical signal, while being stable and providing a variety of functional modes. They jokingly refer to it as playing atomic pinball, since the two moving parts resemble the flippers in a pinball machine—unfortunately they haven't yet got a ball to play with.

"We have demonstrated the stimulated and controllable mobility of an atomic-scale mechanical device," says Harold J.W. Zandvliet. "This atomic-scale variant of pinball machine flippers exhibits a variety of dynamic modes that are exclusively excited by an external electrical signal. Our work is an important advance in atomic-scale engineering since it shows that even on the scale of a few atoms, a device can be constructed that operates only if an external stimulus is applied."

Zandvliet, a professor at the MESA + Institute for Nanotechnology, University of Twente in The Netherlands, together with his team, discovered this atomic pinball machine by accident since they did not set out to design or assemble a NEMS device.

"Our initial aim was to study the physical properties of self-organizing dimerized atomic chains," explains Zandvliet. "These 1-D systems can exhibit a myriad of exotic physical properties such as a Peierls instability, quantization of conductance, and Luttinger liquid behavior."

The elementary building block of the structure that the Dutch researchers have studied is a pair of platinum atoms, referred to as a *dimer*, on a germanium surface.

Zandvliet explains that, unlike an individual atom, a dimer can exhibit a rotational mode that leads to additional distinguishable and often energetically equivalent configurations.

"While our initial experiments did not provide any evidence for dynamic behavior of the dimer pairs, we came across one special case where the atomic structure of a dimer pair deviated from the periodic structure that is observed elsewhere in the chains," he says. "In this special case, the two center atoms of the dimer pair rearranged frequently while the scanning tunneling microscope (STM) tip was scanning across the chain." On closer examination, the team found that one atom in the pair acts as a pivot while the other swings back and forward, like the flipper in a pinball machine. But they did this only when they were exposed to the electrons from the STM. Increasing the strength of the current increased the flipping frequency.

Although this experiment nicely demonstrates the feasibility of controlling the mobility of an atomic-scale mechanical device, there are several basic questions and issues that need to be resolved before this can be applied to functional nanomachinery. According to Zandvliet, the two crucial questions that remain to be answered regarding the mobility of the dimer pair are why the particular dimer configuration that they have investigated is mobile, and how it differs structurally from other dimer pairs in the chains.

"We have been analyzing the current time traces in detail," he says. "Out of a billion electrons that tunnel between tip and pinball machine, only one induces a flip–flop event. Our analysis reveals that this one-electron process is entirely stochastic."

The scientists believe that the most likely explanation for the flipping dimer is that the substrate atom underneath the two revolving atoms has been replaced by a different atom, leading to reduced binding of the dimers with the substrate and a sideways displacement of the dimers as they attach to neighboring substrate atoms. "Unfortunately, it is not possible for STM to identify the chemical nature of this subsurface atom and no other available chemical characterization technique has the required resolution, meaning that the chemical identification of this atom cannot be performed experimentally," says Zandvliet.

A more practical complication is that, although the current of the STM tip can be used to reliably influence the rate at which the dimer pair switches from one mechanical configuration to another, using an STM is a very complicated way of applying the external signal. It would be much nicer if one could simply apply a voltage to external contacts and subsequently adjust the flipping frequency of the flippers.

> *Featured scientist:* Harold J.W. Zandvliet
> *Organization:* MESA + Institute for Nanotechnology, University of Twente (The Netherlands)
> *Relevant publication:* Amirmehdi Saedi, Arie van Houselt, Raoul van Gastel, Bene Poelsema, Harold J. W. Zandvliet: Playing pinball with atoms, *Nano Lett.*, online, DOI: 10.1021/nl8022884.

6.8   Plastic Motors Driven by Light Alone

A fast-growing body of nanotechnology research is dedicated to nanoscale motors and molecular machinery. The results of these studies are spectacular—

well-designed molecules or *supramolecules* show various movements upon exposure to various stimuli, such as molecular shuttles, molecular elevators, and molecular motors. However, nobody has yet been able to utilize the mechanical work done by these molecular machines on the macroscale.

Making progress towards this goal, an international group of researchers has succeeded in amplifying the minuscule change in structures at a molecular level caused by an external stimulus (light) to a macroscopic change through a cooperative effect of liquid crystals. Using liquid-crystalline elastomers (LCEs)—unique materials having properties both of liquid crystals (LCs) and of elastomers—the scientists have successfully developed new photomechanical devices, including the first light-driven plastic motor. In other words, with this novel material the energy from light can be converted directly into mechanical work without the aid of batteries, electric wires, or gears.

"A motor device is one of the most useful energy conversion systems that can convert input energy directly into a continuous rotation," explains Tomiki Ikeda. "Although chemomechanical motors and light-switchable molecular machines that can move objects by light have been demonstrated, light-driven plastic motors converting light energy directly into a rotation have not yet been realized."

Ikeda, a professor of polymer chemistry in the Chemical Resources Laboratory at Tokyo Institute of Technology, and his group focus on the fundamental understanding of the interaction between light and polymer materials from the viewpoint of innovative photonic applications. In the process, they create photofunctional polymer materials with precisely controlled molecular alignment.

Ikeda's team, in collaboration with Yanlei Yu's group at Fudan University in Shanghai and Christopher J. Barrett's group at McGill University in Montreal, have demonstrated new sophisticated motions of LCEs and their composite materials, which resulted in a plastic motor driven only by light.

The researchers prepared a continuous ring of LCE film which incorporated photochromic molecules such as an azobenzene. Upon simultaneous exposure to UV light from the downside right and visible light from the upside right, the ring rolled intermittently toward the light source, resulting in an almost 360° roll at room temperature.

"This is the first example of this kind of photoinduced motion in a single-layer film, although the rolling of the LCE ring was slow and stopped when the ring was broken by irradiation," says Ikeda.

He describes a plausible mechanism for the rotation: "Upon exposure to UV light, a local contraction force is generated at the irradiated part of the belt near the right pulley along the alignment direction of the azobenzene mesogens, which is parallel to the long axis of the belt. This contraction force acts on the right pulley, leading it to rotate in the counterclockwise direction. At the same time, the irradiation with visible light produces a local expansion force at the irradiated part of the belt near the left pulley, causing a counterclockwise rotation of the left pulley. These contraction and expansion forces, produced simultaneously at different places along the long axis of the belt, give rise to the rotation of the pulleys and the belt in the same direction. The rotation then brings new parts of the belt to be exposed to UV and to visible light, which enables the motor system to rotate continuously."

The size of the samples used in the experiments is in the millimetre range, but is not in principle material-limited, so numerous applications even on the nanoscale are possible, especially where efficient power supply to mechanical systems needs to be battery-free and non-contact. Photomobile polymers function with a minimum of moving parts, which minimizes the friction and surface contact difficulties that exist on a very small scale.

Ikeda points out that various complex 3-D movements such as rolling and rotation of cross-linked liquid-crystalline polymers with azobenzene dyes can be induced upon irradiation with light. Light can be handled remotely, instantly and precisely, especially with lasers, and these photomobile plastic materials can work as the main driving parts of light-driven actuators without the aid of any other external power source.

"Few scientists believe that one could build mechanical systems only with organic materials like the ones that make up the human body," Ikeda says. "I believe that our work has demonstrated one possible way of building mechanical systems with organic materials. Ultimately, all-plastic cars driven solely by exposure to sunlight might be possible. They will have two sets of wheels covered with belts of photomobile polymer materials. By filtering the sunlight with plastic sheets, one can irradiate specific parts of the plastic belts with either UV or visible light to enable the plastic car to move. Today, this is just a dream. But one day it might become reality."

Apart from the fact that photomobile materials do not require batteries, electric wires, or gears, another intriguing aspect of these polymers is that they can be controlled remotely just by manipulating the irradiation conditions. By controlling the area of irradiation, the wavelength, and the intensity of the light, one can choose the way in which a film or fiber made of these materials in driven, thus enabling them to be used as a wide range of photoactuators.

> *Featured scientist:* Tomiki Ikeda
> *Organization:* Chemical Resources Laboratory, Tokyo Institute of Technology, Tokyo (Japan)
> *Relevant publication:* Munenori Yamada, Mizuho Kondo, Jun-ichi Mamiya, Yanlei Yu, Motoi Kinoshita, Christopher J. Barrett, Tomiki Ikeda: Photomobile polymer materials: towards light-driven plastic motors, *Angew. Chem., Int. Ed.,* **47**, 4986–4988.

6.9 Nanotechnology Gets a Grip

Nanotechnology researchers are actively working on the beginnings of various nanorobotic systems that could one day lead to automated, assembly-line type nanofabrication processes. One such device is a nanogripper, a kind of a robotic 'hand' some 10 000 times smaller than a human hand. This 'pick-and-place' device uses a silicon gripper which is controlled by a nanorobotic arm and is capable, for example, of picking up a carbon nanofiber and fixing it on to the tip of an AFM cantilever. One of the problems that is vexing researchers is that the nanoscale miniaturization of these grippers comes at the cost of reduced strength—the smaller the gripper, the weaker it becomes. What is needed is a gripper design that is strong enough yet sufficiently flexible and small enough to handle tough materials like carbon nanotubes.

"Finding the right shape of the gripper 'muscles'—the actuator—is a serious problem," says Peter Bøggild. "We previously found that electrothermal actuation—heating up doped silicon to make it expand—enables us to make grippers that are highly compact, yet fairly strong. After years of trying and failing, we have found that grippers that are delicate enough to manipulate nanotubes and nanowires in a practical way are simply not strong and robust enough. We simply could not make them strong and small at the same time."

Bøggild is an associate professor and leader of the Nanointegration Group at the Department of Micro and Nanotechnology at the Technical University of Denmark (DTU). Together with collaborators from Denmark and Germany, he has been working on designing and fabricating microgrippers for nanoscale manipulations. Their latest design is an electrothermal microgripper that can be integrated into a nanorobotic system and can be used for serial device assembly.

Bøggild explains that one of his colleagues at DTU, Ole Sigmund, is doing fascinating work on topology optimization, and this turned out to be a very powerful technique for addressing his team's nanogripper design problem.

"We took an earlier gripper design, a dual (open/close action) actuator, and applied the topology optimization algorithm—it basically figures out the shape that best maximizes certain goals, while obeying the boundary conditions. We just have to tell it: we apply currents here and here, and that point over there should have a large actuation and a large force." The algorithm ignores aesthetic and engineering traditions and habits; it just optimizes the shape to maximal performance. Strangely, the design resembles a human bone structure, even incorporating a wrist-like structure to amplify the movement of the end-effector (the fingers) of the gripper.

As a result, Bøggild and his colleagues now have a gripper that is ~50 times stronger than the previous design, but has similar actuation range and is similar in size. This new design provides a viable route to fast prototyping and even small-scale manufacturing of nanotube-based devices, resembling industrial macroscale robotic assembly lines.

Topology optimization is a finite-element-based method which relies on the redistribution of a given amount of material in a well-defined design domain, using gradient-based deterministic optimization algorithms, which often leads to topologies that are vastly different than conventional approaches, and with superior performance. The two-dimensional (2-D) topology optimization procedure used by Bøggild's team is based on an approach described by

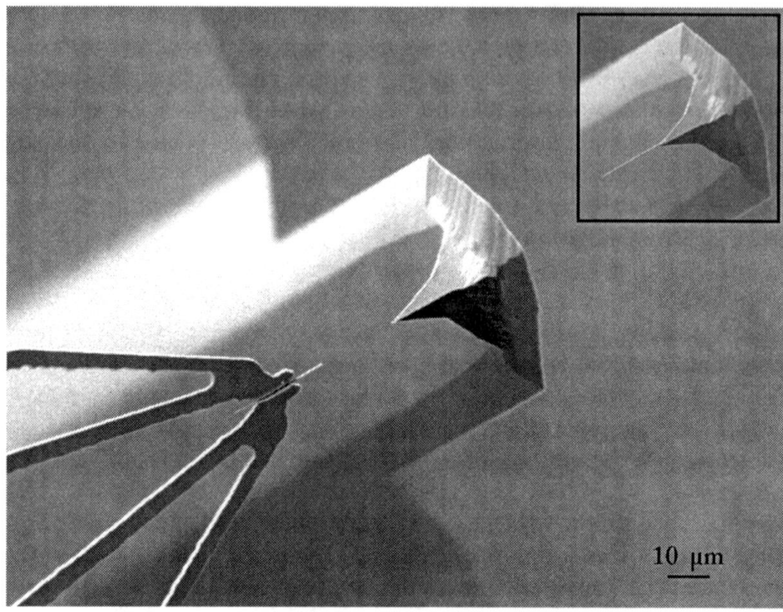

Figure 6.3   Reproducible assembly of CNT-enhanced AFM super-tips using topology-optimized microgrippers. (Image: Özlem Sardan, DTU)

Sigmund.[4] Applying this optimization methodology, the team improved their design and demonstrated that the gripper is capable of picking and placing CNTs with great predictability.

Bøggild tells us that, using the topology-optimized microgrippers in their nanorobotic system, the DTU group together with nanorobotics experts at the Division of Microrobotics and Control Engineering at Oldenburg University in Germany, were able to demonstrate reproducible and consecutive nanomanipulation and assembly of CNTs for different purposes: on to AFM cantilevers (for high aspect ratio AFM tips), on TEM grids (for characterization in transmission electron microscopes), and on electrodes. "These are three useful and pretty generic nanoassembly tasks. This is the first attempt at serial fabrication of CNT devices using nanomanipulation."

This work is done in the framework of a European project, NANOHAND, which ambitiously aims not just to make nanoassembly possible, or to facilitate it, but actually to build equipment and methodology that does nanoassembly, *i.e.* to build a prototype nanoassembly line.

The scientists' aim is to be able to assemble CNTs on AFM tips at least semi-automatically. The project combines 3-D electron microscopy, image recognition, state-of-the-art robotics, and nanogrippers, and unites European companies with interests in nanotechnology and nanomanufacturing with leading European research institutions in the field. Figure 6.3

---

[4] Sigmund O.: Design of multiphysics actuators using topology optimization—Part I: One-material structures, *Comput. Meth. Appl. Mech. Eng.*, **190**, 6577–6604.

"The nanorobotics people in the consortium are taking fascinating approaches," says Bøggild. "They are building two types of set-up—a 'nano-fab' which is a commercial stationary robot system inside a SEM aiming at serial fabrication and prototyping; and a more experimental 'nano-lab' that, among others, features cubic centimetre sized mobile robots from EPFL in Lausanne that 'drive around' on a surface and can collaborate, some of them equipped with our nanogripper. These are really robots in a traditional sense—they look like small tanks with arms."

He points out that mechanical testing is another obvious application area. "Using the topology optimization design approach, we can relatively easily design the tools to perform according to a range of requirements—in other words, we can customize them."

What is so exciting about the NANOHAND project approach is that it shows a real path from lab to fab: the project results are immediately relevant for prototyping of nanotube, nanowire, and nanomaterial devices, perhaps even small-scale fabrication.

Featured scientist: Peter Bøggild

Organization: Nanointegration Group, Department of Micro and Nanotechnology, Technical University of Denmark, Lyngby (Denmark)

Relevant publication: O. Sardan, V. Eichhorn, D. H. Petersen, S. Fatikow, O. Sigmund, P. Bøggild: Rapid prototyping of nanotube-based devices using topology-optimized microgrippers, *Nanotechnology*, **19**, 495503.

PART II:
NANOMEDICINE—DEATH OF THE SLASHER

CHAPTER 7

The Radically Different Medicine of the Future

7.1 Nanotechnology's Medical Applications

For centuries, humans have searched for miracle cures to end suffering caused by disease and injury. Many researchers believe nanotechnologies may be our first giant step toward this goal. Whether this belief is based on fact or hope, many corporations and governments are willing to invest a great deal of resources to find out what happens when nanotechnology is used for medical applications—the emerging field of nanomedicine. Hundreds of millions, if not billions of dollars have been invested by governments and the private sector in nanomedicine research and nanotechnology-related life sciences ventures. The 2008 budget of the U.S. National Nanotechnology Initiative provides more than $200 million for the National Institutes of Health. The European Union, particularly Germany and the UK, and Japan are also investing heavily in this field. It is difficult to find fault with a technology that promises to cure cancer almost before it starts, and to prevent the spread of AIDS and other infectious diseases. Scientists around the globe are searching for ways to exploit nanoparticles to improve human health. However, there are toxicological concerns and ethical issues that come with nanomedicine and they have to be addressed alongside the benefits.

Nanotechnology promises us a radically different medicine compared to the cut, poke, and carpet bomb (think chemotherapy) medicine of today. The two major differences of nanomedicine will be the tools it uses—the main workhorse will be multifunctional nanoparticles—and the fact that it will enable a perfectly targeted and individual treatment: organs and bones, or any body tissue, will one day be diagnosed and treated on a cell by cell basis with precise dosing and monitoring through the use of biomolecular sensors.

RSC Nanoscience and Nanotechnology No. 8
Nano-Society: Pushing the Boundaries of Technology
By Michael Berger
© Michael Berger 2009
Published by the Royal Society of Chemistry, www.rsc.org

The medical advances that may be possible through nanotechnologies range from diagnostic to therapeutic, and everything in between.

Diagnostics

In the past few decades, imaging has become a critical tool in the diagnosis of disease. The advances in the form of magnetic resonance imaging (MRI) and computed tomography (CT) are remarkable, but nanotechnology promises sensitive and extremely accurate tools for *in vitro* and *in vivo* diagnostics far beyond the reach of today's state-of-the-art equipment.

As with any advance in diagnostics, the ultimate goal is to enable physicians to identify a disease as early as possible. Nanotechnology is expected to make diagnosis possible at the cellular or even the subcellular level.

Quantum dots in particular have finally taken the step from pure demonstration experiments to real applications in imaging. In recent years, scientists have discovered that these nanocrystals can enable researchers to study cell processes at the level of a single molecule. This may significantly improve the diagnosis and treatment of cancers. Fluorescent semiconductor quantum dots are proving to be extremely beneficial for medical applications, such as high-resolution cellular imaging. Although their use could revolutionize medicine, unfortunately most quantum dots are toxic. However, studies have shown that protective coatings may eliminate the toxicity.

Therapy

In terms of therapy, the most significant impact of nanomedicine is expected to be realized in drug delivery and regenerative medicine. Nanoparticles enable physicians to target drugs at the source of the disease, which increases efficiency and minimizes side effects. They also offer new possibilities for the controlled release of therapeutic substances. Nanoparticles are also used to stimulate the body's innate repair mechanisms. A major focus of this research is artificial activation and control of adult stem cells.

Magnetic nanoparticles and enzyme-sensitive nanoparticle coatings that target brain tumors; smart nanoparticle probes for intracellular drug delivery and gene expression imaging; and quantum dots that detect and quantify human breast cancer biomarkers are just a few of the advances researchers have already made in the laboratory.

Interestingly enough, there could be massive shifts in economic value among pharmaceutical companies. While the new nanomedicines open up enormous market and profit potentials, entire classes of existing pharmaceuticals such as chemotherapy agents worth billions of dollars in annual revenue would be displaced.

Toxicological Concerns

Nanomedicine, as with nanotechnology in general, is new and we have little experimental data about unintended and adverse effects. The lack of knowledge about how nanoparticles might affect or interfere with the biochemical pathways and processes of the human body is particularly troublesome. Scientists are primarily concerned with toxicity, characterization, and exposure pathways.

Despite these concerns, over 130 nanotechnology-based drugs and delivery systems and 125 devices or diagnostic tests have entered pre-clinical, clinical, or commercial development since 2005.

The issue of safety is a global concern. In the United States the National Institutes of Health (NIH) is evaluating several safety issues, including particle pathways in the human body; the length of time nanoparticles remain in the body; the effects on cellular and tissue functions; access to systemic circulation through dermal exposure; and unanticipated reactions *in vivo*. The National Cancer Institute's Nanotechnology Characterization Laboratory is working to develop standards for advancing the new class of molecular-sized cancer drugs through clinical trials.

In Europe, the *SCENIHR Report*[1] and the white paper *Nanotechnology Risk Governance*[2] published by the International Risk Governance Council address these issues. Both reports emphasize the lack of data on potential risks associated with nanomedicine and nanotechnology with regard to the human-health and ecological consequences of nanoparticles accumulating in the environment.

Other than the obvious potential risks to patients, there are toxicological risks associated with nanomedicine. There are concerns over the disposal of nanowaste and environmental contamination from the manufacture of nano-medical devices and materials (see Chapter 15).

Ethical Issues

Beyond the issue of safety lies the question of society's ethical use of nano-technology. Informed consent, risk assessment, toxicity, and human enhancement are just a few of the ethical concerns voiced in what has become a passionate debate.

Another issue is the fine line between medical and non-medical uses of nanotechnology for diagnostic, therapeutic, and preventive purposes. The question of whether nanotechnology should be used to make intentional changes in or to the body when the change is not medically necessary is yet another hot topic in the long list of concerns.

The good news is that these questions are being asked, but there is still much work to be done. The European Union has taken the lead in addressing ethical aspects of nanomedicine and it appears that Europeans in general, and in particular when it comes to emerging technologies, are more sympathetic to the precautionary principle than is the case in the United States.

---

[1] This can be downloaded from the European Commission website: http://ec.europa.eu/health/ph_risk/committees/04_scenihr/scenihr_opinions_en.htm
[2] http://www.irgc.org/Nanotechnology.html

7.2 Mathematical Engines of Nanomedicine

The process of bringing a major new drug to market, from discovery to marketing, takes ~10–12 years and costs an average of $500–$800 million in industrialized countries. Most drugs fail before they even make it on to a pharmacy shelf: ~80% of drugs never make it through their clinical trials. Of the medications that actually enter consumer use, an average of just 60% provide therapeutic benefits to patients.

For a pharmaceutical company, the results of the process of designing new drugs lead to a library of novel compounds that are created with a specific goal and a given set of criteria. Often these criteria include selectivity for a particular known receptor. A new drug treatment can be discovered by testing those drugs on other receptors by trial and error. Since this is a very expensive approach, pharmaceutical companies have developed sophisticated computer models that help reduce the risk and uncertainty inherent in the drug-development process. Here, one starts with a computer model of the structure of a receptor and a drug. The goal is to predict by simulation how a drug will dock (interact with a receptor), or how the receptor will fold. Drug design based on mathematical models will also become a massive task within the emerging field of nanomedicine. Although nanotechnology offers great visions of improved, personalized treatment of disease, at the same time it renders the problem of selecting the candidates for biological testing astronomically more complex.

The new notion of 'design maps' for nanovectors—similar to the concept of the periodic table for chemical elements—could provide guidance for the development of optimized injectable nanocarriers through mathematical modeling.

"The number of potential combinatorial variations that can be developed by choosing different nanoparticle core materials, targeting moieties, and payload molecules is very large," says Mauro Ferrari. "It is easy to compile a catalogue of 100 realistic choices in each of these entry categories. Combining them results in 1 million possible candidates. Choosing just 10 different particle diameters in the range of say 10–500 nm, the total number rises to 10 million. To give the fullest power to the method, each of the 1 million molecules in an *a priori*, 'conventional' combinatorial library may be used as payload: molecular docking for specificity becomes a less stringent criterion if the target selectivity is accomplished by the nanovector, and not solely by the drug. The total number of candidates is now 100 billion. Even more awe-inspiring numbers are obtained by considering variations in shape, multiple targeting moieties, biological barriers, avoidance mechanisms, and multiple payloads."

Ferrari, the Director of the Research Center for NanoMedicine at the Brown Foundation Institute of Molecular Medicine for the Prevention of Human Diseases at the University of Texas Health Science Center at Houston, points out that the problem—and, with it, the great opportunity for the mathematically inclined—is that there exists no consensus on nanovector selection criteria, as there is for drug libraries.

Ferrari's group is the first to use mathematics to guide nanotechnology development in drug delivery and nanomedical formulations, in effect bringing nanomedicine closer to the pharmaceutical world, where computer modeling has been used for a long time.

"Rather than focusing on molecular docking, we started by considering three fundamental processes in the journey that takes an intravascularly injected nanovector from convective transport in the bloodstream to its cellular target on and beyond the vascular endothelial wall," says Ferrari. "These processes are margination, firm adhesion by specific and nonspecific means, and cellular uptake."

By individually analyzing these three processes as a function of nanovector design parameters (size, shape, surface properties, bulk properties, surface density of targeting moieties) and the biological characteristics of the target lesion in the body (receptor density, blood-flow descriptors, wall permeability) the researchers achieved results and methods that can be used to analyze and design nanovectors of specific overall margination, adhesion, and cell-uptake properties (endocytosis).

Analyzing these results, the University of Texas scientists developed the concept of design maps for nanovectors to be used as a preliminary reference for choosing the properties of the nanoparticle as a function of physiological parameters at the site of desired adhesion within the target.

Ferrari explains that this approach is somewhat similar to the concept of the periodic table of chemical elements, but more akin to phase diagrams in its graphic presentation. "The design maps provide guidance for the development of optimized injectable nanocarriers and are for the time being limited to the combination of adhesion and endocytosis. These maps, and their more sophisticated successors that will evolve over time, will allow for therapeutic regimens to be personalized to individual cases, resulting in greater efficacy in the fight against disease and a reduced burden of undesired and harmful side effects.

"The design maps are but the first steps of a long journey, yet they are along a path charted by the power of mathematics to act as a compass, pointing the way to the right location, deep within a forest of extraordinary complexity, where each buried treasure corresponds to advances against human disease and suffering," he says.

Led by Paolo Decuzzi, a visiting associate professor, Ferrari's group looked into ways to mathematically model nanovector selection and to identify suitable selection criteria. Not surprisingly, they found that there is no one design for a nanoparticle that is generally best; but they did find that there is an "absolute worst".

"We discovered that spherical nanoparticles with diameters of $\sim 100\,nm$ have the very worst margination properties; that is, they have the lowest likelihood of encountering the conjugate antigen on the target endothelium, among all spherical nanoparticles," Ferrari says. "Similarly, they have the absolute worst likelihood of penetrating through the vascular fenestrations that make the cancer neovasculature more permeable, and thereby provide the mechanism for differentially favorable accumulation of the drugs into the target tumor. Furthermore, spherical particles perform the absolute worst in terms of adherence to the endothelial wall—a requirement for the targeted delivery of a therapeutic payload."

The bad news here is that these results indicate that almost all the nano-carriers that are in the clinical or in the preclinical pipeline are basically the worst possible size and shape for their intended purpose. The good news is that these nanovectors have nevertheless already proved quite successful in practice. According to Ferrari, the pleasant nature of his team's unpleasant discoveries is that even small improvements in nanoparticle design, fueled by mathematical analysis, will lead to greater success in the fight against human disease.

> *Featured scientist:* Mauro Ferrari
> *Organization:* Research Center for NanoMedicine in the Brown
>   Foundation Institute of Molecular Medicine for the Prevention of
>   Human Diseases, University of Texas Health Science Center,
>   Houston, TX (USA)
> *Relevant publication:* Mauro Ferrari: The mathematical engines of
>   nanomedicine, *Small*, **4**, 20–25.

7.3   How Medicine can Learn from Materials Science

Can a major component of a catalytic converter or a fullerene derivative lead to an eventual treatment for Parkinson's disease or arthritis? Research to date certainly hints at this possibility. In chemistry, radicals (often referred to as *free radicals*) are atomic or molecular species with unpaired electrons on an otherwise open shell configuration. These unpaired electrons are usually highly reactive, so radicals are likely to take part in chemical reactions. Radicals play an important role in physiology but, because of their reactivity, they also can participate in unwanted side reactions resulting in cell damage. Free radicals damage components of the cell's membranes, proteins, or genetic material by oxidizing them—the same chemical reaction that causes iron to rust. This is called *oxidative stress*. Many forms of cancer are thought to be the result of reactions between free radicals and DNA, resulting in mutations that can adversely affect the cell cycle and potentially lead to malignancy. Oxidative stress is believed to play a role in neurodegenerative diseases such as Alzheimer's and Parkinson's. Some of the symptoms of aging, such as arteriosclerosis, are also attributed to free-radical-induced oxidation of many of the chemicals making up the body.

Despite the broad role that oxidative stress plays in human disease, medicine has been limited in its development of treatments that counteract free radical damage and the ensuing burden of oxidative stress. In contrast, in the field of engineering, considerable effort has been developed to counter the effects of oxidative stress at the materials science level. Nanotechnology has provided numerous constructs that reduce oxidative damage in engineering applications with great efficiency. Beverly A. Rzigalinski, a professor at the Virginia College of Osteopathic Medicine and Virginia Polytechnic and State University, looks at how these nanoengineering concepts could be applied to biomedical problems, ultimately leading to nanotechnology-based therapeutic treatments for oxidative stress-induced diseases.

"The chemical and physical processes involved in three-way catalysis for improved combustion and removal of environmental contaminants from engine exhausts have similarities with biological redox reactions and antioxidants, from a chemical and physical standpoint," she explains. "Likewise, the role of coatings in the reduction of metal oxidation involves chemical principles similar to those associated with the prevention of oxidation in biomolecules. Nanotechnology has provided dramatic improvements in controlling or eliminating oxidation reactions in materials applications. This may provide a new basis for pharmacological treatment of diseases related to oxidative stress."

Three of the most-studied nanoparticle redox reagents at the cellular level are rare earth oxide nanoparticles (particularly cerium), fullerenes, and carbon nanotubes (CNTs).

"Our initial results suggest that cerium oxide nanoparticles extend cell and organism longevity through their actions as regenerative free-radical scavengers," Rzigalinski summarizes the core findings of her work so far. "Additional studies suggest that these nanoparticles are also potent anti-inflammatory agents. Although much work remains to be done in this realm, ceria nanoparticles hold high promise for future development of nanopharmacological agents to treat age-related neurodegenerative and inflammatory disorders."

Research has already shown that nanoparticles composed of cerium oxide or yttrium oxide protect nerve cells from oxidative stress and that the neuroprotection is independent of particle size.

Several studies also suggest that ceria nanoparticles are potent anti-inflammatory agents. Most intriguingly, Rzigalinski's research shows that ceria nanoparticles directly added to their food increased the maximum and average lifespan of fruit flies. Of course, it is a long way from fruit flies to humans, but this research indicates that these particles might have antioxidative properties that, once their mechanism of action is fully understood, one day also could benefit therapeutic applications for humans.

"Our work brings future potential for nanopharmacology closer to reality," says Rzigalinski. "By designing additional ceria constructs, we hope to be able to direct and control free radical scavenging activity."

So far, the chemistry and physics of ceria nanoparticles support the hypothesis that the biological actions of ceria are related to a regenerative free-radical scavenging ability. During the experiments with fruit flies it was clearly shown that a single low dose of ceria nanoparticles protected cells from free-radical damage over an extended period of time.

Several reports also describe the free-radical scavenging capabilities of fullerene derivatives and CNTs. Their antioxidant activity is hypothesized to be related to the large electronegative center of these constructs.

However, several studies also showed detrimental effects of certain functionalized fullerenes in the treatment of oxidative stress. The potential toxicity of CNTs is much debated but, nevertheless, they also have been reported to have free-radical scavenging properties. Much more research needs to be done on carbon nanomaterials to determine their antioxidant properties.

Rzigalinski points out that the case for nanoparticles as free-radical scavengers holds great promise for future pharmacotherapy of diseases in which oxidative stress is a component. However, these studies, as with much of nanomedicine, are in their infancy. "Our knowledge of the physiological behavior of nanoparticles is scant and we are just starting down the road from bench to bedside—a road that will no doubt require numerous adaptations to our traditional concepts of pharmacology."

> *Featured scientist:* Beverly A. Rzigalinski
> *Organization:* Virginia College of Osteopathic Medicine, Blacksburg, VA (USA)
> *Relevant publication:* Beverly A. Rzigalinski, Kathleen Meehan, Richey M. Davis, Yang Xu, William C. Miles, Courtney A. Cohen: Radical nanomedicine, *Nanomedicine*, **1**, 399–412.

7.4 Next-Generation Tissue and Cell Engineering

There is a huge demand for tissue regeneration technologies, covering a wide range of potential applications in such areas as cartilage, vascular, bladder, and neural regeneration. Just consider the need for bone and dental implants. Each year, almost 500 000 patients receive hip implants worldwide, about the same number need bone reconstruction as a result of injuries or congenital defects, and 16 million Americans lose teeth and may require dental implants. The market for medical implant devices in the United States alone is estimated to be $23 billion per year and it is expected to grow by about 10% annually for the next few years. Unfortunately, medical implant devices have been associated with a variety of adverse reactions, including inflammation and fibrosis. It has been suggested that poor tissue integration is responsible for loosening of implants and mechanical damage to the surrounding host tissues.

On the basis of the expanding body of nanobiomedical research, there is a growing consensus among scientists that nanostructured implant materials may have many potential advantages over existing, conventional ones. The key, as indicated in a number of findings, seems to be that physical properties of materials, especially with regard to their surface's nanostructure, affect cell attachment and eventually the tissue response to the implant. Although nanotopography-mediated cell responses have been shown in previous work, the mechanism of these responses is mostly undetermined. A group of researchers led by Liping Tang, a professor of bioengineering at the University of Texas at Arlington, has now tried to determine the influence of nanopore size on cellular responses. Interestingly, these studies have revealed that larger nanopores (200 nm) trigger DNA replication and cell proliferation via various signal transduction pathways.

"Because of the direct interactions with host cells and tissue, the surface properties of medical implants play a critical role in determining the host responses and reactivity," says Tang. "Several lines of evidence reveal that surface topography can influence cellular responses and activities at tissue–implant interfaces both *in vitro* and *in vivo*. Although the effects of surface topography on cellular responses have been widely reported, little is known about the molecular mechanism of nanotextured material-mediated cellular responses. Therefore the aim of our study was to evaluate the cellular and molecular responses such as cell adhesion, morphology, and proliferation as well as the early gene expression of cells cultured on large (200 nm) and small (20 nm) nanotextured surfaces."

Tang points out that his team's research has shown that nanotopography affects not only cell morphology and proliferation but also gene profiles of vascular smooth muscle cells. "In spite of the relationship between cell responses, including cell proliferation, and gene expression, relatively little work has been done to investigate the effects of surface topography on the gene expression of exposed cells," he says.

"We found that several genes involved in cell adhesion, cell morphology, cell cycle, DNA replication, cell proliferation, and signaling transduction pathways are dependent on the surface nanotopography," says Tang. "Exposure to larger pores induced genes involved with cytoskeletal elements such as villin, myosin, cofilin, and caveolin that play an important role in cell morphology. In addition, cell proliferation is greater in cells exposed to 200 nm pore membranes compared to 20 nm pore membranes since exposure to larger pore surfaces induced expression of genes involved in the cell cycle, DNA replication, and cell proliferation while at the same time it reduced expression of genes involved in cell apoptosis."

The results from this University of Texas study seem to confirm the increasingly popular belief among nanomedical researchers that nanotopography coatings and surface structuring will improve the biocompatibility and safety of medical implant devices.

"The future of tissue and cell engineering depends on the development of next-generation biomaterials, the nanostructured materials," says Tang. "These materials must provide control over cell attachment and cell development into tissue. Since surface topography influences many aspects of cellular and molecular responses, surfaces of implanted devices can be developed so that they allow the engineering of the desired cell shape and cell responses to the points of interest."

Tang believes that such a topographic modulation of cell response may be one of the most important considerations during the design and manufacture of medical devices. "There is no doubt that a nanodimensional surface has great potential for the design of next generation of tissue-engineered replacements."

7.5   Improving the Tools for Single-Cell Nanosurgery

Nanosurgery holds the amazing promise of studying, manipulating, and repairing individual cells without damaging them. For instance, nanosurgery could remove or replace certain sections of a damaged gene inside a chromosome; sever axons to study the growth of nerve cells; or destroy an individual cell without affecting neighboring cells. Already, scientists routinely transplant the cell nucleus from one cell to another during cloning using micropipette technologies, but these methods are too crude for other subcellular structures.

First steps towards cell nanosurgery have been made using so-called 'optical tweezers', where the energy of laser light is used to trap and manipulate nanoscale objects, for instance the nucleus of a cell, without mechanical contact. Combined with a laser scalpel—the use of a laser for cutting and ablating biological objects—optical tweezers have been used to study cell fusion or DNA cutting. Unfortunately, while optical tweezers offer exquisite sensitivity in their ability to position micro- and nanoparticles, they suffer from one important disadvantage: the trapped particle is localized at the laser focus where light intensity is the highest. As a result, the laser light that is used to trap an object also has a propensity to photobleach and photodamage it, especially when the object is fragile and small (*e.g.* a fluorescently labeled subcellular organelle). Minimizing this drawback, Daniel T. Chiu, professor of chemistry at the University of Washington, developed the use of polarization-shaped optical vortex traps for the manipulation of particles and subcellular structures.

An optical vortex, or Laguerre–Gaussian (LG) beam, is laser light that has a helical phase distribution at its wave front. Optical vortices are characterized by the presence of a dark core, *i.e.* a point of zero intensity. Because of the spiral phase distribution at the wave front, the phase at the center of the beam is undefined and thus a singularity is formed where the light intensity has a zero value. In simple terms, light can be twisted like a corkscrew around its axis of travel. This corkscrew of light, with darkness at the center, is called an optical vortex. Because of the twisting, the light waves at the axis itself cancel each other out. On a flat surface, an optical vortex looks like a ring of light, with a dark hole in the center.

"The dark core of an optical vortex takes on different appearances upon tight focusing through a high numerical aperture objective, depending on the polarization state of the incident beam," explains Chiu. "We take advantage of these polarization effects to fashion vortex traps for the trapping of particles of different sizes while minimizing photodamage to the trapped particles and subcellular organelles."

Chiu's group focuses on the development of new tools that combine ultrasensitive, laser-based detection and manipulation methodologies with micro- and nanofabrication techniques for interfacing with biological systems at the nanoscale. In their experiments, Chiu and his colleagues found that light damage was always less when they used a polarization-shaped vortex trap. Similarly, by comparing the time course of bleaching of mitochondria, the researchers found that the mitochondria trapped in the optical tweezers were rapidly bleached within the first few seconds, but the ones held by the vortex trap displayed a relatively stable fluorescent signal over the duration of the observation.

Applying these findings to single-cell nanosurgery, Chiu demonstrated that the use of polarization-shaped optical vortex traps minimizes photobleaching of fluorescent tagging and damage to the cell caused by optical tweezers.

"The ability to optically manipulate submicrometer subcellular structures while minimizing photodamage has several practical applications," says Chiu. "In particular, we are developing a droplet nanolab platform for single-organelle assays and chemical analysis, because many diseases are caused by organelle malfunction where a given cell can contain a heterogeneous population of healthy and defective organelles. To achieve single-organelle analysis, it requires the optical manipulation and encapsulation of the select organelle within a femtolitre-volume aqueous droplet. Here, it is pertinent the organelle remains functionally intact during optical manipulation so subsequent biological or chemical analysis reflects the true physiological state of the organelle."

Chiu's research points to a new strategy based on shaping intensity variations within the diffraction-limited laser focus to create the next generation of optical tweezers. He says that the availability of such tools will offer unprecedented control in active nanoscopic manipulation, from subcellular structures to nanoparticles, such as metal colloids and quantum dots.

Featured scientist: Daniel T. Chiu
Organization: Department of Chemistry, University of Washington, Seattle, WA (USA)
Relevant publication: Gavin D. M. Jeffries, J. Scott Edgar, Yiqiong Zhao, J. Patrick Shelby, Christine Fong, Daniel T. Chiu: Using polarization-shaped optical vortex traps for single-cell nanosurgery, *Nano Lett.*, **7**, 415–420.

7.6 Microbotics—Nanoparticles Hitching a Ride on Bacteria

Vaccination has resulted in the eradication of smallpox and the control of measles, rubella, tetanus, diphtheria, and other infectious diseases in many areas of the world (at least where vaccines are available and affordable; providing vaccines to many parts of the developing world still is one of the basic medical needs that is far from being met). The basic idea of vaccination is to inject weakened or killed forms of pathogens such as bacteria or viruses into the body in order for the immune system to develop antibodies against them. If the same types of microorganisms enter the body again, they will be recognized and destroyed by the antibodies.

About 25 years ago, the basic idea of vaccination gave rise to *bactofection*—the technique of using bacteria as nonviral gene carriers into target cells. The DNA cargo is transported inside the bacteria and, once it arrives at the target location, the bacteria are broken up in order to release the therapeutic gene or protein. A novel technique takes advantage of the invasive properties of bacteria for delivery of nanoparticles into cells. Here, the gene or cargo is not carried inside the bacteria, but rather remains on the surface conjugated to nanoparticles. Consequently, this approach does not require bacterial disruption for delivery, or any genetic engineering of the bacteria for different cargo.

"Our bacteria-mediated nanoparticle and cargo delivery approach, which we term *microbotics*, promises excellent potential for nonviral gene delivery and unique capabilities for biomedical nanorobotics and nanomedical therapy," explains Rashid Bashir. "Although more than one gene can be delivered by means of bactofection, many more copies of a target cargo can be carried with one bacterium using our method. We also show that nucleic-acid-based model drugs loaded on the nanoparticles can be released from the carriers and eventually find their way into the nucleus, with subsequent transcription and translation of their respective proteins, for both *in vitro* and *in vivo* conditions. Such bacteria, which we call *microbots,* can potentially be used to carry proteins, small molecules, and even synthetic objects like sensors and therapeutic moieties into different types of cells."

Bashir, professor of electrical and computer engineering at Purdue's Birck Nanotechnology Center, together with colleagues from various departments at Purdue University, describes a clever use of microbiology and nanotechnology to employ bacteria to deliver nanoparticles into cells. Bashir's team coupled polystyrene nanoparticles loaded with plasmid DNA to the surface of weakened *Listeria monocytogenes* bacteria.

According to Bashir, three steps were necessary to make their microbots: "First we treated the bacteria with a biotin-carrying antibody that acts against—and will therefore attach to—proteins on the bacterial surface called muraminidase. Next, we mixed the treated bacteria with nanoparticles, ranging from 40 nm to 200 nm, coated with streptavidin, a protein that binds strongly to biotin. Finally, the nanoparticle-loaded bacteria were mixed with plasmid DNA carrying biotin, which binds to the free streptavidin sites on the surface of the nanoparticles."

Because the nanoparticles are linked to the bacteria by means of an antigen–antibody interaction, the cargo and the bacteria can readily separate in the lower pH environment of the subcellular compartments. Bashir says that other factors, such as intracellular enzymatic processing or destabilization of antigen–antibody binding or a reduction in the biotin–streptavidin interactions can also be involved in the release mechanisms of the DNA, and all of these possibilities can potentially be used for endowing microbots with smart cargo release ability.

In their experiments, the researchers at Purdue found that microbots successfully delivered their cargos of nucleic acid-based model drugs, plasmid DNAs for firefly luciferase, and SEAP enzymes into multiple organs of live mice, and the delivered genes also resulted in functional protein expression by 3 days after treatment.

"The delivered plasmid DNAs were able to escape from intracellular entrapment and were targeted to the nuclei of the cells, resulting in transcription and expression of the enzymes," says Bashir. "Hence, our novel technology can be used to deliver these reporter molecules for whole-animal live imaging agents (luciferase) or for non-invasive *in vivo* reporter assays."

Advances in bactofection have so far been mostly incremental but the Purdue team's approach of combining microbiology with nanotechnology could be a big step forward. As well as having scientific potential, the microbots also show what can be achieved when scientists from different disciplines get together, and the Purdue microbiology/nanotechnology work should lead the way for other interdisciplinary research projects.

Bashir says that his team's studies will concentrate on the development of an attenuated *Listeria* strain, microbot-mediated delivery of artificial biohybrid nanostructures, delivery of larger particles and functional proteins, and investigation of solid organ tumor penetration by microbots for applications in diagnostics and therapy at the single cell level and up to a few cells.

Featured scientist: Rashid Bashir
Organization: School of Electrical and Computer Engineering, Purdue University, West Lafayette, IN (USA)
Relevant publication: Demir Akin, Jennifer Sturgis, Kathy Ragheb, Debby Sherman, Kristin Burkholder, J. Paul Robinson, Arun K. Bhunia, Sulma Mohammed, Rashid Bashir: Bacteria-mediated delivery of nanoparticles and cargo into cells, *Nat. Nanotechnol.*, **2**, 441–449.

7.7 The Modest Pipette Becomes a Nanosurgery Tool

Pipettes are one of those ubiquitous tools you find in every chemical, medical, or biology lab. Originally made of glass, the pipette works by creating a partial vacuum above the liquid-holding chamber and selectively releasing this vacuum to draw up and dispense liquid. With the advance of molecular biology, researchers required pipettes that were able to confine smaller and smaller amounts of specimen, not only for probing but also for injecting drugs, DNA, *etc.* into cells—without damaging the cells, of course.

In 2007, researchers at the Brookhaven National Laboratory developed what is thought to be the world's smallest pipette. Made of a carbon-coated germanium nanowire it can hold a volume of only a few zeptolitres (a zeptolitre is a billionth of a trillionth of a litre, or $1000\,nm^3$). Although today it is possible to process the tip of a glass pipette to have an inner diameter as small as a few tens of nanometers, this involves considerable problems in terms of the processing accuracy and the operability of the pipette, particularly in terms of locating the tip. It is generally considered that probes made from CNTs and nanoscale carbon pipes offer an attractive alternative to glass pipettes because of their small size, high mechanical strength, and high electrical conductivity. Researchers at the University of Pennsylvania have developed a manufacturing technique for carbon nanopipettes (CNPs) that does not require cumbersome nanoassembly and is amenable to mass production.

"We are ultimately interested in developing nanosurgery tools to probe cells, monitor their processes, and control or alter their functions," says Haim H. Bau. "At our Micro-Nano Fluidics Laboratory, CNPs are mass produced and can inject reagents into cells without damage. We feel CNPs will help scientists gain a better understanding of how a cell functions and help develop new drugs and therapeutics."

Bau is a professor of mechanical engineering and applied mechanics at the University of Pennsylvania in Philadelphia. He conducts research on micro- and nanofluidic phenomena with applications in biology and medicine.

"We have developed a fabrication process, amenable for mass production, to produce CNPs without any assembly," explains Michael G. Schrlau. "Our CNPs consist of a carbon film deposited inside quartz micropipettes to form a continuous carbon channel along its length. A unique property of CNPs is the existence of this electrically conductive, hollow, interior carbon film running the entire length of the quartz pipette, lending itself to performing cell physiology measurements during cell injection." Schrlau is a PhD candidate in Bau's group and works with him on developing the CNPs.

The researchers found that their CNP tips are flexible and elastic, yet strong enough to spear cells readily. "Unlike glass pipettes, which would shatter, we observed how CNPs buckle without breaking when pushed against a solid surface, then recover their original shape when the force is removed," says Schrlau. "Yet the CNP is rigid enough to penetrate into smooth muscle cells. CNPs have successfully penetrated various cells such as oral squamous carcinoma cells and nerve cells."

Several research groups have developed CNT and carbon pipe-based tools with similar intentions, although these tools are difficult to manufacture. In contrast, the scalable fabrication method of the University of Pennsylvania team allows for hundreds of probes with consistent nanoscale dimensions to be fabricated concurrently. Schrlau describes the process:

"The inner bore of quartz capillaries is filled with a catalyst solution, allowed to air dry, and then pulled into fine-tipped micropipettes. Carbon is selectively deposited on the catalyzed surface by chemical vapor deposition (CVD). The thickness of the carbon film is controlled by varying the CVD time. The tip of the micropipette is then wet-etched to remove the quartz exterior and expose a short length of the interior carbon pipe. The wet etching time and temperature dictate the length of the exposed carbon pipe. Further reduction in tip outer diameters can be achieved by etching the outer diameter of the carbon pipe by plasma oxidation, resulting in carbon pipes with outer diameters ranging from tens to hundreds of nanometers. The final product is a glass capillary lined with a continuous carbon film along its interior and a nanoscale carbon pipe extending from its end."

Apart from the unique advantage of being able to use the CNP concurrently as a nanoinjector and a nanoelectrode, Schrlau points out that there are several other promising potential applications for them: (1) nanoscale biosensors—carbon can be functionalized with proteins, *etc.* for specific detection of analytes; (2) sample holders for electron microscopy—the electron-transparent properties of CNPs enable them to hold biological samples in scanning or transmission EMs.

In future work the researchers will focus on measuring cell membrane potentials concurrently with microinjection and on developing arrays of CNPs to simultaneously interact with a large number of cells.

"I believe the biggest challenge in the nanosurgery tool field will be to go beyond the proof-of-concept, tool development stage into the utilization stage," says Bau. "This includes finding the appropriate collaborations across engineering, life science, and medical disciplines to, among other things, study how cells respond to their use (*i.e.* toxicity studies) and compare the tools to current (or traditional) technologies."

Featured scientists: Haim H. Bau, Michael G. Schrlau
Organization: Mechanical Engineering and Applied Mechanics, University of Pennsylvania, Philadelphia, PA (USA)
Relevant publication: Michael G. Schrlau, Erica M. Falls, Barry L. Ziober, Haim H. Bau: Carbon nanopipettes for cell probes and intracellular injection, *Nanotechnology*, **19**, 015101.

CHAPTER 8

Improved Diagnostics—Key to Effective Prevention

"Just as different ethical issues exist for preventive medicine *versus* curative or therapeutic medicine, very different kinds of ethical issues arise out of diagnostic nanomedicine *versus* therapeutic nanomedicine," Bawa and Johnson write in an article on the ethical aspects of nanomedicine.[1] "Interventions based on nanotechnologies likely will resurrect old questions about human enhancement, human dignity, and justice that have been asked many times before in the context of pharmaceutical research, stem cell research, and gene therapy."

Bawa argues that diagnostic nanotechnologies will eventually provide the ability to detect and characterize individual cells, subtle molecular changes in DNA, or even minor changes in blood chemistry—scenarios that will likely make us pause and reconsider of what it means to be a 'healthy person' *versus* a 'person who has a disease'. He writes that, "in a 'nanoworld', we might have to reconsider how to diagnose someone who has, say, cancer. Is the presence of a genetic mutation known to have a predisposition for causing cancer in a single cell a diagnosis? Or is it simply a risk factor? How many cells from the body must be of a cancerous nature for it to be defined as cancer? 1? 50? 1000?"

Once diagnostic technologies have reached this stage it will require reconceptualizing our understanding of disease. In some cases, more information might just be too much information. Nevertheless, the balance of information processed and disseminated *versus* benefit to society and individual health is a significant consideration for the ethics of nanotechnology-based diagnostic technologies.

There are numerous engineered constructs, assemblies, and particulate systems used for nanomedical diagnostics, whose unifying feature is their nanoscale size range. This chapter looks at some examples.

---

[1] R. Bawa and S. Johnson, The ethical dimensions of nanomedicine, *Med. Clin. N. Am.*, 2007, **91**, 881–887.

RSC Nanoscience and Nanotechnology No. 8
Nano-Society: Pushing the Boundaries of Technology
By Michael Berger
© Michael Berger 2009
Published by the Royal Society of Chemistry, www.rsc.org

8.1 Nanostoves for Superfast Blood Analysis and Detection of DNA Defects

Walk into an intensive care unit and you're likely to see many of the patients sporting a 'central line'—a plastic tube placed in a large vein that goes to the heart. A central line is a very efficient way of pumping nutrients, antibiotics, or other drugs directly into the bloodstream. Unfortunately it is also a cause for bloodstream infections (sepsis), and central lines remain an important cause of hospital deaths. An estimated 200 000 bloodstream infections occur each year in the United States alone, most of them associated with the presence of an intravascular catheter. The Institute for Healthcare Improvement (IHI) estimates the yearly death toll from blood infections related to intravenous lines to be as high as 28 000. Numerous pathogens can cause sepsis, and the death toll could be reduced if the specific pathogen could be identified more quickly than it is usually done today using blood cultures (the laboratory examination of a blood sample to detect the presence of disease-causing microorganisms). One solution would be to determine the pathogen's DNA, requiring a rapid DNA assay with the potential for mutant identification and multiplexing. Current DNA assays are based on thermal dehybridization or melting of the DNA duplex helix—a process that can take up to an hour. Researchers in Germany have developed a novel technique that allows for DNA analysis in the millisecond range. Their method has great potential to vastly improve the speed of pathogen detection.

Apart from the 'frontline' blood analysis requirements in hospital environments, DNA analysis is a huge and growing field in the area of DNA defect analysis, *e.g.* in researching hereditary diseases. Many diseases are caused by defective DNA sequences. One of the side effects of these defects is that they reduce the melting temperature of the DNA.[2] Researchers use this to identify DNA defects by measuring its melting curve.

The way this has been done so far is by combining DNA sequences with gold nanoparticles and then slowly—usually in a water bath—increasing the temperature of the DNA-containing solution. During the heating process the optical absorption is monitored by laser. Aggregates consisting of gold nanoparticles linked *via* DNA change their absorption properties when the DNA strands dehybridize, since this leads to a break-up of the entire aggregate. The melting temperature is determined by gently heating the entire solution, always keeping the entire system in thermodynamical equilibrium. Since this is a protracted process, taking up to 1 hour, it is not applicable in high-throughput DNA analyses.

"So far, laser-induced heating of gold nanoparticles has mainly been applied to destroy tissue or cell membranes," says Jochen Feldmann, a professor of photonics and optoelectronics at the Ludwig Maximilians University of

---

[2] The melting temperature of DNA is determined by its base sequence. At defect locations the two DNA strands do not match perfectly and therefore separate at a lower temperature than is required to separate matching base pairs.

Munich. "We have used gold nanoparticles as optically controllable and pulsed nanostoves. This local and temporally limited way of heating gold nano-particles provides the possibility of inducing thermally driven reactions on the nanoscale within a limited time window. Instead of waiting for equilibrium conditions on the macroscopic scale, we just measure the temporal behavior of the system on its natural nanoscale."

Feldmann's group at the University of Munich, together with Roche Diagnostics and the group of Thomas Klar from the University of Ilmenau, all in Germany, have developed a technique that allows the analysis of DNA in milliseconds. In contrast to destructive applications of gold nanoparticle-mediated optical heating, the German researchers apply gold nanoparticles as nanoscopic stoves for gentle, controlled, and reversible heating.

"This allows us for the first time to use gold nanoparticles for optically induced DNA melting," says Feldmann. "We use DNA-bound gold nano-particle aggregates as light absorbers to locally convert optical energy from 300 ns laser pulses into thermal energy. This thermal energy is used to melt double-stranded DNA. Subsequently, the aggregates disintegrate on a milli-second time scale."

In this technique, the aggregates of nanoparticles serve both as converters of optothermal energy to heat the DNA and as spectral reporters of DNA melting, allowing for the discrimination of different targets even when they are mixed in one and the same solution.

The perfectly matched DNA can be clearly distinguished from single-base-pair mismatched DNA even in a 1:1 mixture of both targets. The millisecond observation window provides the fastest time frame for a DNA melting analysis to date, with potential for multiplexing and mutant identification.

With this new technique it is possible to adjust the intensity of the laser so that defective DNA melts but healthy DNA does not. This allows us to check with a single, millisecond measurement if a sample of DNA is defective or not. Given its unprecedented speed, this novel technique allows for a new high-throughput analysis technology.

Gold nanoparticles have unique chemical and physical properties. They are non-toxic and can easily be attached to almost any macromolecule or biomo-lecule. They also have unique optical and electrical properties. Apart from their use as optically controllable nanostoves, they can act as nanosized amplifiers for spectroscopic signals such as fluorescence and Raman. At the same time, they can act as nanosized manipulators, sensors, amplifiers, and transmitters. This makes them very interesting for many scientific and technological applications in materials science, biotechnology, cell biology, pharmacy, and medicine.

Feldmann is convinced that his team's findings will stimulate further optothermal manipulation experiments on the nanoscale. "Many important molecular and biomolecular reactions are controlled or initiated by tempera-ture," he says. "Attaching gold nanoparticles allows for external and optical control of such reactions. The non-destructive nature of our optothermal control method is clearly of great importance for *in vivo* thermal manipulation experiments."

Featured scientist: Jochen Feldmann
Organization: Department of Physics and CeNS, Ludwig-Maximilians-Universtität München, Munich (Germany)
Relevant publication: Joachim Stehr, Calin Hrelescu, Ralph A. Sperling, Gunnar Raschke, Michael Wunderlich, Alfons Nichtl, Dieter Heindl, Konrad Kürzinger, Wolfgang J. Parak, Thomas A. Klar, Jochen Feldmann: Gold nanostoves for microsecond DNA melting analysis, *Nano Lett.*, **8**, 619–623.

8.2　Electronic Noses that 'Smell' Cancer

Biomarkers are of increasing importance in modern medicine for the early detection and diagnosis of disease. Biomarkers are mostly protein molecules that can be measured in blood, other body fluids, and tissues to assess the presence or state of a disease. The presence of an antibody may indicate an infection or an antigen, for instance prostate-specific antigen (PSA) might indicate the presence of prostate cancer cells. Although protein-based approaches to early detection and diagnosis of cancer have a clear advantage over other, more invasive methods, protein detection is a challenging problem owing to the structural diversity and complexity of the target analytes. State-of-the-art detection methods have limited application because of their high production cost and instability. Another limitation of current proteomic diagnostics is the limitation of arrays to only one or a few markers; in other words, you can only test for the specific markers that you are looking for, and not generally measure levels of proteins in your blood in order to detect anomalies. A novel nanotechnology-based protein detector array could change that.

"One of the key finding of our research is that nanoparticles can be used to create selective receptors for proteins," says Vincent M. Rotello. "Creation of an array of these receptors can result in a 'chemical nose'—a system where proteins can be identified."

Rotello is professor in the Department of Chemistry at the University of Massachusetts. Together with members from his group and the School of Chemistry and Biochemistry at the Georgia Institute of Technology, led by Uwe Bunz, he describes how assemblies of gold nanoparticles with fluorescent PPE polymer provide efficient sensors of proteins, achieving both the detection and identification of analytes.

"To my knowledge, this is the first use of nanoparticles in array-based sensing," says Rotello. "We were motivated by the need for a general method for sensing and identifying proteins. While array-based sensing of proteins has been done before, our system is much more sensitive, and much more effective, *i.e.* we use fewer receptors to identify more proteins."

There are two key challenges for the development of effective protein sensors: (1) the creation of materials featuring appropriate surface areas for binding protein exteriors, coupled with the control of structure and functionality required for selectivity; (2) the signal transduction of the binding event.

"Nanoparticles provide versatile scaffolds for targeting biomacromolecules that have sizes commensurate with proteins; a challenging prospect with small-molecule-based systems," Rotello points out. "Our strategy for the creation of protein sensors is to use the nanoparticle surface for protein recognition, with displacement of a fluorophore generating the output."

To create the protein sensors, the scientists used six gold nanoparticle–fluorescent polymer conjugates. These particles serve both as selective recognition elements and as quenchers for the polymer. The subsequent binding of protein analytes displaces the dyes, thereby regenerating the fluorescence.

By modulating the nanoparticle–protein and/or nanoparticle–dye association, distinct signal response patterns can then be used to differentiate the proteins. The fluorescent indicator displacement assay does not require special instruments and its sensitivity—due in large part to the high surface area provided by the nanoparticles—and speed facilitate protein detection.

Rotello explains that real-world applications require identification of proteins at varying concentrations and of unknown identity. "Varying protein concentrations would be expected to lead to the drastic alteration of fluorescence response patterns for the proteins, making identification of proteins with both unknown identity and concentration challenging. To enable the detection of unknown proteins, we have designed a protocol combining laser Doppler anemometry (LDA) and ultraviolet (UV) measurements."

In their experiments, the researchers also used a series of unknown protein solutions for quantitative detection. The unknown protein solutions were submitted to the testing procedures: of the 52 unknown protein samples, only 3 were incorrectly identified.

Rotello says that the team's overall goal is to develop a more holistic approach to detecting diseases. "The plan is to monitor levels of proteins in the body, as opposed to looking for just a single marker. Of course, the key question is whether disease can effectively be 'smelled'—are there detectable differences in the relative ratios of proteins that can be used for diagnostic purposes?"

Featured scientist: Vincent M. Rotello
Organization: Department of Chemistry, University of Massachusetts at Amherst, MA (USA)
Relevant publication: Chang-Cheng You, Oscar R. Miranda, Basar Gider, Partha S. Ghosh, Ik-Bum Kim, Belma Erdogan, Sai Archana Krovi, Uwe H. F. Bunz, Vincent M. Rotello: Detection and identification of proteins using nanoparticle–fluorescent polymer 'chemical nose' sensors, *Nat. Nanotechnol.*, **2**, 318–323.

8.3 A Biosensor for *Salmonella* Detection

The U.S. Food and Drug Administration has published a handbook called the *Bad Bug Book*[3] which provides basic facts regarding food-borne pathogenic microorganisms and natural toxins. It contains all you always wanted to know about *Salmonella, E. coli*, parasitic protozoa, worms, viruses and natural toxins that, when they get into your food, as they do from time to time, can make you quite sick, even kill you. The U.S. Centers for Disease Control and Prevention (CDC) keep some pretty scary statistics and estimate that food-borne pathogens cause approximately 76 million illnesses, 325 000 hospitalizations, and 5000 deaths in the United States annually.

Three pathogens combined—*Salmonella, Listeria*, and *Toxoplasma*—are responsible for 1500 deaths in the United States each year. *Salmonella* is the most common cause of food-borne deaths and is responsible for millions of cases of food-borne illness a year. Sources of food poisoning are raw and undercooked eggs, undercooked poultry and meat, dairy products, seafood, fruits, and vegetables—basically, more or less everything we eat. Early detection of food-borne pathogenic bacteria, especially *Salmonella*, is therefore an important task in microbiological analysis to control food safety. Several methods have been developed in order to detect this pathogen; however, the biggest challenges remain detection speed and sensitivity. A novel nanotechnology-based biosensor is showing great potential for food-borne pathogenic bacteria detection with high accuracy.

"Early detection of food-borne pathogenic bacteria is critical to prevent disease outbreaks and preserve public health," says Bosoon Park. "Current detection techniques such as ISO method 6579, fluorescent antibody (FA), enzyme-linked immunosorbent assay (ELISA), or polymerase chain reaction (PCR) are time-consuming, cumbersome, and have limited sensitivity."

Park, an agricultural engineer at the U.S. Department of Agriculture, was part of a team that included scientists from the University of Georgia and the Korea Food Research Institute, which developed a novel and effective food-borne bacteria detection method.

"Our nanotechnology-based biosensor is very promising for the detection of proteins, viruses, and bacteria, with high sensitivity and high resolution," says Yiping Zhao, an associate professor of physics at the University of Georgia. "This biofunctional hetero-nanorod detection method has great potential in the food safety industry as well as in biomedical diagnostics."

The research team, which also included Ralph A. Tripp from the Department of Infectious Diseases at the University of Georgia, fabricated a hetero-structured silicon/gold nanorod array by the glancing angle deposition (GLAD) thin film method and functionalized it with anti-*Salmonella* antibodies and organic dye molecules. Because of the high aspect ratio of the silicon nanorods, dye molecules attached to the nanorods produce enhanced fluorescence upon capture and detection of *Salmonella*.

---

[3] http://www.cfsan.fda.gov/~mow/intro.html

Park explains that traditional microbiological techniques for detecting food-borne pathogens take up to 5 days to obtain a positive result, including pre-enrichment, selective enrichment, and confirmation of colonies, procedures that are time-consuming and labor-intensive. Another downside of culture methods is that they show poor sensitivity when there is only a low level of contamination in the samples.

A number of investigators have used the FA technique for *Salmonella* detection. Although FA procedures offer considerable time savings, a large number of the pathogenic organisms needs to be present in samples in order to observe detectable fluorescent signals. This usually meant that enrichment culture techniques were required prior to immunofluorescence microscopy. Consequently, the FA procedure for *Salmonella* detection has not been routinely used.

Now, though, nanotechnology is playing an increasing role in building sensors that can reliably detect food-borne pathogens. Zhao lists some of these nanotechnology-enabled techniques: detections by luminescence using quantum dots; localized surface plasmon resonance of metallic nanoparticles; enhanced fluorescence; or dye-immobilized nanoparticles.

According to Zhao, the nanostructures used for biosensing applications have two characteristics: "First, they contain certain recognition mechanisms specific to the analyte such as antibodies or enzymes. Second, they are able to generate a distinguishing signal from the analyte and this signal could be generated by the nanostructures themselves or produced by signaling molecules immobilized or contained in the nanostructures."

The scientists point out that for single-component nanostructures, it can be difficult to immobilize the recognition molecules and signaling molecules simultaneously. Hetero-nanostructures provide a promising platform to solve this problem. Thus, different functional molecules can be immobilized to the different components of the hetero-nanostructure to enhance selectivity and specificity of detection.

In their experiment, Zhao, Park, Tripp and collaborators managed to capture a single *Salmonella* bacterium with the antibodies conjugated on the gold nanorod array and detected by thousands of dye molecules immobilized on the silicon nanorods. In principle, the protocol developed in this study could be used for detecting other food-borne pathogenic bacteria such as *E. coli, Staphylococcus, Campylobacter,* and food toxins such as ricin, abrin, or *Clostridium botulinum* if the proper antibody is selected for conjugation with nanorod substrates. Additionally, the fluorescent detection dye can also be replaced by other types of dyes or potentially by quantum dots that may allow for multiplex detection.

This novel nanobiosensor could have broad appeal to the food industry, food safety inspection agencies, government agencies overseeing food safety, and researchers focusing on safety and biosecurity research.

Featured scientists: (a) Bosoon Park, (b) Yiping Zhao
Organizations: (a) U.S. Department of Agriculture, Agricultural
Research Service, Athens, GA (USA); (b) Department of Physics
and Astronomy, University of Georgia, Athens, GA (USA)
Relevant publication: Junxue Fu, Bosoon Park, Greg Siragusa, Les
Jones, Ralph Tripp, Yiping Zhao, Yong-Jin Cho: An Au/Si
hetero-nanorod-based biosensor for Salmonella detection,
Nanotechnology, **19**, 155502.

The Fight Against Cancer

9.1 Carbon Nanotubes as Effective Cancer Killers

Curing cancer is one of the many promises of nanotechnology. Although scientists have been making amazing progress in this area, there are still significant challenges that need to be overcome before highly selective, targeted anticancer therapies become available for everyday clinical use.

A nanotechnology-based system to eradicate cancer requires four elements: (1) molecular imaging at the cellular level so that even the slightest over-expressions can be monitored; (2) effective molecular targeting after identifying specific surface or nucleic acid markers; (3) a technique to kill the cells that are identified as cancerous based on molecular imaging, simultaneously by photodynamic therapy or drug delivery; and (4) a post-molecular-imaging technique to monitor the therapeutic efficacy.

Today, these four techniques are used separately or ineffectively, resulting in an overall poor therapeutic outcome. In what could amount to a quantum leap in cancer nanomedicine, researchers have integrated these techniques simultaneously *in vitro* and shown that this results in higher therapeutic efficacies for destroying cancer cells. Their demonstration of multicomponent molecular targeting of surface receptors and subsequent photothermal destruction of cancer cells using single-walled carbon nanotubes (SWCNTs) could lead to a new class of molecular delivery and cancer therapeutic systems.

Scientists usually develop and refine their techniques for individual components of the above-mentioned four elements, but cancer nanomedicine can only become part of clinical practice if these ideas are combined into a coherent therapy with the single goal of eradicating cancer. A group of U.S. researchers have managed to target more than one class of receptor, show molecular

RSC Nanoscience and Nanotechnology No. 8
Nano-Society: Pushing the Boundaries of Technology
By Michael Berger
© Michael Berger 2009
Published by the Royal Society of Chemistry, www.rsc.org

imaging using simple microscopy techniques, improve selectivity by making carbon nanotubes (CNTs) penetrate the cells for drug delivery, and finally show cell death using photodynamic therapy.

"We have shown that CNTs can be made to enter cells, using receptor-specific antibodies, to a much greater degree than was previously thought possible," explains Balaji Panchapakesan. "Secondly, by targeting several classes of receptors simultaneously, we could improve the selectivity of killing cancer cells. Finally, we showed that molecular imaging using multicomponent targeting can be achieved using simple optical microscopy techniques and is therefore easier to implement in a clinical setting. While there are thousands of papers in cancer nanotechnology, none of these papers have attempted to integrate imaging, drug delivery, and therapy simultaneously."

Panchapakesan, an assistant professor in electrical and computer engineering at the University of Delaware, together with collaborators from the Kimmel Cancer Center at Thomas Jefferson University in Philadelphia, functionalized SWCNT with HER2 and IGF1R specific antibodies and showed that they display selective attachment to breast cancer cells compared to SWCNT functionalized with non-specific antibodies. Photothermal laser treatment resulted in the death of all cancer cells that had antibody/SWCNT hybrids attached (while more than 80% of the cells with SWCNT/non-specific antibody hybrids remained alive).

The researchers used SWCNTs with an average size of 1.4 nm in diameter and 500–1000 nm in length. "One of the most intriguing observations that we made is how the SWCNTs become internalized into the cells," says Panchapakesan. "Optical and confocal images obtained after the incubation of SWCNT–antibody conjugates with the cells demonstrated that the SWCNTs were readily internalized into the cells over large areas."

Panchapakesan says it is reasonable to think that the cells may be acting as a suction pump for internalization of SWCNTs. When antibodies attach to their corresponding receptors in cancer cells, stresses are generated by the release of free energy and this may create pressure differences across the membrane pores, thereby allowing the internalization of the SWCNTs.

Antibodies incubated with the cells acted as biological transport carriers to realize the endocytosis of SWCNTs. Shining near-infrared laser light heated the nanotubes inside the cells. The localized photothermal effect produced heat to destroy the cells completely.

"Light causes interesting properties in nanotubes such as photoconductivity due to exciton generation, light-induced elastic deformation, electrostatic charge separation, and explosions such as SWCNT nanobombs," explains Panchapakesan. "These effects are highly important for biomedical applications. For example, the explosions can be used not only for cancer therapeutics but also to generate acoustic waves at the nanoscale for the next generation of high-efficiency ultrasound imaging applications. Stresses generated by light on the surface of the SWCNTs can also be used as a nanoscale delivery mechanism for proteins from the surface of the SWCNTs into the cells."

The researchers are now working to see if they can integrate all these techniques—targeting, imaging, and therapy—*in vivo*. "If we can," says Panchapakesan, "then we have a clear winner and we can improve clinical outcome by more than 70–80%, which is a quantum leap over the chemo and radiation therapies that are being used today."

The challenge of course is that the researchers do not know whether these techniques will work *in vivo* as well as they did in the test tube. At a minimum, it will take a fair amount of testing. But Panchapakesan is confident that this is the way to do it. "Unless we integrate imaging, drug delivery, and therapy all in one package, clinical outcome might still be poor," he says. "My vision is to package all these things into one 'magic pill' which you can take to get rid of cancer."

> *Featured scientist:* Balaji Panchapakesan
> *Organization:* Electrical and Computer Engineering, University of
>   Delaware, Newark, DE (USA)
> *Relevant publication:* Ning Shao, Shaoxin Lu, Eric Wickstrom, Balaji
>   Panchapakesan: Integrated molecular targeting of IGF1R and
>   HER2 surface receptors and destruction of breast cancer cells using
>   single wall carbon nanotubes, *Nanotechnology,* **18**, 315101.

9.2 Boosting Cancer Drug Efficacy by Targeting Specific Sites Inside Tumor Cells

The ability of a drug molecule to reach its intended target is key among several aspects that determine how effective a pharmaceutical drug is. This need for target-specific delivery of drugs is well accepted in modern drug therapy. Many research efforts are geared towards improving not just the tissue accumulation, but also the cell-specific accumulation of drug molecules in the hope of improving their efficacy.

The ability to target nanoparticles to specific types of cancer cells is one of the main reasons why nanoparticles have gained favor as a promising drug delivery vehicle. By increasing the amount of an anticancer agent that reaches tumor cells, as opposed to healthy cells, researchers hope to minimize the potential side effects of therapy while maximizing therapeutic response. A group of scientists has taken this approach one step further by targeting the specific location inside a tumor cell, where a cancer drug then exerts its cell-killing activity.

Volkmar Weissig, an associate professor of pharmacology at the Midwestern University College of Pharmacy Glendale, explains that efforts aiming at the development of pharmaceutical nanocarriers suitable for the targeted delivery of biologically active agents to specific organs, tissues, and cells date back to the late 1970s. Then, Gregory Gregoriadis discussed the enormous potential of liposomes as a colloidal drug and DNA delivery system for biomedical applications for the first time in two prescient papers published in 1976[1]—almost three decades before the term 'nano' started to replace the term 'colloidal' in the pharmaceutical literature.

During the 1980s and early 1990s, problems and issues linked to the biomedical applications of nanoparticles that are nowadays widely discussed, such as biocompatibility, biodegradability, biodistribution, toxicity, surface modification, drug encapsulation, drug retention, and release, were addressed and largely solved in the field of liposome technology. This culminated in 1995 in the marketing of Doxil® (liposomal encapsulated doxorubicin) as the first FDA-approved nanomedicine.

However, despite the progress made in using nanocarriers to increase tissue accumulation of drug molecules in order to improve efficacy and to reduce unwanted side effects, the successful subcellular targeting of drugs specifically to cell organelles has only recently gained broader recognition. Many drugs have target sites inside the cell, at specific cell organelles or even inside organelles such as mitochondria or lysosomes. This means that to dramatically increase the efficiency of such intracellularly acting drugs, they not only need to be delivered selectively to organs, tissues, and cells, but also to targets inside cells, to organelles, and even to the interior of organelles such as the mitochondrial matrix, which is surrounded by two cell membranes. The subcellular, organelle-specific delivery of drugs has emerged as the new frontier for drug delivery.

Weissig and his former colleagues from Northeastern University have clearly demonstrated that pharmaceutical nanocarriers can be targeted to subcellular compartments.

"These nanocarriers offer a significant benefit because they allow the specific delivery of drugs to subcellular targets without the need for chemical modification of the drug molecules," says Weissig. "Most importantly, we have shown that such organelle-specific drug-loaded nanocarriers can significantly enhance therapeutic effect. With suitable ligands, our strategy could be applied to other organelle targets, thereby offering improved therapy for a number of diseases associated with organelle dysfunction."

Weissig's team developed a lipid-based nanoparticle and decorated its surface with a molecule known as triphenylphosphonium cation, which is known to be taken up specifically by mitochondria, the cell's energy-producing organelles. The investigators then loaded this nanoparticle with ceramide, a drug that forms holes in the mitochondrial membrane, which in turn triggers cell death by a process known as apoptosis. Ceramide also has been shown to work in concert with other anticancer agents to overcome the multiple drug resistance that develops in many tumors.

Using this formulation, Weissig and his colleagues observed that targeted cells, and only targeted cells, treated in culture accumulated ceramide in their mitochondria. Moreover, the scientists found that only the targeted nanoparticle was able to deliver ceramide to the mitochondria and trigger apoptosis.

---

[1] G. Gregoriadis, The carrier potential of liposomes in biology and medicine. *N. Engl. J. Med.*, 1976, **295**, 704–710; 761–770.

Untargeted nanoparticles loaded with ceramide, as well as free ceramide, did not accumulate in mitochondria, nor did they trigger apoptosis.

Weissig notes that an accidental discovery led him to start this work. "In the mid-1990s, as a postdoc in Tom Rowe's lab at the University of Florida in Gainesville, I was screening drugs known to accumulate in mitochondria for their ability to inhibit a certain—at that time still putative—enzyme in *Plasmodium falciparum*, the causative agent of malaria. When trying to make a stock solution of such a drug, dequalinium chloride, I realized this drug is able to self-assemble into tiny small vesicles, just like phospholipids form liposomes. In other words, by accident, I made what today we would call nanocarriers completely composed of molecules with high affinity for mitochondria, which in addition bear a positive surface charge." For lack of a better name, Weissig named these vesicles DQAsomes—for *DeQAlinium derived liposome-like vesicles.*

"Being well aware of the use of cationic liposomes for the delivery of DNA to the nucleus (as nonviral transfection vectors) it struck me that 'my' DQAsomes might have the potential to deliver DNA to mitochondria," Weissig recounts. "The only problem I had at that time was the question of why anybody would want to deliver DNA to mitochondria. Going back to the literature, I quickly learned about breathtaking discoveries made just a few years earlier revealing the link between mutations of mitochondrial DNA and human diseases, and with astonishment I read about the requirement for a mitochondria-specific transfection vector."

Weissig's realization that his accidental discovery about the self-assembly of mitochondria-specific cationic molecules might actually meet this requirement totally changed the direction of his research, and during the past 8 years he and his colleagues have demonstrated that DQAsomes do indeed meet all requirements for mitochondria-targeted DNA delivery (Figure 9.1).

Besides the application of DQAsomes—and similar mitochondria-targeted DNA delivery systems—as mitochondrial transfection vectors, a particularly

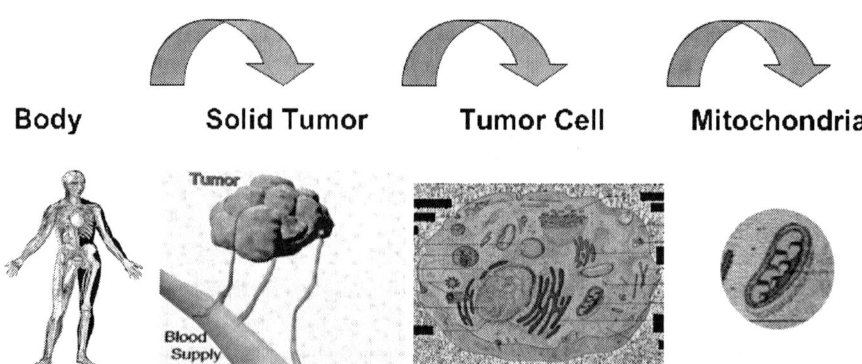

Figure 9.1   Three levels of targeting cancer. (Image: Dr V. Weissig, Midwestern University College of Pharmacy)

intriguing aspect of this work involves the central role mitochondria play in apoptosis, *i.e.* programmed cell death.

Weissig points out that several clinically used drugs and many experimental ones are known to act directly on mitochondria in order to exert their apoptotic activity, *i.e.* to make the cell commit suicide. However, sufficiently high drug concentrations at the site of mitochondria usually require the administration of drug doses which can easily reach the toxic range. By selectively delivering these drugs to their intracellular target site the drug dose can be reduced significantly. In other words, subcellular, organelle-specific drug delivery may increase the therapeutic efficiency of a drug.

Furthermore, nanocarrier-mediated mitochondria-specific delivery of pro-apoptotic drugs may even have two additional major effects, explains Weissig. "First, cancer cells have developed the ability to avoid apoptosis by a variety of ways of 'ignoring' the body's command to commit suicide for the benefit of the whole. The specific delivery of drugs known to trigger apoptosis by acting at mitochondrial target sites directly to tumor cell mitochondria, however, would literally bypass defective apoptotic mechanisms 'upstream' of mitochondria and subsequently trigger apoptosis in tumor cells which otherwise would be resistant.

"Second, selective targeting to and selective release of drug molecules at their site of action, *i.e.* at the surface of mitochondria, may constitute a novel strategy to bypass the p-glycoprotein, which resides in the cell membrane and which is one major cause for multidrug resistance. The carrier would literally hide the drug from the membrane-bound pump, just as the Trojan horse hid the warriors inside it from the gatekeepers."

The ultimate goal of research such as Weissig's is the design of nanocarriers, which combine all levels of drug targeting: systemic administration to specific organs, then to specific tissues, then to specific cells, and subsequently to organelles.

> *Featured scientist:* Volkmar Weissig
> *Organization:* Department of Pharmacology, Midwestern University College of Pharmacy, Glendale, AZ (USA)
> *Relevant publication:* Sarathi V. Boddapati, Gerard G. M. D'Souza, Suna Erdogan, Vladimir P. Torchilin, Volkmar Weissig: Organelle-targeted nanocarriers: specific delivery of liposomal ceramide to mitochondria enhances its cytotoxicity *in vitro* and *in vivo*, *Nano Lett.*, **8**, 2559–2563.

9.3   Inhaling Bubbles to Fight Lung Cancer

Inhalation (or respiratory) therapy is a traditional medical treatment that dates back to ancient times—and not always for purely therapeutic effects; witness the hookah. In the late 18th century, earthenware inhalers became popular for the inhalation of air drawn through infusions of plants and other ingredients, and about 50 years ago the first pressurized metered dose inhaler was put on the market. People suffering from asthma are very familiar with inhalers—devices that help deliver a specific amount of medication to the lungs. The delivery of drugs via the pulmonary route is a potentially effective form of therapy not only for asthma but also for patients with other chronic diseases, including cystic fibrosis, type I diabetes (insulin is absorbed well through the lungs), and recently, lung cancer. In inhalation therapy, the drugs are delivered in aerosol form, meaning that very small particles of the drug are suspended in air (liquid particles make mist, solid particles make fume or dust).

Unfortunately, state the-of-the-art aerosol delivery technologies do not make it possible to target aerosols to specific regions of the lung. They can only be targeted to the central regions or periphery of the lung. Researchers in Germany have shown that aerosols containing magnetic nanoparticles can be guided inside the lungs, thereby offering a potential new route for lung cancer treatment. Their research effort was designed to provide a potentially new technology to offer gentler treatment options for patients suffering from severe lung diseases such as lung cancer. At first glance it might seem strange that lung cancer, which is often caused by inhaling carcinogenic particles found in polluted air and tobacco smoke, can be treated just by inhaling some more particles. Although aerosol therapy does work to some degree, there are side effects (as is the case with all other current cancer drug therapies) caused by the administered drugs damaging healthy tissue.

"Being able to target specific areas in the lung with cancer drug aerosols would avoid potential drug-related site effects in healthy tissue," explains Carsten Rudolph, a researcher at the Ludwig Maximilians University of Munich. "We were able to demonstrate theoretically by computer-aided simulation and for the first time experimentally in mice, that targeted aerosol delivery to the lung can be achieved with aerosol droplets made up of superparamagnetic iron oxide nanoparticles (SPIONs), so-called nanomagnetosols, in combination with a target-directed magnetic gradient field."

Rudolph, together with colleagues from various universities in Munich and Berlin, and other institutions across Germany, show that with their technique higher doses of drugs can be delivered to the cancerous region without increasing side effects. The German research team suggests that nanomagnetosols may advance the field towards a patient tailor-made aerosol treatment, *e.g.* for treating localized lung diseases, such as foci of bacterial infection or tumor nodules.

Rudolph explains that in traditional aerosol therapy the inhaled particles are carried through the respiratory tract with tidal air and are deposited in the lungs when their trajectories differ from the inspired air stream. This difference is induced by forces such as gravity and impulse transfer from gas molecules acting on the particles, and inertia.

Although it has previously been demonstrated[2] that aerosol deposition can be magnetically targeted *in vitro*, these particles could not be moved easily

---

[2]J. Ally, B. Martin, M. B. Khamesee, W. Roa, and A. Amirfazli, Magnetic targeting of aerosol particles for cancer therapy, *J. Magn. Magn. Mater.*, **293**, 442–449.

because of their size and small magnetic momentum. The novel aspect in the German researchers' work is that they assembled a large number of SPIONs into microdroplets, which led to an increased magnetic moment. This resulted in aerosols that are guidable by medically compatible magnetic fields. By including one or several drugs within the aerosol droplets, the nanomagnetosols become guided drug delivery vehicles. In medical practice, drug dosing will be as high as required to secure an effective drug level at the target site, unless side effects become unacceptable, which is quite frequently the case.

"To reduce deposition in the untargeted lung lobe, on the basis of our calculations we suggest the positioning of a bypass magnet close to the main bifurcation of the trachea," says Rudolph. "Our calculations suggest that this is a realistic scenario."

"The challenge of the technology will be to scale up the appropriate conditions from mice to human patients," says Rudolph. "For instance, the non-linear fall in magnetic force with distance is a significant issue. When we positioned the magnetic tip 1 mm above the lung lobe, the relative SPION concentration was eightfold higher than in a lung lobe not exposed to a magnetic field. When we moved the tip to 2 mm away we observed a much reduced relative accumulation, only 2.5-fold higher. This will be an issue with human patients, where the magnetic field has to penetrate a much larger distance. Scaling up the magnetic gradient field to address the size of the human lung represents the major challenge in translating this technology to humans."

> *Featured scientist:* Carsten Rudolph
> *Organization:* Klinikum der Universität München, Munich (Germany)
> *Relevant publication:* Petra Dames, Bernhard Gleich, Andreas Flemmer, Kerstin Hajek, Nicole Seidl, Frank Wiekhorst, Dietmar Eberbeck, Iris Bittmann, Christian Bergemann, Thomas Weyh, Lutz Trahms, Joseph Rosenecker, Carsten Rudolph: Targeted delivery of magnetic aerosol droplets to the lung, *Nat. Nanotechnol.*, **2**, 495–499.

9.4 Nanoparticle Photodynamic Therapy for Deep Cancer Treatment

Photodynamic therapy (PDT) is a cancer treatment that combines a chemical compound, called a photosensitizer, with a particular type of light to kill cancer cells. The treatment works like this: the photosensitizing agent is injected into the bloodstream. The agent is absorbed by cells all over the body, but stays in cancer cells longer than it does in normal cells. One to three days after injection, when most of the agent has left normal cells but remains in cancer cells, the tumor is exposed to light. The photosensitizer in the tumor absorbs the light and produces an active form of oxygen (singlet oxygen) that destroys nearby cancer cells. PDT has been used for the past 30 years and is a treatment that works. It takes very little time, is often done as an outpatient procedure, can be accurately targeted to the affected area, can be repeated, and has no known long-term side effects. Also, it is not as expensive or invasive as some other cancer treatment options. The limitation of this form of cancer treatment is that the light needed to activate most photosensitizers cannot pass through more than 1 cm of tissue. For this reason, PDT is usually used to treat tumors on or just under the skin, or on the lining of internal organs or cavities. It is not effective for treating large or deep tumors.

Researchers have proposed a new PDT system in which the light is generated by X-ray scintillation[3] nanoparticles with attached photosensitizers. When the nanoparticle–photosensitizer conjugates are targeted to tumors and stimulated by X-rays, the particles generate visible light that can activate the photosensitizers for photodynamic therapy.

"I have been working on nanotechnologies for 15 years," says Wei Chen. "My original work was trying to use quantum dots for *in vivo* imaging. I was facing the challenge of light penetration. I also have experience with the design and synthesis of scintillation nanoparticles. I knew that light delivery was also a challenging issue for PDT, just as in *in vivo* optical imaging. Then, I came up with the idea of combining PDT with radiation therapy through scintillation nanoparticles for deep cancer treatment."

Chen, an assistant professor of nanobiophysics at the University of Texas at Arlington, points out that PDT is not new, and radiation therapy is not new; but the combination of the two through scintillation nanoparticles is new and potentially important for deep cancer treatment. He originally introduced this concept in 2006 and has since worked on refining it.

Chen explains that, when the nanoparticle–photosensitizer conjugates are targeted to a tumor and stimulated by X-ray or other radiation sources during radiation therapy, the particles will generate energy in the form of light to activate the photosensitizers. With this novel therapeutic approach, no external light is necessary to activate the photosensitizing agent within tumors. Tissue thickness would therefore no longer be a limiting factor for PDT.

"Effectively, the radiation and photodynamic therapies are combined and occur simultaneously, and the tumor destruction will be more effective," he says. "More importantly, it can be used for deep tumor treatment as X-rays can penetrate deep into the tissue and only an extremely low dose of radiation is needed for the treatment. Therefore, this provides a simple but more efficient modality for cancer treatment. We called this new modality *nanoparticle self-lighting photodynamic therapy*."

Working with Chen's group were Shaopeng Wang and Yuanfang Liu, senior research scientists at ICx/Nomadics Inc.; Alan G. Joly, an optical physicist and a senior scientist at Pacific Northwest National Laboratory; and Carey Pope, Regents Professor and Head Sitlington Chair in Toxicology at the Center for Veterinary Health Sciences, Oklahoma State University. Their pilot studies

---

[3]Scintillation is a flash of light produced by an ionization event.

indicate that water-soluble scintillation nanoparticles, with a particle size of ~15 nm, can potentially be used to activate PDT as a promising deep cancer treatment modality.

For practical applications, the nanoparticle–photosensitizer conjugates must be delivered to the tumor cells in vehicles such as antibodies, peptides, liposomes, or other functional molecules. Chen and his team used folic acid to target folate receptors at tumor cells. Their results indicate that folic acid has no effect on the quantum yield of singlet oxygen production in the nanoparticle conjugates, making this system practical for photodynamic activation applications.

Initial results of the studies have been promising. But before nanoparticle self-lighting photodynamic therapy becomes a clinical reality, the researchers must overcome two main challenges: (1) they need to develop a class of water-soluble scintillation nanoparticles with very high quantum efficiencies of X-ray luminescence, and (2) they need to improve the targeting capabilities of the nanoparticle–photosensitizer compound—but this is a challenge that applies to all drug-based cancer treatments.

Featured scientist: Wei Chen
Organization: The University of Texas at Arlington, TX (USA)
Relevant publication: Yuanfang Liu, Wei Chen, Shaopeng Wang,
  Alan G. Joly: Investigation of water-soluble X-ray luminescence
  nanoparticles for photodynamic activation, *Appl. Phys. Lett.*, **92**,
  043901.

9.5 Highly Effective Radiofrequency Absorbers for Cancer Therapy

Another treatment for cancer is radiofrequency ablation (RF ablation) which works by inserting a thin needle guided by CT or ultrasound through the skin and into a tumor. Electrical energy is then delivered through a number of electrodes deployed through the needle, causing a zone of thermal destruction that encompasses the tumor.

Although minimally invasive, this procedure requires the surgeon to insert the RF probes into the tumor and typically creates a zone of non-specific tissue destruction 3–5 cm in size, with both malignant and normal tissues around the needle electrode undergoing thermal injury. Non-complete tumor destruction occurs in 5–40% of cases and complications arise in 10% of patients. In clinical use today, RF ablation is limited to liver, kidney, breast, lung, and bone cancers.

Researchers have experimented with noninvasive radiowave thermal ablation of cancer cells that uses nanoparticles as a novel approach to treat cancer. The idea is that RF treatment of malignant tumors at any site in the body should be feasible if it were possible to get agents that release heat under the influence of the RF field to the specific tumor site. Most RF ablation research so far has worked with gold nanoparticles—these are easily prepared, and binding of cancer-targeting agents to the nanoparticles is easily achieved—though successful results have been demonstrated even with CNTs.

A novel nanomaterial has proved to be a very strong RF absorber and provides high enough thermal-ablation ability in order to generate localized thermally-driven damages inside the cancer cells. Even more, these nanoparticles have shown relatively low cytotoxicity and they absorb low-frequency RF radiation, which has significant penetration depth inside living organisms. Although several research groups presented similar concepts, the nanomaterials used in this particular study proved to be superior at very low RF frequencies.

"Our technique uses magnetic nanoparticles that are covered with graphitic layers as RF absorbers and localized heat providers," says Alexandru S. Biris. "Moreover, we tried to understand the mechanisms that are responsible for the death of cancer cells, such as DNA fragmentation, protein denaturation, development of holes in various membranes, *etc*. Our work was motivated by the desire to find a method that can provide enough thermal energy into a tumor and destroy it with RF energy provided from outside the body."

In their work, Biris and his collaborators have demonstrated that graphitic carbon-coated ferromagnetic cobalt nanoparticles were efficiently taken up by cervical cancer HeLa cells (an immortal cell line used in medical research) and acted as high-efficiency RF absorbers. After a short exposure time of 2–30 min at low frequency (350 kHz) of RF radiation, the nanoparticles were found to be responsible for inducing cell death in over 98% of the cancer cells. The scientists believe that the RF-driven heating of the nanoparticles is responsible for the generation of extremely localized damage inside the cells.

Interestingly, most of the thermally induced biological modifications were found to happen inside the cellular subcompartments. According to Biris, chief scientist and assistant professor at the University of Arkansas Nanotechnology Center, these results indicate that the cancer cells do not die because of an overall temperature increase, but rather as a result of the internal localized damage produced by the nanostructures. The research team, which also included scientists from the University of Arkansas for Medical Sciences, the Electrostatics and Surface Physics Laboratory at NASA, the National Institute for Research and Development of Isotopic and Molecular Technologies in Romania, and the AgCenter at Louisiana State University, determined that these nanoparticles clearly have the ability to cross the various intracellular membranes and to reach the nucleus. They observed that after the nanoparticles are taken up into the cytoplasm they tend to agglomerate around the nuclear membrane, and a small number crossed the nuclear membrane into the nucleus.

"Because of the localized RF heating provided by the nanoparticles, we found that the cells were going through an accelerated apoptotic process and consequently cellular decomposition," says Biris. "We believe that the disintegration of nanolocalized cell environments such as the nucleus, nuclear membranes, and DNA is the main effect of the RF inducement of heat in the cobalt nanoparticles."

To further prove the function of the ferromagnetic metal nanoparticles, Biris and his team used SWCNTs as a comparison. Here, under identical conditions, only 3.1–6.6% of the cells were found to be dead as against 75.2–90.1% with cobalt nanoparticles.

The results of this work indicate that magnetic nanoparticles can become strong RF absorbers and therefore can generate thermally localized cellular damage which can induce cell death and necrosis of cancerous tissue.

A main problem of noninvasive RF thermal ablation is the need to provide enough thermal energy to the tumor site from outside the body.

"In conclusion, we have demonstrated that, using a combination of a low frequency, low power RF radiation—which has a high penetration ability in human tissue—with graphitic–magnetic composite nanoparticles could prove an excellent means of raising the temperature at the cellular level above the threshold required for DNA fragmentation or protein denaturation, which will ultimately be responsible for the death of the cells," Biris sums up the team's results.

The group envisages several main research directions growing out of this particular work:

(1) use of these nanomaterials *in vivo* as high efficiency agents for the thermal ablation of cancer cells and tumors
(2) specific delivery of the carbon–magnetic nanomaterials to the tumors and their heating by single or combinations of electromagnetic sources (RF, lasers, X-rays, magnetic, *etc.*)
(3) development of hybrid materials that can perform complex functions such as delivery of biopharmaceutical agents that can further enhance their thermal ablation properties, in order to accomplish an efficient killing of all the cancer cells
(4) targeting, detection, and killing of single cancer cells *in vivo* in real time.

According to Biris, the main problem that still needs to be addressed is the binding of their novel nanomaterials to antibodies, growth factors, amino acids, *etc.* for their efficient delivery to the tumor site.

Featured scientist: Alexandru S. Biris
Organization: University of Arkansas Nanotechnology Center, Little Rock, AK (USA)
Relevant publication: Yang Xu, Meena Mahmood, Zhongrui Li, Enkeleda Dervishi, Steve Trigwell, Vladimir P. Zharov, Nawab Ali, Viney Saini, Alexandru R. Biris, Dan Lupu, Dorin Boldor, Alexandru S. Biris: Cobalt nanoparticles coated with graphitic shells as localized radio frequency absorbers for cancer therapy, *Nanotechnology*, **19**, 435102.

9.6   Cancer Treatment with Diamonds

Developed originally for the surface finishing industry, diamond nanoparticles are finding new and far-reaching applications in modern biomedical science and biotechnologies. Because of its excellent biocompatibility, diamond has been called the biomaterial of the 21st century and medical diamond coatings are already heavily researched for implants and prostheses. Nanoscale diamond is being discussed as a promising cellular biomarker and a nontoxic alternative to heavy-metal quantum dots. Further extending the nanomedical use of diamond, researchers have demonstrated a nanodiamond-embedded device that could be used to deliver chemotherapy drugs locally to sites where cancerous tumors have been surgically removed.

"Our work has shown that nanodiamonds serve as versatile platforms that can be embedded within polymer-based microfilm devices," Dean Ho tells us. "The nanodiamonds are complexed with a chemotherapeutic agent, and subsequently enable sustained, slow release of the drug for a minimum of 1 month, with a significant amount of drug in reserve. This opens up the potential for highly localized drug release as a complementary and potent form of treatment with systemic injection towards the reduction of continuous dosing, and as such, attenuation of the often powerful side effects of chemotherapy."

Ho is an assistant professor in the Departments of Biomedical and Mechanical Engineering at the Robert R. McCormick School of Engineering and Applied Science at Northwestern University, and in developing this flexible microfilm he has worked with collaborators from Nanocarbon Research Institute, Asama Research Extension Center, at Shinshu University in Japan, and the Robert H. Lurie Comprehensive Cancer Center of Northwestern University.

One of the big advantages of this micropatch, which resembles clingfilm and can be customized easily into different shapes, is that it prevents a significant downside of conventional therapies—the often observed large-scale initial release of toxic drugs, which also precludes long-term drug release.

"Slow and sustained drug release that eliminates the 'burst' release effect was an important component of our work," explains Ho. "Furthermore, our research has resulted in a highly scalable technology in that the chemical vapor deposition capabilities of the polymer microfilm, coupled with its batch synthesis, purification, and functionalization of the nanodiamonds with the drug, allow for massively parallel microfilm fabrication capabilities."

Nanodiamonds are therefore potentially quite economical, enabling the use of these devices in the treatment of a broad spectrum of disorders, *e.g.* serving as a local chemotherapeutic patch, or as a pericardial device to suppress inflammation after open heart surgery, a key need in cardiothoracic medicine.

Ho and his colleagues embedded millions of tiny drug-carrying nanodiamonds in the FDA-approved polymer parylene. Currently used as a coating for implants, parylene is a biostable, flexible, and versatile material. A substantial amount of drug can be loaded on to clusters of nanodiamonds, which have a high surface area. The nanodiamonds then are placed between extremely thin films of parylene, resulting in a device that is minimally invasive.

To build the biomedical device, the researchers developed a streamlined approach where a double layer of parylene was fabricated with the nanodiamond–drug complexes sandwiched in between. The bottom layer, ~ 20–$30\,\mu m$ thick, serves as the backbone of the device, allowing it to be easily handled. For the top layer, the research team created a thinner semi-porous film that allows the drug to release slowly from the device.

To test the device's drug release performance, the researchers used doxorubicin, a chemotherapeutic agent used to treat many types of cancer. They found the drug was slowly and consistently released from the embedded nanodiamond clusters for 1 month, with more doxorubicin in reserve, indicating that a more prolonged release (several months and longer) was possible.

"Our nanodiamond patch could be used to treat a localized region where residual cancer cells might remain after a tumor is removed," says Ho. "If a surgical oncologist, for example, was removing a tumor from the breast or brain, the device could be implanted in the affected area as part of the same surgery. This approach, which confines drug release to a specific location, could mitigate side effects and complications from other chemotherapy treatments."

Another example that Ho mentions is the important need in cardiothoracic medicine for a device that can slowly release anti-inflammatories or anti-scarring agents after open heart surgery while only having a small footprint. The micropatch with its micron-scale dimensions fits this bill and can be considered only minimally invasive.

Ho notes that after open heart surgery, a large percentage of patients experience scarring or inflammation that can lead to congestive heart failure, arrhythmia, *etc.* "We envision that a thin strip of the microfilm device can be placed near the heart to slowly release anti-inflammatory drugs, which can reduce serious infections and postoperative complications. Overall, a key benefit of this technology in the aforementioned applications, as well as other medical applications, is that it can serve multiple functions—*e.g.* a shield against scar tissue, a multidrug release device—in one system while also being scalable and scalably produced."

The research team is also investigating the use of nanodiamonds with a broad range of therapeutic classes, including additional small molecules, proteins, therapeutic antibodies, RNAi,[4] *etc.* Ho says that they are also exploring animal model work to assess the *in vivo* therapeutic efficacy of their system.

An area of paramount importance in realizing the translational significance of this work will involve determining how the nanodiamonds are processed by the body after drug release, and developing strategies to ensure that they are removed after drug delivery. Preliminary studies have indicated that the

---

[4] RNA interference (RNAi) is a type of interaction between RNA molecules that helps to determine the activity of genes within living cells.

nanodiamonds are amenable to biological surroundings, and important continued work in this area will explore their long-term interaction with animal models.

Featured scientist: Dean Ho

Organization: Departments of Biomedical and Mechanical Engineering, Robert R. McCormick School of Engineering and Applied Science, Northwestern University, Evanston, IL (USA)

Relevant publication: Robert Lam, Mark Chen, Erik Pierstorff, Houjin Huang, Eiji Osawa, Dean Ho: Nanodiamond-embedded microfilm devices for localized chemotherapeutic elution, *ACS Nano*, **19**, 435102.

CHAPTER 10

Next-Generation Implants, Prostheses, and other Cures

10.1 Improved Implant Devices

Implantable cardioverter defibrillators, cardiac resynchronization therapy devices, pacemakers, tissue and spinal orthopedic implants, hip replacements, phakic intraocular lenses, and cosmetic implants are among the top sellers of the medical implant devices market. Devices like orthopedic implants and heart valves are made of titanium and stainless steel alloys, primarily because these materials are biocompatible. Unfortunately, in many cases these metal alloys, with a lifetime of 10–15 years, may wear out within the lifetime of the patient. They also may not achieve the same fit and stability as the original tissue, and in the worst case, the host organism may reject the implant altogether. Although available implants can alleviate excruciating pain and allow patients to live more active lives, there are often problems getting bone to attach to the metal devices. Small gaps between natural bone and the implant can increase over time, requiring the need for additional surgery to replace the implant.

In the quest to make bone, joint, and tooth implants almost as good as Nature's own version, scientists are turning to nanotechnology. Researchers have found that the response of host organisms to nanomaterials, even at the protein and cellular level, is different than the observed response to conventional materials. While this new field of nanomedical implants is in its very early stage, it holds the promise of novel and improved implant materials.

"We are continually finding more and more positive attributes of using nanomaterials as improved orthopedic implants," says Thomas J. Webster. "Our results point to the idea that no matter what material chemistry one is interested in—metals, ceramics, polymers, composites—increased bone formation can be achieved by nanostructuring these materials. This can be done by using a number

RSC Nanoscience and Nanotechnology No. 8
Nano-Society: Pushing the Boundaries of Technology
By Michael Berger

Published by the Royal Society of Chemistry, www.rsc.org

of manufacturing techniques including the use of nanoparticles themselves, electron-beam evaporation, chemical etching, or lithography. In some cases, we see a three to five times faster regeneration of bone on materials of the same chemistry but with nano instead of micron grain sizes."

Webster is an associate professor in the Division of Engineering and Orthopaedics at Brown University. Together with Ganesan Balasundaram, also at Brown, he discusses recent studies that have been conducted to determine the efficacy of nanostructured materials as orthopedic implants.

Webster outlines the requirements for successful bone implants. They should not only temporarily replace missing bone but provide a framework into which the host bone and vascular network can regenerate and heal; they should act as a scaffold to support newly formed bone, blood vessels, and soft tissue as they grow to connect fractured bone segments; and, ideally, the implant should also interact with the host tissue, recruiting and even promoting differentiation of osteogenic cells, rather than acting as a passive stage for the performance of any itinerant cells.

Webster says that there are clearly two important but not unrelated factors that determine the selection of orthopedic implant material: the correct chemistry to support or stimulate an appropriate host response, and the geometric engineering of an appropriate scaffold structure.

He points out that there has been a history of 'trial and error' approaches for orthopedic material chemistries, which has not greatly improved the success of bone implants. Frequently, implant materials promote the formation of undesirable soft connective tissue by other cells such as fibroblasts.

"It has become clear that, in order to design better orthopedic implant materials, we need to concentrate on cellular processes that lead to efficient new bone growth," says Webster. "The efficacy of bone regeneration is determined mainly by surface characteristics such as the chemical composition and physical properties of the implant that control initial protein adsorption."

Rather than just experimenting by trial and error, scientists are aiming to intelligently design implant surfaces to control protein interactions important for subsequent cell adhesion. This may provide answers to the problems that have plagued current orthopedic implants.

Three factors have become key in the development of improved orthopedic devices: topography, where nanoscale surface structuring would optimize cell colonization; surface chemistry, where scientists attempt to control and optimize the chemical surface properties of an implant material; and wettability, because cell adhesion and subsequent activity are generally better on hydrophilic surfaces.

Some studies imply that cell responses might be more sensitive to changes in surface roughness features in the nanometer range (< 100 nm), compared with conventional ranges (> 100 µm), and sensitivity may vary with cell type.

Surface structuring and chemistry modification require nanoscale processes, and engineered nanomaterials could play a role in increasing wettability. "For example," says Webster, "we have demonstrated that aqueous contact angles were three times smaller (*i.e.* the material was more wettable) when alumina grain size was decreased from 167 nm to 24 nm."

Although nanostructured implant materials may have many potential advantages in the context of promoting bone cell responses, it is important to remember that studies on nanophase materials have only just begun.

Webster cautions: "There are still many other issues that must be answered. Most importantly, influences of nanoparticulates on human health are not well understood, whether exposure occurs through the manufacturing or through the implantation of nanophase materials. Clearly, detailed studies in this context are required if nanoparticles are to be used in implant systems."

For instance, nanoparticles could become loose through the degradation of implanted polymeric materials, or they could be generated at artificial joints where friction between two surfaces is high. Although the outcome of micron-sized wear debris on bone health has been well studied, the influence of nanoparticulate wear debris, or nanoparticles in general, on bone cell health is only just beginning to be understood.

Despite the challenges that lie ahead, Webster is optimistic: "There is now substantial evidence that nanophase materials represent an important growth area of research that may improve bonding between an implant and surrounding bone. It has proved to be a versatile approach that can increase bone cell functions on a wide range of orthopedic implant chemistries."

> *Featured scientist:* Thomas J. Webster
> *Organization:* Division of Engineering, Brown University, Providence, RI (USA)
> *Relevant publication:* Ganesan Balasundaram, Thomas J. Webster: A perspective on nanophase materials for orthopedic implant applications, *J. Mater. Chem.*, **16**, 3737–3745.

10.2 Diamond Ice Coatings could Improve Knee Prostheses and Solar Cells

With recent advances in the industrial synthesis of diamond and diamond-like carbon film bringing prices of these materials down significantly, researchers are increasingly experimenting with diamond coatings for medical implants. On the upside, the wear resistance of diamond is dramatically superior to that of titanium or stainless steel. On the downside, because it attracts coagulating proteins, its blood clotting response is slightly worse, and the possibility has been raised that the nanostructured surface features of diamond might abrade tissue (although some of the currently used implant materials cause problems as well). To find solutions for these problems, researchers have run simulations showing that thin layers of ice could persist on specially treated diamond coatings at temperatures well above body temperature. The soft and hydrophilic ice multilayers might enable diamond-coated medical devices that reduce abrasion and are highly resistant to protein absorption.

"Low-temperature and short-range interfacial ordering of water has been previously predicted and observed on a variety of other planar substrates, including muscovite mica, platinum, chlorine-terminated silicon, quartz, and graphite," Alexander Wissner-Gross explains. "However, in order to stabilize ice coatings with the nanoscale thicknesses relevant to macromolecular adsorption at temperatures suitable for *in vivo* application, a novel substrate is needed."

Wissner-Gross, Environmental Fellow at Harvard University who worked on this research with Efthimios Kaxiras, the Gordon McKay Professor of Applied Physics at Harvard University, found that chemically modified diamond is the strongest surface stabilizer of ice yet discovered. According to the scientists' simulations, diamond with an atomically thin treatment of sodium can prevent ice films a few nanometers thick—the size of a very small protein—from melting at temperatures beyond human body temperature. You might call it 'warm ice'.

"We think the result is exciting from a nanotechnological perspective because it opens the door to a new class of atomically engineered surfaces with novel physical properties, such as long-range ice stabilization," says Wissner-Gross.

The physicists arrived at their result by using a computer simulation based on molecular dynamics. Direct simulations of the freezing phase transition are challenging, says Wissner-Gross, so the focus was instead on understanding the stability of initially crystalline ice layers. Many hypothetical models have been developed in order to discover the structure of water, and Wissner-Gross and Kaxiras chose to use the TIP4P/Ice model of water in order to reproduce the melting temperature of ice.

Modeling the motion of water atoms sitting on top of a sodium–diamond surface at different temperatures over long time periods resulted in the surprising conclusion that ice layers could persist on the treated diamond up to temperatures of 42 °C, and in some circumstances could remain frozen beyond the boiling point of water.

The researchers think that the main application of their research will be for coating medical implants. "In particular, we have demonstrated a new, potentially biocompatible diamond nanocoating that could make human joints completely scratch- and wear-resistant," says Wissner-Gross. "This is a major finding for patients who suffer from arthritis and other diseases that require joint replacement surgery. Diamond has long been an attractive material for scratch-free coatings and it is becoming inexpensive enough for biomedical applications such as bone and enamel coatings. Our new research demonstrates that, with the proper chemical modification, diamond might be made extremely bio-friendly by shielding itself with a thin layer of ice at human body temperature."

In addition to the medical application, this novel physical effect may prove useful in making solar energy more economically viable. In particular, the effect may enable solar thermal collectors to compensate for variability in the insulation supply by locally buffering their thermal output using hot water sequestered on such diamond surfaces. Because diamond can elevate the solid–liquid transition of water close to its liquid–vapor transition, 50% more latent heat could be stored near the melting point than with a traditional heat exchanger based on the solid–liquid transition. This water-based approach to small-scale solar energy storage and buffering would have the additional advantage of being more environmentally friendly than lithium ion batteries or organic phase-change materials.

"I see surface physics as a tremendously interesting field going forward," says Wissner-Gross. "The main trend I see is toward adding more functionality to everyday surfaces, whether by nanostructuring or other mechanisms. Synthesizing all the surfaces we can imagine will always be a challenge, but I think the field is up to it."

A short film that the researchers made from of some of their simulations was a finalist in the 2007 Materials Research Film Festival.[1] The work also earned Wissner-Gross the 2007 Dan David Prize Scholarship from Tel Aviv University and the 2007 Graduate Student Silver Award from the Materials Research Society.

> *Featured scientist:* Alexander Wissner-Gross
> *Organization:* Department of Physics, Harvard University, Cambridge, MA (USA)
> *Relevant publication:* Alexander D. Wissner-Gross, Efthimios Kaxiras: Diamond stabilization of ice multilayers at human body temperature, *Phys. Rev. E*, **76**, 020501(R).

10.3 Break a Bone—and Fix it with Carbon Nanotubes

Carbon nanotubes (CNTs) have shown promise as an important new class of multifunctional building blocks and innovative tools in a large variety of nanotechnology applications, ranging from nanocomposite materials through nanoelectronics to biomedical applications. The exploration of CNTs in biomedical applications is well under way and exploratory uses have included CNT-coated implants, drug delivery, and CNTs as components of biosensors. Notwithstanding the issue of toxicity—still not satisfactorily addressed—properties such as high strength, high electrical and thermal conductivities, and high specific surface area render CNTs particularly useful in the fabrication of nanocomposite-derived biomedical devices.

In one particular area—biomaterials applied to bone—CNTs are anticipated to improve the overall mechanical properties for applications such as high-strength arthroplasty prostheses expected to remain in the body for a long time, or fixation plates and screws that will not fail or impede healing of bone. In addition, CNTs are expected to be of use as local drug delivery systems or

---

[1] Watch it on YouTube: http://www.youtube.com/watch?v=LrsOXSas4bU

scaffolds to promote and guide bone tissue regeneration. A study by Japanese scientists clearly demonstrates that multi-walled CNTs (MWCNTs) have good bone-tissue compatibility, permitting bone repair and becoming closely integrated with bone tissue. Furthermore, under certain circumstances, their results indicate that MWCNTs even accelerate bone formation.

"Bone-tissue compatibility is extremely important for using CNTs in biomaterials placed in contact with bone, but no studies have characterized this property or another very important one: the effects of CNTs on bone healing and bone regeneration," explains Naoto Saito. "Our study is the first to clarify the bone-tissue compatibility of CNTs and their influence on new bone formation to determine whether and how CNTs might perform in biomaterials in contact with bone or as scaffolding for bone regeneration."

Saito is a professor in the Department of Applied Physical Therapy at Shinshu University School of Health Sciences in Matsumoto, Japan. "We were able to clearly demonstrate that MWCNTs have good bone-tissue compatibility, permitting bone repair and becoming closely integrated with bone tissue," he says. "Furthermore, our results indicate that MWCNTs accelerate bone formation in response to bone morphogenetic protein."

In their study, the Japanese scientists tested highly crystalline MWCNTs with an average diameter of 80 nm and length 10–20 µm, prepared by catalytic chemical vapor deposition and subsequent thermal treatment above 2800 °C in argon. They note that the high-temperature treatment was a critical step to improve the structural integrity of the their high-purity MWCNTs (~98%) and also to remove any remaining metallic impurities to less than 100 ppm.

Saito and his collaborators basically tested four issues: (1) Are MWCNTs compatible with bone tissue? (2) What is their influence on bone healing? (3) What is their influence on new bone formation? and (4) Do MWCNTs promote crystallization of hydroxyapatite (the inorganic mineral that makes up 70% of bone)?

With regard to compatibility, the experiments indicated that MWCNTs did not cause strong inflammatory reactions and had no influence on bone even when in contact with it, representing good bone-tissue compatibility.

To examine their influence on bone healing, the researchers implanted 100 nL of MWCNTs in defects (0.7 mm in diameter and 2 mm deep) created in the shin bones of mice. After 4 weeks they found that the cortical bone and the bone-marrow cavity were completely restored. MWCNT particles were present not only in the bone marrow but also in the bone matrix. Observation of the interface between the MWCNT particles and bone matrix disclosed that MWCNTs adhered directly to the bone itself.

For the third issue, new bone formation, the scientists used bone morphogenetic proteins (BMPs), a group of proteins known for their ability to induce the formation of bone and cartilage. They made a composite consisting of BMP, MWCNTs, and collagen, and implanted it in the dorsal musculature of mice. Again they found that MWCNT particles were integrated entirely into new bone and bone marrow, but also that, compared to a control group using only BMP–collagen, the MWCNTs seemed to accelerate new bone formation in response to BMP.

Finally, Saito's team confirmed that hydroxyapatite was formed and crystallized on the MWCNT surface in simulated body fluid remarkably quickly, and that the MWCNTs acted as the core for the initial crystallization of hydroxyapatite.

Saito points out that, while further studies are required to elucidate the mechanism by which MWCNTs affect bone formation, the observation that bone formation is accelerated by the addition of MWCNTs is extremely important for the medical application of MWCNTs to biomaterials concerning bone. "Our finding should facilitate development of new drug delivery systems or scaffold materials for bone regeneration using MWCNTs," he says. "Furthermore, including MWCNTs in implants for the treatment of fractures, such as plates and screws, may promote bone repair and thus facilitate rapid fracture healing."

He also notes that it is necessary that other structures and dimensions of CNTs should be assessed as well. "Further studies are necessary to determine the characteristics of bone containing CNTs, including remodeling properties. In addition, the metabolism of CNTs *in vivo* should be studied in detail. As is true generally, safety-related testing involving carcinogenesis and other toxicity is absolutely imperative before CNTs are used in humans."

Featured scientist: Naoto Saito
Organization: Department of Applied Physical Therapy, Shinshu
  University School of Health Sciences, Matsumoto (Japan)
Relevant publication: Yuki Usui, Kaoru Aoki, Nobuyo Narita,
  Narumichi Murakami, Isao Nakamura, Koichi Nakamura, Norio
  Ishigaki, Hiroshi Yamazaki, Hiroshi Horiuchi, Hiroyuki Kato,
  Seiichi Taruta, Yoong Ahm Kim, Morinobu Endo, Naoto Saito:
  Carbon nanotubes with high bone-tissue compatibility and bone-
  formation acceleration effects, *Small*, **4**, 240–246.

10.4 Nanoparticles make Better Eye Drops

For the treatment of eye conditions, conventional eye drops have three major disadvantages: they must be applied frequently; their ocular bioavailability is low (on average, less than 5% of the administered active is absorbed or becomes available at the site of physiological activity); and their use is often associated with high systemic exposure to active free acid. Ophthalmic inserts, the common alternative option, achieve sustained drug delivery but suffer from other limitations: they are difficult to insert (especially for elderly or visually impaired people); they are easily expelled from the eye; patient compliance is low (discomfort and blurring of vision, difficulty of insertion, need for removal at the end of their useful life); and they are costly to manufacture. Researchers in the UK believe that biodegradable polymer nanoparticles show great promise as drug delivery devices for the eye. They have developed well-tolerated systems that combine the sustained release characteristics of inserts with the patient acceptability of conventional eye drops.

With the new systems, biodegradable polymers can be combined with drugs in such a way that the drug is released from the material into the eye in a predesigned manner. Drug release can be constant or cyclic over a long period of time, or triggered by the environment or a chemical signal. At the end of its useful life, the drug-delivering polymer can be broken down naturally by the body.

John Tsibouklis, an expert in biomaterials and drug delivery from the School of Pharmacy and Biomedical Sciences at the University of Portsmouth, UK, explains this new drug delivery system: "We have developed nanoparticle-based ocular formulations that can release actives over extended time scales (days). These are administered into the conjunctival cul-de-sac and, at the end of their useful life, disappear into the lachrymal fluid. Among other advantages, the materials are well tolerated and the rate of their bioerosion/dissolution can be controlled."

An ideal polymeric drug carrier should have a loading capacity that can ensure therapeutic doses, be able to penetrate to and reside at the desired site of action, and release the active in a controlled manner. It should also be non-toxic, biocompatible, and biodegradable. For most ophthalmic applications the carrier should not impede vision, since tolerability and acceptance by the patient are critical.

Polymeric vehicles for controlled drug release have been classified as either reservoir type, where the polymer essentially coats the drug core, or matrix type, where the active is homogeneously mixed with the polymer, or is bound to it through covalent or hydrogen/donor–acceptor interactions. The vehicles may assume a variety of forms, ranging from solutions or gels to colloidal systems or solid inserts.

"The formulation of biodegradable or bioerodible polymers as water-based colloidal nanosystems holds significant promise for ophthalmic drug delivery," says Tsibouklis. "A colloidal system for poorly water-soluble drugs would allow dropwise administration while maintaining drug activity at the site of action."

Although several synthetic methods and drug loading techniques have been reported to be safe and reproducible, no procedure for the formulation of drug-loaded nanoparticles has yet been standardized. Formulation stability, particle size uniformity, control of drug release rate, and the large-scale manufacture of sterile preparations are major developmental issues that need to be addressed before commercialization.

Tsibouklis points out that the nanoparticulate drug carriers developed at the University of Portsmouth's Biomaterials and Drug Delivery Group have been formulated and tested with drugs for glaucoma (pilocarpine, atropine) and with antibiotics (chloramphenicol, norfloxacine).

As drugs can now be coupled to nanocarriers that are specific for cells and/or organs, nanotechnologies will play a vital role in the future of ophthalmic medication.

Featured scientist: John Tsibouklis
Organization: School of Pharmacy and Biomedical Sciences, University
   of Portsmouth (UK)
Relevant publication: Eugen Barbu, Liliana Verestiuc, Thomas G.
   Nevell, John Tsibouklis: Polymeric materials for ophthalmic drug
   delivery: trends and perspectives, *J. Mater. Chem.*, **16**,
   3439–3443.

10.5   Contributing to the Fight against Alzheimer's

There is currently no cure for Alzheimer's disease, and its ultimate cause is still unknown. The disease affects millions of people around the world. Experts predict that this figure is expected to rise dramatically as the population, especially in developed countries, ages. Genetic factors are known to be important in causing the disease and dominant mutations in different genes have been identified that account for both early-onset and late-onset Alzheimer's. For a number of years, researchers have been working on alleviating neurodegenerative disorders such as Alzheimer's or Parkinson's disease through gene therapy. In this type of treatment, a gene's DNA is delivered to the neurons in individual cells, allowing them to produce their own therapeutic proteins. Gene therapy typically aims to supplement a defective mutant allele (the location of DNA codings on a chromosome) with a functional one. Currently, the most common carrier vehicles to deliver the therapeutic genes to the patient's target cells are viruses that have been genetically altered to carry normal human DNA. These viruses infect cells, deposit their DNA payloads, and take over the cells' machinery to produce the desirable proteins. One problem with this method is that the human body has developed a very effective immune system that protects it from viral infections.

Thanks to advances in nanotechnological fabrication techniques, the development of nonviral nanocarriers for gene delivery has become possible. This is attractive because of the potential for improved safety, reduced ability to provoke an immune response, ease of manufacturing and scale up, and the ability to accommodate larger DNA molecules than is possible with virus-based delivery tools.

A number of people working in nonviral gene therapy used to work with viruses as delivery systems. They had different reasons for turning to nonviral methods, but they have one thing in common: they saw problems with viral gene therapy. In the early days, even retroviruses were used for gene therapy (HIV is a retrovirus) but later on researchers mostly used adeno-associated viruses, thought to be safe because they are not associated with human disease. Nevertheless, some scientists think that viruses are inherently unsafe and their use should be avoided.

Jon A. Wolff, professor of pediatrics and genetics at the Waisman Center at the University of Wisconsin, Madison, used to work with retroviruses, which use

RNA instead of DNA as their genetic material, as a way to deliver genes. But then he realized that viruses had some limitations. "I thought nonviral might be a simpler and easier way of [effecting delivery] and avoid the problems with viral therapy," he says. The main problem is the immunogenicity of the viruses. "Viruses have evolved over a billion years to efficiently deliver DNA into our cells, but our immune system has also evolved over that time," he says.[2]

This is where nonviral gene delivery comes in. By developing nonviral synthetic nanocarriers from scratch, researchers can design a system that simultaneously achieves high efficiency, prolonged gene expression, and low toxicity. A number of these carriers fulfill one or two of those criteria, but it is difficult to meet all three. Besides the consistent problem of toxicity, synthetic systems such as polymer-, lipid- and peptide-based gene carriers are typically much less efficient in delivering genes to primary neurons compared to viruses.

Research undertaken at Johns Hopkins University in Baltimore has begun to offer a systematic approach to understanding the gene delivery process in neurons and the intracellular barriers to nonviral gene delivery, and possible ways to improve their effectiveness with nanotechnology.

"Much work has been done already to improve gene delivery into central nervous system (CNS) neurons to treat neurodegenerative disorders using various gene delivery systems with several therapeutic agents," explains Justin Hanes, a professor of chemical and biomolecular engineering who leads the Advanced Drug and Gene Delivery Group at Johns Hopkins. "Nevertheless, so far there have been no systematic studies investigating the individual intracellular barriers, which should be the guideline for the design of future gene vectors for CNS therapy."

Hanes and his group have systematically examined several potential intracellular barriers that may limit the effectiveness of polymeric gene carriers. Jung Soo Suk, one of the scientists working with Hanes, elaborates: "In general, potential barriers include cellular uptake, intracellular transport, endosome escape, and nuclear import. On the basis of our mechanistic studies, we identified cellular uptake as one major barrier that hampers efficient gene delivery in differentiated neurotypic cells."

To overcome this cellular uptake barrier, the researchers incorporated targeting peptides (*i.e.* RGD or HIV Tat-1 peptide) to their polymeric gene carriers to endow them with some desirable 'virus-like' properties. They found that attachment of either of these peptides improves cellular uptake substantially.

Polycationic polymers have shown significant promise as nonviral gene delivery vectors. Polyethyleneimine (PEI) is currently the most popular polymer used to deliver genes into various cell types, including neurons. PEI is able to condense genes into small nanoparticles and protect the DNA from degradation by nucleases. In addition, the cationic nature of PEI facilitates entry of these gene vectors into cells.

---

[2]Quoted from C. M. Henry, Gene delivery—without viruses, *CENEAR*, 2001, **79**(48), 35–41.

"Surprisingly," says Hanes, "RGD peptide attachment to our gene-carrying nanoparticles improved both the uptake by neurotypic cells and the escape of the particles from lysosomes. Lysosomes typically degrade gene carriers and prevent them from delivering their DNA payload to the cell nucleus, the target site within the cell for gene therapy."

Hanes cautions that this work is still in its early stages and many more obstacles must be overcome. Suk points out that in order to accomplish the ultimate goal of successful clinical application, a thorough investigation of CNS gene delivery mechanisms *in vitro* and *in vivo* must be conducted.

While this research at Johns Hopkins deals with intracellular barriers, gene vectors in practical applications need first to successfully overcome several barriers outside the cells before they reach the target cell itself. According to Suk, future work along the lines of this study will include the selection of the optimal administration method and the investigation of extracellular barriers that gene vectors may encounter. "Extensive and systematic approaches toward the field will ultimately guide us to design virus-like gene vectors without the drawbacks of viruses including safety issues."

Hanes sees the future research in this field focusing on improved biomaterials that are more efficient while remaining safe to use in the CNS; improved basic understanding of the remaining bottlenecks to efficient gene delivery, both inside the cell and outside; and studying the most promising particles in animal models of disease.

Featured scientists: Justin Hanes, Jung Soo Suk
Organization: Institute for NanoBioTechnology, Johns Hopkins
    University, Baltimore, MD (USA)
Relevant publication: Jung Soo Suka, Junghae Suha, Kokleong Choya,
    Samuel K. Laib, Jie Fub, Justin Hanes: Gene delivery to
    differentiated neurotypic cells with RGD and HIV Tat peptide
    functionalized polymeric nanoparticles, *Biomaterials*, **27**, 5143–5150.

10.6   Nanotechnology-based Stem Cell Therapies for Damaged Heart Muscles

Regenerative medicine is an area in which stem cells hold great promise for overcoming the challenge of limited cell sources for tissue repair. Stem cell research is being pursued vigorously in the hope of achieving major medical breakthroughs. Scientists are striving to create therapies that rebuild or replace damaged cells with tissues grown from stem cells, offering hope to people with cancer, diabetes, cardiovascular disease, spinal cord injuries, and many other disorders.

Embryonic stem cells are pluripotent. That means that during normal embryogenesis—the process by which the embryo is formed and develops—

human embryonic stem cells can differentiate into all derivatives of the three primary germ layers: ectoderm, endoderm, and mesoderm. Researchers have also found undifferentiated cells (adult stem cells) in children and adults. Unlike embryonic stem cells, the use of adult stem cells in research and therapy is not controversial because the production of adult stem cells does not require the creation or destruction of an embryo.

Often, adult stem cells are not pluripotent but multipotent. That means they can differentiate only into a limited variety of cell types. One example is mesenchymal stem cells (MSC)—adult stem cells found in bone marrow that can be differentiated into bone, cartilage, fat, and connective tissues. These offer tremendous potential for the repair and or regeneration of damaged tissues and organs.

An area of particular interest is differentiation of MSC into cardiomyocytes for damaged heart muscle tissue. During a heart attack, part of the heart muscle loses its blood supply and cells in that part of the heart die, thereby damaging the muscle. This reduces the ability of the heart to pump blood around the body. Considering that coronary heart disease is the leading cause of death in most developed countries, stem cell therapy that repairs heart muscle cells and restores the viability and function of the area already damaged could have a tremendous impact on modern medicine.

"Recently, CNTs have been generating great excitement in the fields of bioengineering and drug delivery research. However, very little is known about the affect of CNTs on MSC response," says Valerie Barron. "Therefore, the main goal of one of our research studies was to investigate the effect of CNTs on human MSC (hMSC) biocompatibility, proliferation, and multipotency."

Barron, a senior researcher at the National Centre for Biomedical Engineering Science at National University of Ireland (NUI), together with collaborators from NUI's Regenerative Medicine Institute and Department of Anatomy, investigated a range of different types of CNTs, including single-walled nanotubes (SWCNTs), multiwalled nanotubes (MWCNTs), and functionalized CNTs.

The scientists revealed that at low concentrations of SWCNTs functionalized with carboxylic acids (COOH), the CNTs had no significant effect on cell viability or proliferation. In addition, by fluorescently labeling the COOH-functionalized SWCNTs, the CNTs were seen to migrate to a nuclear location within the cell after 24 h, without adversely affecting the cellular ultrastructure. Moreover, the nanotubes had no affect on adipogenesis (the development of fat cells), chondrogenesis (the development of cartilage), or osteogenesis (the development of bone).

Previous research had shown that CNTs migrate into cancer cells and can therefore be used for delivery of biomolecules directly into the cells. Barron's study is the first to examine the effect of CNTs on hMSC and as such is important for new and emerging technologies in drug delivery, tissue engineering, and regenerative medicine. It appears that at low concentrations, CNTs have minimal affect on MSC viability and multipotency. Therefore, they have great potential to advance the field in a number of ways, including

1. Manipulation of MSC differentiation pathways
2. Development of nanovehicles for delivering biomolecule-based cargoes to mesenchymal stem cells
3. Creation of novel biomedical applications for electroactive carbon nanotubes in combination with mesenchymal stem cells.

In a previous position at Trinity College Dublin, Barron had worked in Werner Blau's Molecular Electronics and Nanotechnology group where she gained a tremendous appreciation for CNTs. "As a biomaterials scientist, I could see their potential in biomedical applications," she says. At NUI Galway she teamed up with Mary Murphy, a principal investigator in the Orthobiologics Group at the university's Regenerative Medicine Institute, to examine the effect of CNTs on MSC differentiation. Both researchers were aware that, since no clinical therapy is available for the repair of damaged heart muscle, there exist tremendous opportunities for the creation of novel nanotechnology-based therapies.

CNTs are electrically conductive and this provides the potential for manipulating MSC differentiation pathways to create electroactive cells such as those found in the heart. In particular, specific applications could result in novel MSC-based cell therapies for electroactive tissue repair; novel biomolecule delivery vehicle for manipulation of MSC differentiation pathways; and electroactive CNT scaffolds for damaged electroactive tissues.

"At present, we are developing a novel electrophysiological environment to promote MSC differentiation towards a cardiomyocyte lineage," says Barron. "In the short term, we plan to focus on optimizing this approach to develop nanotechnology-based cell therapies. In the longer term we hope to use the CNTs as delivery vehicles for a range of different biomolecules for the manipulation of MSC differentiation pathways towards a range of different cell types."

Featured scientist: Valerie Barron
Organization: National Centre for Biomedical Engineering Science at National University of Ireland, Galway (Ireland)
Relevant publication: Emma Mooney, Peter Dockery, Udo Greiser, Mary Murphy, Valerie Barron: Carbon nanotubes and mesenchymal stem cells: biocompatibility, proliferation and differentiation, *Nano Lett.*, **8**, 2137–2143.

CHAPTER 11

Concepts a Bit Further Out

11.1 Cell-like Nanofactories Inside the Body

Notwithstanding the huge amount of research going into this field, nanomedicine by and large is still in the basic research stage. Fundamental problems like the targeting of nanoparticles *in vivo*, the transport of unstable drugs, and the dosage control of drug-carrying nanoparticles lead some scientists to think even further, much further, ahead. Rather than delivering external drugs into the body, they conceptualize artificial-cell nanofactories that work with raw ingredients already in the body to manufacture the proper amount of drug *in situ* under the control of a molecular biosensor.

"This approach draws its inspiration from the ability of the human body to self-medicate by actively adapting molecular production in response to its intrinsic biochemistry," explains Philip R. LeDuc, an assistant professor of mechanical engineering at Carnegie Mellon University. "It proposes that molecular machinery could, in principle, be introduced into the body to convert pre-existing materials into therapeutic compounds, or to change molecules that a patient is unable to process, owing to some medical condition, into other compounds that the body can process."

What would such a nanofactory look like, and what would it do?

"We would need six essential components for realizing a biochemical/biomaterials-based pseudo-cell factory," says Michael S. Wong, assistant professor in chemical and biomolecular engineering at Rice University who works with LeDuc: "(1) a structural shell or scaffold; (2) transport to convey biomolecules to and from the environment; (3) sensing functionality; (4) encapsulation of biochemical machinery; (5) targeting of the factory within the body; and (6) an externally triggered 'kill switch' to terminate a treatment in a controlled fashion."

RSC Nanoscience and Nanotechnology No. 8
Nano-Society: Pushing the Boundaries of Technology
By Michael Berger

Published by the Royal Society of Chemistry, www.rsc.org

One of the most challenging aspects of the nanofactory will be the mechanism that facilitates the transport of the required molecules into the artificial cell, and the outward transport of the products.

"Biological cells closely regulate the transport of materials across their wall through a variety of complex mechanisms including channels, endocytosis, and lipid rafts," says Wong. "There is no obvious solution to this challenge at present, but research into synthetic approaches to cellular-based transport provides some insight. Potential solutions include the development of active artificial carrier proteins or the use of passive channels that allow transport across the membrane."

The outer structure of the artificial cell could be designed on the basis of a variety of encapsulation systems that are already in the experimental stage, such as liposomes or hybrid organic/inorganic composites. Whatever its composition, a key task of the shell structure will be its ability to evade the body's immune system for a sufficiently long period of time. Part of the outer shell would also have to incorporate some kind of sensing mechanism that recognizes the required biomolecules.

Once biomolecules enter the nanofactory, they need to be modified in order to create the desired end product. This could be done using encapsulated modifiers such as enzymes and vesicles as the compartment material.

The ability to localize the factories to a specific area *in vivo* would be essential for their success. Although targeting on the nanoscale has been investigated intensively, especially for drug delivery, the scientists point out that the success of this component will require advances on many fronts.

The final component, the 'kill switch' would allow an external operator to stop the operation of the nanofactory, for instance through ultrasonic stimulation, and have it break down into smaller parts. The goal here is not only to stop the production of the nanofactory's output, but also prevent the disassembled parts from causing any unwanted side effects.

The artificial cell concept has a long history, including research into developing systems that mimic living cells as well as using cells in an artificial environment or device. Efforts to mimic living cells focus on how to generate vesicles with cellular functionality. Such structures would be considered 'living' if they could replicate, self-heal, and evolve.

"When comparing the use of an artificial cell system with living cells, there are benefits and risks to both," says Wong. "Living cells are wonderfully adapted to performing a wide variety of tasks. Yet, the ability to engineer control of a cell to produce billions of molecules for a specific function has limitations."

"So, rather than reverse engineering an extremely complex system, generating an artificial cell provides a robust platform where researchers can add functionality in a component-by-component process," adds LeDuc. "Furthermore, artificial approaches may be more controllable as living cells can respond in unanticipated ways."

There are many challenges in developing this type of system, including issues with fabrication, immunological response, and safety. If successful, though, this could become an important platform technology that could benefit a variety of therapeutic approaches. Not only would it allow us to cure cancer by repairing or destroying malignant cells one by one, but it would revolutionize gene therapy as well. Even NASA is looking into the nanofactory concept for the prevention of radiation-induced cancer in astronauts during long space missions.

Featured scientists: (a) Philip R. LeDuc, (b) Michael S. Wong
Organizations: (a) Department of Mechanical Engineering, Carnegie Mellon University, Pittsburgh, PA (USA); (b) Chemical and Biomolecular Engineering, Rice University, Houston, TX (USA)
Relevant publication: Philip R. LeDuc, Michael S. Wong, Placid M. Ferreira, Richard E. Groff, Kiryn Haslinger, Michael P. Koonce, Woo Y. Lee, J. Christopher Love, J. Andrew McCammon, Nancy A. Monteiro-Riviere, Vincent M. Rotello, Gary W. Rubloff, Robert Westervelt, Minami Yoda: Towards an *in vivo* biologically inspired nanofactory, *Nat. Nanotechnol.*, **2**, 3–7.

11.2 Nanoscale Power Plants

For the more than 100 million people worldwide who have diabetes, testing their blood glucose level is the only way to be sure that it is within normal range and to adjust the insulin dose if it is not. The current method for monitoring blood glucose requires poking a finger to obtain a blood sample. The equipment needed to perform the blood test includes a needle device for drawing blood, a blood glucose meter, single-use test strips, and a log book. Now imagine this scenario: your doctor implants a tiny device the size of a rice grain under your skin. This device automatically and accurately measures your blood glucose levels at any desired interval, even constantly if required. It transmits the data to an external transceiver. If any abnormality is detected, the device warns you and automatically transmits the data to your doctor's computer. This scenario is one of the many promises of nanomedicine where *in situ*, real-time, and implantable biosensing, bio-medical monitoring, and biodetection will become an everyday fact of normal health care.

Nanosensors are already under intensive development in laboratories around the world. One of the important components for implantable nanosensors is an independent power source, either a nanobattery or a nanogenerator that harvests energy from its environment, so that the sensor can operate autonomously. Not only has such a nanogenerator now been developed, but a new prototype has been demonstrated to effectively generate electricity inside biofluid, *e.g.* blood. This is an important step toward self-powered nanosystems.

Huge research efforts go into the development of nanoscale sensing devices for applications ranging from medical and biosensing to environmental monitoring to military use. In comparison, until recently, innovations for delivering nanoscale energy sources to power these devices have been almost nonexistent. The energy to be fed into a nanogenerator is likely to be mechanical energy that is converted into electric energy that will then be used to power nanodevices without using a battery. Examples of mechanical energy are body movement or muscle stretching, vibration energy such as acoustic/ultrasonic waves, and hydraulic energy such as body fluid and blood flow.

Let us revisit Zhong Lin Wang and his nano power generators, whom we met in Chapter 6.1. Wang's group has also developed a DC nanogenerator that is driven by ultrasonic waves. "The basic principle is to use piezoelectric and semiconducting coupled nanowires, such as zinc oxide, to convert mechanical energy into electricity," explains Wang. "This nanogenerator has the potential to convert hydraulic energy in the human body, such as blood flow, heartbeat, and contraction of blood vessels, directly into electric energy. Our nanogenerator is able to generate electricity in biocompatible fluid as driven by ultrasonic waves."

"In order for this nanogenerator to have practical uses in real-life applications, however, we had to come up with an innovative design to drastically improve its performance with regard to the following aspects," says Wang: "First, we must eliminate the use of an atomic force microscope for the mechanical deformation of the nanowires, so that the power generation can be achieved by an adaptable, mobile, and cost-effective approach on a larger scale. Secondly, all of the nanowires are required to generate electricity simultaneously and continuously, and all the electricity must be effectively collected and output. Finally, the energy to be converted into electricity has to be provided in the form of waves or vibration from the environment, so the nanogenerator can operate independently and wirelessly."

Using their innovative approach, the Georgia Tech scientists have made their nanogenerator work in biofluid. By generating electricity in liquid, this work sets a platform for developing self-powering nanosystems with important applications in implantable *in vivo* biosensing.

As an ultrasonic wave travels through the fluid, it triggers the vibration of the electrode and nanowires to generate electricity. The size of the nanogenerators used in these studies was $\sim 2\,\text{mm}^2$. There are more than 1 million nanowires in each of these generators. Wang and his team kept the ultrasonic wave on for 4 h without interruption and the nanogenerator remained active and generated electricity continuously. "We expect the lifetime of the nanogenerator is much longer than the time we have tested," says Wang.

The nanogenerators in this study also show the possibility of integrating multiple nanogenerators in biofluid for receiving high power output. This is a key step towards self-powering biosensing and medical applications.

In future, Wang and the team will try to optimize the growth of the nanowire arrays in terms of size and height uniformity and their distribution on substrate as well as the design of the top electrode, so that most of the nanowires will generate electricity. The second goal will be to raise the output voltage to more than 0.5 V so that it can be used for practical applications. The third goal is to improve the packaging of the nanogenerator to enhance the efficiency of energy generation. The final goal is to test and improve the power generation at low frequency.

> *Featured scientist:* Zhong Lin Wang
> *Organization:* Center for Nanostructure Characterization, Georgia
>   Institute of Technology, Atlanta, GA (USA)
> *Relevant publication:* Xudong Wang, Xudong Wang, Jin Liu, Jinhui
>   Song, Zhong Lin Wang: Integrated nanogenerators in biofluid, *Nano
>   Lett.*, **7**, 2475–2479.

11.3   Nanotechnology Coming to a Brain Near you

If you have seen the movie *The Matrix* you will be familiar with 'jacking in'—a brain–machine neural interface that connects a human brain to a computer network. For the time being, this is a science fiction scenario, but researchers are already working on laying the foundation for future brain–machine interfaces. What is already reality today is something called *neuroprosthetics*, an area of neuroscience that uses artificial microdevices to replace the function of impaired nervous systems or sensory organs. Different biomedical devices implanted in the central nervous system, so-called *neural interfaces*, have already been developed to control motor disorders or to translate willful brain processes into specific actions by the control of external devices. These implants could help increase the independence of people with disabilities by allowing them to control various devices with their thoughts (not surprisingly, the other candidate for early adoption of this technology is the military).

The potential of nanotechnology application in neuroscience is widely accepted. In particular, single-walled carbon nanotubes (SWCNTs) have received great attention because of their unique physical and chemical features, which allow the development of devices with outstanding electrical properties. In a crucial step towards a new generation of future neuroprosthetic devices, a group of European scientists developed a SWCNT/neuron hybrid system and demonstrated that carbon nanotubes (CNTs) can directly stimulate brain circuit activity.

Examples of existing brain implants include brain pacemakers to ease the symptoms of such diseases as epilepsy, Parkinson's disease, and dystonia; retinal implants that consist of an array of electrodes implanted on the back of the retina, a digital camera worn on the user's body, and a transmitter/image processor that converts the image to electrical signals sent to the brain; and most recently, cyberkinetics devices such as the BrainGate™ Neural Interface System[1] which has been successfully used by quadriplegic patients to control a computer by thoughts alone.

---

[1] http://www.cyberkineticsinc.com/content/medicalproducts/braingate.jsp

Thanks to the application of recent advances in nanotechnology to the nervous system, a novel generation of neuro-implantable devices is on the horizon, capable of restoring function loss as a result of neuronal damage or altered circuit function. This research will very soon be mature enough to explore *in vivo* neural implants in animal models.

Italian scientists Laura Ballerini and Maurizio Prato have developed an integrated system that couples SWCNTs to an *ex vivo* reduced nervous system where a mesh of SWCNTs deposited on glass acts as a growing substrate for cultured rat neurons. "We demonstrated that neurons formed functional healthy networks *in vitro* over a period of several days and developed a dense array of connection fibers, unexpectedly intermingled with the SWCNT meshwork, with tight contacts to cellular membranes," Ballerini describes their work.

Ballerini, an associate professor in physiology, and Prato, a professor in the Department of Pharmaceutical Science, both at the University of Trieste, are also involved in the European Neuronano[2] project, an advanced multidisciplinary project designed to develop neuronal nanoengineering by integrating neuroscience with materials science, micro- and nanotechnology. The Neuronano network's major aim is to integrate CNTs with multielectrode array technology to develop a new generation of biochips to help repair damaged CNS tissues.

"For the first time, we can show how electrical stimulation delivered through CNTs activates neuronal electrical signaling and network synaptic interactions," says Michele Giugliano, a researcher at the Brain Mind Institute at the Ecole Polytechnique Fédérale de Lausanne in Switzerland who collaborated with Ballerini and Prato. "We developed a mathematical model of the neuron–SWCNT electrochemical interface. This model provides for the first time the basis for understanding the electrical coupling between neurons and SWCNT."

Over the past few years, there has been tremendous interest in exploiting nanomaterials and nanodevices in basic neurosciences research. So far, the interactions between CNTs and cellular physiology have been studied and characterized as an issue of biochemical mechanisms involving molecular transport, cellular adhesion, biocompatibility, *etc.* The research by the Italian team boosts scientists' understanding of interfacing the nervous system with conductive nanoparticles, at the very fast time scale of electrical neuronal activity which in mammals determines behavior, cognition, and learning.

"Recently, the Neuronano research group pioneered the exploration of CNTs as an artificial way to interact with the collective electrical activity emerging in networks of vertebrate neurons," says Giugliano. "Biocompatibility of CNTs has been described in the literature and several groups have recently attempted to couple neurons to CNTs to probe or elicit electrical impulses. However, specific considerations of the electrophysiological techniques that are crucial for understanding signal transduction and electrical coupling have been underestimated."

---

[2] http://www.neuronano.net/

The researchers achieved direct SWCNT–neuron interactions by culturing rat hippocampal cells on a film of purified SWCNTs for 8–14 days, to allow for neuronal growth. This neuronal growth was accompanied by a variable degree of neurite extension on the SWCNT mat. A detailed scanning electron microscopy analysis suggested the presence of tight interactions between cell membranes and SWCNTs at the level of neuronal processes and cell surfaces.

The scientists point out that their results as a whole represent a crucial step towards future neuroprosthetic devices, exploiting the surprising mechanical and (semi)conductive properties of CNTs. This field is now closer to a quantitative understanding of how precise electrical stimulation may be delivered in deep structures by 'brain pacemakers' in the treatment of brain diseases.

"From current and previous results of our group, it seems that CNTs could functionally interact with electrical nervous activity even in the absence of signal-conditioning integrated electronics and explicit external control," says Ballerini. "In fact, at least to some extent, (semi)conductive properties of the nanotubes might facilitate the emergence of synaptic activity. These achievements offer a promising strategy to further develop next-generation materials to be used in neurobiology."

> *Featured scientists:* (a) Laura Ballerini, (b) Maurizio Prato, (c) Michele Giugliano
>
> *Organizations:* (a,b) University of Trieste (Italy); (c) Brain Mind Institute, Ecole Polytechnique Fédérale de Lausanne (Switzerland)
>
> *Relevant publication:* Andrea Mazzatenta, Michele Giugliano, Stephane Campidelli, Luca Gambazzi, Luca Businaro, Henry Markram, Maurizio Prato, Laura Ballerini: Interfacing neurons with carbon nanotubes: electrical signal transfer and synaptic stimulation in cultured brain circuits, *J. Neurosci.*, **27**, 6931–6936.

11.4   Plugging into the Brain with Carbon Nanofibers

Neural engineering is highly interdisciplinary and relies on expertise from computational neuroscience, experimental neuroscience, clinical neurology, electrical engineering, and signal processing of living neural tissue, and encompasses elements from robotics, computer engineering, neural tissue engineering, materials science, and nanotechnology. In order for neural prostheses to augment or restore damaged or lost functions of the nervous system they need to be able to perform two main functions: stimulate the nervous system and record its activity. To do that, neural engineers have to gain a full understanding of the fundamental mechanisms and subtleties of cell-to-cell signaling *via* synaptic transmission and then develop the technologies to replicate these mechanisms with artificial devices and interface them to the neural system at the cellular level.

Scientists' understanding of distributed neuronal network processing has been facilitated by the advent of microelectrode array (MEA) recording platforms, capable of recording activity from multiple neurons simultaneously.

According to Barclay Morrison, an assistant professor in biomedical engineering at Columbia University in New York, simultaneous recording from multiple locations within a tissue may unlock the neural code underlying higher brain functions. "We have introduced a new type of MEA with electrodes made not of metal but of vertically aligned carbon nanofibers (VACNFs). We have shown that these arrays can perform the standard complement of electrophysiological techniques possible with commercial MEAs," he says.

Researchers have already demonstrated the potential of nanofiber-based electrode architectures for interfacing with excitable cell matrices. Carbon nanofibers are electrochemically active structures that may be integrated into parallel arrays using the conventional tools and approaches of microfabrication. In contrast to conventional planar arrays, however, nanofibers provide a novel, nonplanar, high aspect ratio structure that may provide unique opportunities for probing extra-, inter-, and ultimately intracellular phenomena.

"We have previously demonstrated that these structures could be interfaced to individual cells, but it was not known if they would be robust enough to interface to intact tissue," says Morrison. "Further, it was not known whether our carbon nanofibers could stimulate and record action potentials from neural tissue. We have shown that both were possible with our vertically aligned carbon nanofiber arrays."

Morrison and collaborators from his department at Columbia University and the Oak Ridge National Laboratory, led by Timothy E. McKnight and Milton Nance Ericson, were able to demonstrate that these same carbon nanofiber electrodes may also be used to stimulate and record electrophysiological signals from intact hippocampal slices.

For their experiments, the researchers fabricated devices which consisted of a linear array of 40 individually addressed VACNF electrodes, 10 µm in height and spaced 15 µm apart along a total length of 600 µm. To record electrical activity, a slice of live hippocampal tissue was positioned on the VACNF array, allowing the electrodes to penetrate into the tissue. Morrison and his colleagues were able to record spontaneous activity as well as responses evoked by stimuli applied to the electrodes.

In the field of neural engineering, carbon electrodes have several potential advantages over other electrode materials. Morrison explains that, perhaps most importantly, carbon electrodes are well suited for electrochemical measurements in neuronal environments and have been used to observe dynamic chemical events in and around neuronal tissue. "Combined with their ability to also record electrophysiological signals, this dual sensing capability opens the door to developing neural prostheses that sense both neurotransmitter levels and electrical signals simultaneously," he says.

The long-term goal of this research is to demonstrate the ability to measure both electrophysiological signals and neurotransmitter concentrations using these novel VACNF electrode arrays. Morrison hopes that this ability could ultimately lead to improved neural prosthetics that can interface to tissue with high numbers of connections, while also providing feedback on the local chemical environment, including tissue health or activity over time.

In addition to future neuroprosthetic platforms, there are many applications within the neuroscience and neural engineering fields. "Dual-sensing MEAs could increase our fundamental understanding of neuronal network dynamics—how groups of neurons communicate with each other—and how the network gives rise to emergent properties that underlie how the brain functions," says Morrison. "The dual-sensing electrodes would provide information about not only the output of the neuronal network (electrical signals) but also the inputs (local neurotransmitter concentrations) that drive the electrical activity."

This dual-sensing ability could find use in neural prostheses too. Local neurotransmitter concentration could serve as a feedback channel for devices such as deep brain stimulators so that their output could be modulated with respect to the brain's response.

Featured scientist: Barclay Morrison
Organization: Neurotrauma and Repair Laboratory, Columbia
   University, New York, NY (USA)
Relevant publication: Zhe Yu, Timothy E. McKnight, M. Nance
   Ericson, Anatoli V. Melechko, Michael L. Simpson, Barclay
   Morrison: Vertically aligned carbon nanofiber arrays record
   electrophysiological signals from hippocampal slices, *Nano Lett.*, **7**,
   2188–2195.

PART III:
SIMPLY GREEN—ENVIRONMENTAL APPLICATIONS AND RISK MANAGEMENT

CHAPTER 12
Green Nanotechnology

'Green nanotechnology' is usually defined as a set of technologies that offer the possibility of changing the manufacturing process in two ways: "Incorporating nanotechnology for efficient, controlled manufacturing would drastically reduce waste products; and the use of nanomaterials as catalysts for greater efficiency in current manufacturing processes by minimizing or eliminating the use of toxic materials and the generation of undesirable by-products and effluents."[1]

In a broad sense, this term includes a wide range of possible applications, from nanotechnology-enabled, environmentally friendly manufacturing processes that reduce waste products (in what, as visionaries hope, will ultimately lead to atomically precise molecular manufacturing with zero waste); the use of nanomaterials as catalysts for greater efficiency in current manufacturing processes by minimizing or eliminating the use of toxic materials (green chemistry principles); the use of nanomaterials and nanodevices to reduce pollution (*e.g.* water and air filters); and the use of nanomaterials for more efficient alternative energy production (*e.g.* photovoltaics and fuel cells).

Unfortunately, there is a flip side to these benefits. As scientists experiment with the development of new chemical or physical methods to produce nanomaterials, the concerns about a potential negative impact on the environment are also heightened. Some of the chemical procedures involved in the synthesis of nanomaterials use toxic solvents, could potentially generate hazardous by-products, and often involve high energy consumption, not to mention the unsolved issue of the potential toxicity of certain nanomaterials. This is leading to a growing awareness of the need to develop clean, nontoxic, and environmentally friendly procedures for synthesis and assembly of nanoparticles. Here are a few examples.

---

[1] Quoted from the EPA website: http://es.epa.gov/ncer/nano/research/nano_green.html

---

RSC Nanoscience and Nanotechnology No. 8
Nano-Society: Pushing the Boundaries of Technology
By Michael Berger
© Michael Berger 2009
Published by the Royal Society of Chemistry, www.rsc.org

12.1   Flame Retardants with less Toxic Chemicals

Flame-retardant materials have become a major business for the chemical industry and can be found practically everywhere in modern society. In a countries where houses are mostly built from wood, like the USA, most structural timber and wood elements such as paneling are treated to make them more fire resistant.

Plastic materials are replacing traditional materials like wood and metal—just think of the toys you played with and the ones kids have today. Unfortunately, the synthetic polymeric materials we group under the term 'plastics' are flammable. To decrease their flammability they require the addition of flame-retardant chemical compounds. The plastic casings, circuit boards, and cables of computers, electrical appliances, or cars are flame retardant. So is practically every material in airplanes, trains, and ships, from the fabric of seats to every kind of plastic structure found on board. Name any plastic product and chances are it has been made flame retardant.

Unfortunately, conventional methods for making plastic flame retardant involve a range of not exactly harmless chemicals. Improving the flame retardancy of polymeric materials without the use of toxic chemicals could now become possible thanks to the synergistic effect of carbon nanotubess (CNTs) and clay.

Acrylonitrile–butadiene–styrene (ABS) resins are a good example of an ubiquitous industrial plastic material used in a variety of diverse applications such as appliances (where 23% of all produced ABS is used), transportation (21%), piping (13%), electrical and electronic components (11%), medical applications (4%), and miscellaneous other applications including, toys, luggage, lawn and garden products, shower stalls, furniture, and ABS resin blends with other polymers (28%). ABS is the largest-volume engineering thermoplastic resin. Global demand for ABS resins is estimated to be almost 4 million tonnes per year.

Flame retardants (FRs) can either be incorporated in the manufacture of structural plastics, foams, and textile fibers or impregnated into timber, textile yarns, *etc.* They can also be added as protective coatings. The main chemical families of FRs are inorganic chemicals (including antimony, aluminium, and tin compounds), bromine and chlorine based FRs, phosphorus based FRs, and nitrogen (melamine) based FRs. The use of some of these chemicals in FRs has come under increased scrutiny because of their suspected negative health and environmental impact. In particular, the use of brominated flame retardants in polymer formulations has come under a lot of fire (so to speak). Firefighters, for instance, are at risk from the potentially toxic chemicals that are created when products containing these FRs burn.

In response, individual European countries started to require companies to replace brominated FRs with safer alternatives. To harmonize efforts in Europe, the European Union banned the use of all polybrominated diphenyl ethers (PBDEs) and polybrominated biphenyls (PBBs) in electronic products starting in 2006.

Hoping to provide a possible solution, researchers have been able to modify the flammability properties of polymers with CNTs. Previous research has shown that nanoparticle fillers are highly attractive for the purpose of making a material more flame retardant because they can simultaneously improve both the physical and the flammability properties of the polymer nanocomposite. It also has been shown that CNTs can surpass nanoclays as effective flame-retardant additives if they form a jammed network structure in the polymer matrix, so that the material as a whole behaves rheologically like a gel.[2]

Both CNTs and clay can improve the flame retardancy of plastics. Their flame-retarding mechanisms are different and their effect for improving the flame retardancy is limited when they are used alone. However, when they are used together, a significant synergy takes place.

"The common explanation of the synergistic effects between clay and CNTs on improving flame retardancy of polymers is that CNTs act as a sealing agent in final chars with network structure after combustion," explains Zhengping Fang. The flame retardancy of polymer nanocomposites is strongly affected by the formation of a network structure.

"But how is this network formed?" asks Fang. "Does it form during the first stage or during combustion? How does the network structure affect the flammability properties of nanocomposites?"

Fang, professor and director of the Institute of Polymer Composites at Zhejiang University, Hangzhou, China, together with scientists from his institute, has provided a detailed exploration of the synergistic effects between clay and CNTs.

He points out that in previous studies, a synergistic effect has been found on improving flame retardancy in polymer nanocomposites. "The existence of clay enhances the degree of graphitization of multiwalled CNTs (MWCNTs) during combustion. Clay assists in the elimination of dislocations and defects and the rearrangement of crystallites. Aluminium oxide, one of the components of clay, acts as the catalyst of graphitization."

"The coexistence of clay and MWCNTs in the composites can form a more effective confined space and enhances the network structure, which can be responsible for the improved flame retardancy for polymer nanocomposites in our study," says Fang.

What could be of interest to industrial manufacturers of ABS and other synthetic polymeric materials is that the synergistic effect between clay and CNTs can reduce the amount of FR chemicals required for flame retardant materials. "Because of the excellent barrier properties of clay and tensile strength of CNTs, we expect to obtain FR materials with high performance characteristics and with much less use of potentially toxic chemicals," says Fang.

Traditionally, to improve the dispersion of nanofillers in a polymer matrix, a surfactant was grafted on to the nanofillers. This of course reduces the

---

[2] Takashi Kashiwagi, Fangming Du, Jack F. Douglas, Karen I. Winey, Richard H. Harris and John R. Shields: Nanoparticle networks reduce the flammability of polymer nanocomposites, *Nat. Mater,* **4**, 928–933.

flame-retarding effect of the nanofillers because the surfactants themselves are usually flammable. Thus, one of the future directions of Fang's research could focus on improving the dispersion of nanofillers in polymer matrix without reducing their flame-retarding effect.

> *Featured scientist:* Zhengping Fang
> *Organization:* Institute of Polymer Composites, Zhejiang University, Hangzhou (China)
> *Relevant publication:* Haiyun Ma, Lifang Tong, Zhongbin Xu, Zhengping Fang: Synergistic effect of CNT and clay for improving the flame retardancy of ABS resin, *Nanotechnology*, **18**, 375602.

12.2   Turning Diesel Soot into Carbon Nanotubes

Diesel-burning engines are a major contributor to environmental pollution. They emit a mixture of gases and fine particles that contain some 40 mostly toxic chemicals including benzene, butadiene, dioxin, and mercury compounds. Diesel exhaust is listed as a known or probable human carcinogen by several regulatory agencies. Wouldn't it be nice if diesel soot could be rendered harmless before it is released into the environment? Wouldn't it even be nicer if this soot could be used to manufacture something useful? Well, Japanese scientists have come up not only with a unique technique for effectively collecting diesel soot but also a method for using this soot as a precursor for the production of single-walled CNTs.

As we saw in the quote at the beginning of this chapter, the U.S. Environmental Protection Agency and others look into the future and see optimized 'green' manufacturing processes that reduce pollution at the source; but it will take decades to develop the nanotechnology-enabled, 'green chemistry'-type manufacturing technologies that will have a real impact on the environment, and roll them out across the existing industrial infrastructure. In the meantime, much more relevant to today's environmental problems, we have to deal with the remediation of existing pollutants.

One immediate area of application is filtration. Nanofiltration devices are in (very) limited use already. You might have a nanoparticle-based water filter installed on your kitchen tap, there might be a nanoparticle-containing air filter in your car, or some coal- and gas-fired plants might use nanoparticulate sorbents in their smokestack filters. While filters keep the bad stuff out of the air or water, the questions remains what happens to the filter, and the pollutants in it, after use. If they are just dumped in a landfill than this only shifts the problem from one medium, *e.g.* air, to another, *e.g.* soil or groundwater.

Filtration—nano-enabled or not—and recycling have to go hand in hand. Otherwise the solution is only half-baked. A group of researchers in Japan have come up with an interesting solution for dealing with diesel soot that could serve as an example for novel, short-term solutions to existing environmental problems.

In 2005, Tetsuro Nishimoto of Juon Co. Ltd, a Japanese company specializing in the development and production of environmental equipment, filed a patent application for a unique technique for effectively collecting diesel soot from internal combustion engines. Subsequently, researchers at Yokohama City University and Nissan Arc Ltd have shown that diesel soot can be recycled as a carbon source for the synthesis of single-walled CNTs (SWCNTs).

The researchers first collected the exhaust soot from the diesel engines of an electric generator and a vessel using a technique developed by Nishimoto. Particulate matter from diesel engines is effectively trapped on ceramic filters in the exhaust pipe. This particulate matter, mainly composed of soot, soluble organic fractions, and sulfates, is separated from the filters by ultrasonic washing with water or ethanol. The diesel soot was predominantly collected by Soxhlet extraction of the particulate matter with ethanol. The soot recycled by this method was then subjected to laser vaporization to synthesize SWCNTs. The extracted soluble organic fractions can be recycled as diesel fuel.

"We synthesized SWCNTs by laser vaporization of diesel soot that includes fragments of various fullerenes," explains Masaru Tachibana. "We demonstrated that the fragments in diesel soot are suitable precursors for the synthesis of SWCNTs. The successful synthesis of SWCNTs from this soot shows that diesel soot can be recycled as a carbon source for the synthesis of nanomaterials."

Tachibana is a professor of physics in the Nanoscale Science and Technology Group at Yokohama City University in Japan. The synthesis method developed by him and his collaborators provides an important insight into the growth mechanism of SWCNTs. Tachibana explains: "In electric arc-discharge and laser vaporization methods, graphite can generally be used as a carbon source. However, a few studies have demonstrated the synthesis of SWCNTs from other carbon sources, among them synthesis from fullerenes by a laser vaporization method. The laser decompositions of fullerenes, or their fragments, seemed to be suitable precursors for the synthesis of SWCNTs. When we found fullerenes in diesel soot, this motivated us to try synthesizing SWCNTs and thus find a recycling solution for the soot."

The scientists found that diesel soot contains $C60$, $C70$, and other fullerenes, which are thought to be formed during the combustion of light or heavy oil. They also found that diameters of SWCNTs synthesized from diesel soot are smaller than those of SWCNTs synthesized from graphite.

Although Tachibana and his team did not examine the detailed behavior of the toxic chemicals contained in the soot, they believe that almost all toxic chemicals are destroyed during the laser vaporization process. They concede the possibility that some might remain in the soluble organic fractions, though.

Tachibana hopes that the recycling applications of diesel soot and its modification will prove useful not only for the synthesis of SWCNTs but also for electrodes of fuel cells and gas storage materials.

"Although our work may be a small contribution to the scientific field, we hope it will prove to be very important for its contribution to a cleaner environment," says Tachibana.

> *Featured scientist:* Masaru Tachibana
> *Organization:* Yokohama City University (Japan)
> *Relevant publication:* Takashi Uchida, Ouji Ohashi, Hironori
>   Kawamoto, Hirofumi Yoshimura, Ken-ichi Kobayashi, Makoto
>   Tanimura, Naohiro Fujikawa, Tetsuro Nishimoto, Kazuhiko Awata,
>   Masaru Tachibana, Kenichi Kojima: Synthesis of single-wall carbon
>   nanotubes from diesel soot, *Jpn J. Appl. Phys.*, **45**, 8027–8029.

12.3   Nanocatalyst Improves Important Reaction in the Chemical Industry

Back in the early 1800s, it was observed that certain chemicals can speed up a chemical reaction—a process that became known as *catalysis* and that has become the foundation of the modern chemical industry. By some estimates, 90% of all commercially produced chemical products involve catalysts at some stage in their manufacturing process. Catalysis is the acceleration of a chemical reaction by means of a substance, a catalyst, which is itself not consumed by the overall reaction. The most effective catalysts are usually transition metals or transition metal complexes.

Research with CNTs coming out of Germany contains some implications for catalysis in general. Researchers at the Fritz Haber Institute of the Max Planck Society in Berlin have been working for some time at metal-free catalysis using nanocarbons. While their focus initially has been on ethylbenzene, an aromatic hydrocarbon that plays an important role as an intermediate in the production of various plastic materials, they have also used CNTs to activate butane. The results indicate that the use of CNTs can make certain important chemical reactions more effective, more energy efficient, and safer.

The chemical industry produces butadiene and other alkenes by catalytic dehydrogenation of normal butane. Butadiene is an important basic chemical with a global demand of ~9 million tonnes. Over half of that amount is used to prepare synthetic rubbers such as the ones used in tire production.

"CNTs had been used as catalysts before, for instance in converting ethylbenzene into styrene, a precursor to polystyrene, an important synthetic material," explains Dangsheng Su, a scientist at the Fritz Haber Institute of the Max Planck Society in Berlin. "But butane is much less reactive than ethylbenzene and we were surprised how well it worked—the high selectivities we observed were unexpected."

Conventional synthesis technologies use complex metal oxide catalysts to produce butadiene and other alkenes from butane through oxidative dehydrogenation. These catalysts require high temperatures and lots of oxygen to maintain the catalytic activity but this leads to unwanted product oxidation, which in turn leads to a low selectivity for butadiene. What the Max Planck scientists found is that the selectivity to butadiene in the catalytic dehydrogenation of n-butane can be improved by using modified CNTs as catalyst.

"We demonstrated that CNTs can compete with the best metal oxide-based catalysts on the market for converting butane into butenes—and they were nearly twice as selective for butadiene," says Su. He and his team functionalized surfaces of pristine CNTs with oxygen-containing groups and then additionally modified them by passivating defects with phosphorus. The scientists hypothesize that the phosphorus covers up nanotube defects, preventing them reacting with oxygen and producing electrophilic oxygen species that would destroy the alkene products. With this catalyst they were able to achieve an alkene yield of 13.8%.

According to Su, their CNT catalyst is as selective as the best vanadium–magnesium oxide complexes—the material of choice—developed during the past 20 years.

Another important aspect of using CNT catalysts is improved efficiency. The new process uses less energy because it can run at much lower temperatures—400–450 °C as compared to more than 900 °C for existing processes; and not only does it require far less oxygen (an oxygen/butane ratio of 2) but ordinary air could be used in place of pure oxygen, making the reaction safer. Even when the oxygen/butane ratio was reduced from 2 to 0.5, the CNT sample still exhibited outstanding stability.

In order to assure that the reactivity originated exclusively from metal-free active sites on CNTs and that the residual metals played no positive role, Su and his colleagues took great care in preparing their CNT catalysts.

"One of the key concerns in identifying and describing metal-free catalysis is the suspected influence of metal impurities in the catalyst," says Su. "Through X-ray fluorescence spectrometry and high-resolution transmission electron microscopy we made sure that the residual metals in the CNTs we tested were very low and that those that remained were embedded in the carbon and not exposed to the reactants."

The next challenge for Su's team will be to demonstrate that this process also could work on an industrial scale. If it does, it could have a far-reaching impact on the use of CNTs in catalytic processes in the chemical industry.

Featured scientist: Dangsheng Su
Organization: Fritz Haber Institute of the Max Planck Society, Berlin (Germany)
Relevant publication: Jian Zhang, Xi Liu, Raoul Blume, Aihua Zhang, Robert Schlögl, Dang Sheng Su: Surface-modified carbon nanotubes catalyze oxidative dehydrogenation of n-butane, *Science*, **322**, 73–77.

12.4 Golden Nanoparticles + Sunlight = Greener Chemistry

Scientists have discovered that stained glass windows that are painted with gold purify the air when they are lit up by sunlight. Medieval artists used gold nanoparticles (unknowingly) to achieve the bright red color in church windows—particles of gold at the nanoscale are red, not golden. In effect, the glaziers in their medieval workshops were the first 'nanotechnologists' who produced colors by using gold nanoparticles of different sizes.

"For centuries, people appreciated the beautiful works of art and the long life of the colors but little did they realize that these works of art are also, in modern language, a photocatalytic air purifier with nanostructured gold catalysts," says Huai Yong Zhu, an associate professor in the School of Physical and Chemical Sciences at the Queensland University of Technology in Australia.

One of the great challenges for catalysis is to find catalysts that can work well under visible light. If scientists manage to crack this problem it would mean that we could use sunlight—the ultimate free, abundant, and 'green' energy source—to drive chemical reactions. This is in contrast to today's conventional chemical reactions that often require high temperatures and therefore consume a lot of energy. Almost half of the solar energy that hits the surface of the Earth comes in the form of visible light. Developing photocatalysts that respond to visible light would allow us to utilize much of the solar spectrum, for instance for the production of hydrogen energy by splitting water, purification of water and air (see the church windows), and other catalytic applications.

"Many approaches have been proposed to develop visible-light photocatalysts, including doping titanium dioxide with metal ions or metal atom clusters, incorporating nitrogen and carbon into titanium dioxide, and employing other metal oxides or polymetallates as catalyst materials," explains Zhu. "Research has concentrated mainly on semiconductor oxides. Sulfides have also been studied, but they are not suitable catalysts because of their poor chemical stability. However, searching for catalysts that can work under visible light should not be limited to semiconductor materials with bandgap structure, but can be extended to other materials, such as gold nanoparticles."

What Zhu and his collaborators have found—and described so poignantly with the example of church windows as photocatalytic air purifiers with nanostructured gold catalysts—is that visible light can drive the oxidation of formaldehyde and methanol on gold nanoparticles dispersed on oxide supports in air without external heating.

Gold nanoparticles absorb visible light intensely because of the *surface plasmon resonance* (SPR) effect. Zhu and his collaborators reckoned that the combination of the SPR absorption and the catalytic activity of gold nanoparticles could present an important opportunity: "If the heated gold nanoparticles could activate the organic molecules attached to them to induce oxidation of the organic compounds, then oxidation on gold catalysts can be driven by visible light at ambient temperature," says Zhu.

Since the SPR is a local effect, limited to the noble metal particles, the particles—which generally account for only 2–4 wt% of the overall catalyst mass—are heated up quickly to a temperature at which the organic molecules are activated to react with oxygen. Compared to the conventional catalytic oxidation in which both the gold particles and the support material are heated to high temperatures, this leads to significant saving in energy consumption for catalyzing organic compound oxidation.

The scientists found that a number of factors, such as particle size and morphology and the dielectric constant of the oxide medium, can affect the SPR absorption of gold nanoparticles and thus the activity of the gold catalyst.

"Our findings open up a new direction in photocatalysis as the proposed reaction mechanism is distinctly different from that catalyzed by semiconductor photocatalysts," says Zhu. "Based on our results, we propose a tentative reaction mechanism for light-driven catalytic oxidation. The irradiation of incident light with a wavelength in the range of the SPR band may result in two consequences. The first is that light absorption by the gold nanoparticles could quickly heat these nanoparticles. The second consequence is that the interaction between the oscillating local electromagnetic fields and polar molecules also assists in activating the molecules. The activated polar organic molecules react with oxygen in close proximity."

Zhu points out that this light-driven reaction can proceed at ambient temperature at reaction rates similar to those of the catalytic oxidation under heating and could, for example, find convenient applications in indoor air purification. The findings also reveal the possibility of driving other reactions with abundant sunlight on gold nanoparticles at ambient temperature.

Featured scientist: Huai Yong Zhu
Organization: School of Physical and Chemical Sciences, Queensland University of Technology, Brisbane (Australia)
Relevant publication: Xi Chen, Huai-Yong Zhu, Jin-Cai Zhao, Zhan-Feng Zheng, Xue-Ping Gao: Visible-Light-Driven Oxidation of Organic Contaminants in Air with Gold Nanoparticle Catalysts on Oxide Supports, *Angew. Chem. Int.*, **47**(29), 5353–5356.

CHAPTER 13
Dealing with Pollution

13.1 Novel Nanocomposite Material to Combat White Pollution

Polystyrene, also known under its trademark name Styrofoam, is found in your home, office, local grocery, fast food outlet, and cafeteria. It comes in many shapes and forms—foam egg cartons and meat trays, plates and salad boxes, coffee cups and utensils, CD jewel boxes, and the foam peanuts used in packaging. According to the U.S. Environmental Protection Agency's Municipal Solid Waste statistics, solid waste in the USA in 2005 contained almost 2.6 million tonnes of polystyrene—a material that takes hundreds of years to break down and is not recovered in recycling. Polystyrene is also a principle component of the marine debris floating on the world's oceans.

Motivated by this problem of 'white pollution', scientists have developed a nanocomposite material that not only has superabsorbent capabilities but also utilizes waste polystyrene foam. If commercially successful, this and similar methods of recycling waste polystyrene into new products could go a long way in reducing the worldwide harmful effects of white pollution.

Superabsorbent materials have the ability to absorb many times their own weight in aqueous solutions. They are used not only to improve the performance of a wide range of personal care products but also for industrial products such as plant-friendly water storage to improve soil quality, to regulate moisture in upholstered furniture, and as a water-blocking component in cable sheaths. Have you seen TV commercials for baby diapers where they pour on that blue stuff and it disappears? What you saw was a superabsorbent in action. Superabsorbents are lightly cross-linked functional polymers that can absorb, swell, and retain aqueous solutions up to 1000 times their own weight in a relatively short period of time.

A group of researchers in China have developed a superabsorbent nano structured composite with water absorbency of about 1200 times its own weight.

RSC Nanoscience and Nanotechnology No. 8
Nano-Society: Pushing the Boundaries of Technology
By Michael Berger
© Michael Berger 2009
Published by the Royal Society of Chemistry, www.rsc.org

"Our work has two main aspects," explains Jian Shen: "One is to reduce the production cost of the superabsorbent material and at the same time promote its performance. The other is to find a new way to utilize waste polystyrene foam. The introduction of polystyrene chains into the nanocomposite not only reduces the production cost of the superabsorbent, but also improves the water absorbency rate."

Shen is the director of the Jiangsu Engineering Research Center for Biomedical Function Materials at Nanjing Normal University. He and his collaborators focused on cross-linked (polyacrylic acid), PAA, one of the most effective superabsorbents (and a key substance used in making baby diapers). They used partially neutralized acrylic acid, montmorillonite—a low-cost layered aluminium silicate—and waste polystyrene foam to create a novel superabsorbent nanocomposite. For their experiments they used ordinary polystyrene foam from the packaging of household electric appliances.

"Not only does the use of waste polymer, especially hydrophobic polymer, for synthesizing superabsorbents distinguish our work from other research in this area, we also found that the introduction of polystyrene chain to the composite not only reduced the production cost but improved the water absorbency rate as well," says Shen.

In their experiments, the proportion of waste polystyrene used in the nanocomposite material was between 7.7% and 20%. According to Shen, this could reduce the production cost of superabsorbent materials by 2.9% to 8.7%.

This research by the Chinese scientists creates a new way of utilizing waste polystyrene foam. By converting waste polystyrene into a functional polymer rather than dumping it in a landfill, the resulting materials could be used in water-related areas such as water preservation material, leak prevention, matrices for enzyme immobilization, medicine carriers for controlled release, or everyday personal care products.

If these results could be successfully incorporated into industrial manufacturing processes, this could mean a big step towards eliminating the problem of white pollution.

> *Featured scientist:* Jian Shen
> *Organization:* Jiangsu Engineering Research Center for Biomedical
>   Function Materials, Nanjing Normal University (China)
> *Relevant publication:* Ping-Sheng Liu, Li Li, Ning-Lin Zhou, Jun
>   Zhang, Shao-Hua Wei, Jian Shen: Waste polystyrene foam-graft-
>   acrylic acid/montmorillonite superabsorbent nanocomposite, *J. Appl.
>   Polymer Sci.*, **104**, 2341–2349.

13.2   Capturing and Storing Greenhouse Gases

The greenhouse effect is primarily a function of the concentration of water vapor, carbon dioxide, and other trace gases in the Earth's atmosphere, absorbing the terrestrial radiation that leaves the surface of the Earth. Changes in the atmospheric concentrations of these greenhouse gases can alter the balance of energy transfers between the atmosphere, space, land, and the oceans. The capture and storage of greenhouse gases could play a significant role in reducing the amount of greenhouse gases released into the atmosphere. Carbon dioxide (CO_2) is the most important greenhouse gas and basks in the limelight in most reports on global warming. Although other greenhouse gases make up less of the atmosphere, they account for about 40% of the greenhouse gas radiation sent back to Earth. They can also be much more efficient at absorbing and re-emitting radiation than CO_2, so they are small but important elements in the equation. In fact, molecule-for-molecule, some gases containing lots of fluorine are 10 000 times stronger at absorbing radiation than CO_2. A new systematic computational study shows an interesting approach of how nanotechnology, in this case the use of carbon nanotubes and other nanomaterials, could lead to effective filters for the capture and storage of greenhouse gases.

Tetrafluoromethane (CF_4) is a perfluorocarbon (PFC) gas with an extremely stable molecule whose lifetime in the atmosphere is 50 000 years (compared to an average CO_2 lifetime of 50–200 years). It is a particularly powerful greenhouse gas—also called a super greenhouse gas—that, when present in the troposphere, has a particular ability to absorb the outgoing infrared radiation.

The global warming potential (GWP) index is intended as a quantified measure of the globally averaged relative radiative forcing impacts of a particular greenhouse gas. It is defined as the cumulative radiative forcing—both direct and indirect effects—integrated over a period of time from the emission of a unit mass of gas, relative to CO_2 as a reference gas. The GWP for CO_2 is 1 per 100 years whereas that for CF_4 is 6500 per 100 years. So although total worldwide emissions of CF_4 are small in comparison to those of CO_2, its GWP is vastly higher because it is a much more efficient absorber of infrared radiation than CO_2 is.

PFCs have been extensively used in the microelectronic and semiconductor industry in plasma cleaning of chemical vapor deposition chambers. Although the semiconductor industry is moving away from the use of PFCs toward other less problematic gases, other sources of PFCs are still significant: as an unintended byproduct during aluminium production; as a drop-in replacement of chlorofluorocarbon refrigerants; as potential solvents and co-solvents for supercritical fluid extraction processes; and in the petrochemical industry. New uses for PFCs are being explored, *e.g.* for therapeutic purposes, eye surgery, and modifiers for inhaled anesthetics. As a consequence, it is estimated that global emissions of PFCs will rise by 150% in the next 50 years. That means that super greenhouse gases even with small emissions have the potential to influence climate far into the future and could become a serious environmental problem.

"We have conducted simulations that show that single-walled carbon nanotubes (SWCNTs) can serve as efficient nanoscale vessels for encapsulation of CF_4 at room temperature," says Robert Holyst. "A good filter should keep the CF_4 molecules inside and stay cheap. Carbon seems promising, but in order to efficiently store CF_4 we need well-defined pores inside the carbon material. CNTs are a carbon material that offers well-defined pore sizes for storage applications and they are also most efficient in terms of storage capacity."

Holyst, a professor in the Institute of Physical Chemistry at the Polish Academy of Sciences, together with Piotr Kowalczyk, a postdoctoral research fellow in the Department of Applied Physics at RMIT University in Australia, have found that the amount of the encapsulated CF_4 under ambient external conditions (1 bar, ~ 27 °C) is maximized for well-defined pore sizes of SWCNTs. These pore sizes change with a change in external pressure. They also demonstrate that the high enthalpy of adsorption cannot be used as the only measure of storage efficiency.

Hoyst says that the optimal balance between the binding energy (*i.e.* enthalpy of adsorption) and space available for the accommodation of molecules (*i.e.* presence of inaccessible pore volume) is a key for encapsulation of van der Waals molecules.[1] He explains that CNTs can have a very narrow distribution of pore sizes, which is particularly important in view of the results predicting a maximum adsorption at some pore sizes. "In CNTs we can highly compress the gas reaching the density of a solid phase. We also point out that the optimal structure of CNTs for volumetric storage capacity is different from the structure for the optimal mass storage capacity, so it is important whether we consider optimal adsorbent for mass or for volumetric storage."

In their models, the two researchers show that two elements have to be taken into account in the search for optimal adsorbents: heat of adsorption and pore size (it is not true, as is commonly believed, that high adsorption enthalpy is the sole condition for high adsorption capacity).

"The key for optimizing the amount of CF_4 trapped in nanotubes at assumed external operating conditions is the size of the internal cylindrical pores and interstitial channels of an idealized bundle of SWCNTs," explains Kowalczyk. "Because of the large molecular size of CF_4, the internal pores play a major role in the process of encapsulation of CF_4 via the physical adsorption mechanism."

This work shows that an optimized structure of SWCNT bundles seem to be very promising for the encapsulation of CF_4, superior to the currently used activated carbons and zeolites. The efficiency of encapsulation in nanotubes can be explained by their intermediate properties in comparison to these other materials.

Holyst makes the point that, in practice, CF_4 exists as a gas mixture (for example, a mixture with nitrogen that can mimic the air mixture). "So the question arises about the transferability of our simulation results to the selective adsorption of CF_4 from a gas mixture," he says. "Previous studies suggest that our current simulation results of CF_4 adsorption in CNTs are transferable for the problem of CF_4–N_2 mixture adsorption. Our results, as well as previous

---

[1] A van der Waals molecule is a stable cluster consisting of two or more molecules held together by van der Waals forces or by hydrogen bonds.

results, show that optimal adsorption is achieved only when the distribution of pore sizes is sharp."

Based on these model simulations, experimental investigations into the capture and storage of CF_4 in a real bundle of SWCNTs are now needed in order to develop this concept as a practical solution that works on an industrial scale.

> *Featured scientists:* (a) Robert Holyst, (b) Piotr Kowalczyk
> *Organizations:* (a) Institute of Physical Chemistry, Polish Academy of Sciences (Poland), (b) Department of Applied Physics, RMIT University, Melbourne (Australia)
> *Relevant publication:* Piotr Kowalczyk, Robert Holyst: Efficient adsorption of super greenhouse gas (tetrafluoromethane) in carbon nanotubes, *Environ. Sci. Technol.*, **42**, 2931–2936.

13.3 Novel Material Addresses Water Pollution and Oil Spills

Research conducted at MIT describes a novel nanostructure membrane that can be switched on demand between superhydrophilic (strongly absorbing) and superhydrophobic (repelling) behavior. One use for such materials could be as a 'paper towel' for cleaning up oil spills that absorbs only the oil but not the water. Given the global scale of severe water pollution arising from oil spills and industrial organic pollutants, this nanomaterial may prove particularly useful in the design of recyclable absorbents with significant environmental impact.

"There are many materials that can preferentially absorb oil from water, but none has the ability to fully discriminate in the way our new nanoporous membranes do," says Francesco Stellacci, Finmeccanica associate professor of materials science and engineering at MIT. "We have developed a self-assembly method for constructing thermally stable, free-standing nanowire membranes that exhibit controlled wetting behavior ranging from superhydrophilic to superhydrophobic. These membranes can selectively absorb oils up to 20 times the material's weight in preference to water through a combination of super-hydrophobicity and capillary action. Moreover, the nanowires that form the membrane structure can be re-suspended in solutions and subsequently re-form the original paper-like morphology over many cycles."

Many methods for fabricating various types of membrane materials for water filtration have limitations for practical use because of the need to withstand harsh conditions, the need for multistep procedures for implementation, or limitations in substrate size. New and recyclable membranes like the one developed by Stellacci and his collaborators from National University of Singapore, NIMS in Japan, and the University of Connecticut, which overcome these fundamental limitations, could significantly reduce material waste and operating costs.

"The thermally stable membrane material that we developed is based upon self-assembled, free-standing, paper-like structures of cryptomelane-type manganese oxide nanowires," explains Stellacci. "The nanowire membrane, composed of 3-D porous nanostructures, exhibits a superhydrophilic character. When coated with a thin layer of hydrophobic molecules, the membrane becomes superhydrophobic, as is made evident by its high water contact angle of more than 170°. These two extreme wetting characteristics are completely switchable upon coating with or removal of the hydrophobic molecules at elevated temperatures."

In effect, the researchers have created a surface that repels water while selectively allowing oil to spread. They demonstrated the efficient absorption and high level of selectivity of this superwetting nanowire membrane not only for oil but also for a broad range of organic solvents such as toluene or the highly toxic benzene.

The scientists first synthesized free-standing nanowire membranes of arbitrary size exhibiting a uniform surface morphology. This self-assembled membrane has a pore size distribution centered at 10 nm and a surface area of $44 \, m^2/g$. The wetting time for a water droplet ($\sim 2 \, \mu l$) added to the surface was found to be 0.05 s.

"To obtain superhydrophobic surfaces, we silicon-coated the membrane using a vapor deposition technique that provides a coating over the entire surface of the porous material," Stellacci describes the process. "As anticipated, the modified membrane becomes superhydrophobic, as is made evident by its water contact angle of $\sim 172°$. We also found that the new material appears to be completely impervious to water—the superhydrophobicity of the silicone-coated membrane remains unaltered after being immersed in water at ambient temperature for more than 3 months."

An important feature of this nanomaterial is that the hydrophobic coating can be easily removed by heating the nanowire membrane to elevated temperatures (390 °C), resulting in switchable wetting behavior between its superhydrophilic and superhydrophobic states.

The membrane, with a calculated density of $0.286 \, g/cm^3$, demonstrated uptake capacities up to 20 times its weight for a collection of organic solvents and oil. Stellacci points out that $1 \, m^3$ of the material could absorb about 14 tonnes of motor oil (it can selectively absorb emulsified oil suspensions in water with a remarkable uptake capacity of 9 tonnes/m^3), making it an ideal candidate for oil absorption.

Furthermore, the membrane can be regenerated after each use by ultrasonic washing and autoclaving (at $\sim 130 °C$ for 20 min), making recycling schemes for both the membrane and the absorbed liquid possible.

Although the use of manganese oxide in synthesizing the membranes might pose a potential toxicological hurdle for large-scale use, the design principles used by Stellacci and his collaborators can clearly serve as a blueprint for fabricating other novel nanomaterials that can help address the widespread water pollution problems that many regions are confronted with.

> *Featured scientist:* Francesco Stellacci
> *Organization:* Department of Materials Science and Engineering, MIT, Cambridge, MA (USA)
> *Relevant publication:* Jikang Yuan, Xiaogang Liu, Ozge Akbulut, Junqing Hu, Steven L. Suib, Jing Kong, Francesco Stellacci: Superwetting nanowire membranes for selective absorption, *Nat. Nanotechnol.*, **3**, 332–336.

13.4   From Waste to Power in One Step

An interesting, but not yet commercially viable, type of fuel cell is the microbial fuel cell (MFC) where bacteria oxidize compounds such as glucose, acetate, or wastewater. This revolutionary new environmental biotechnology turns the treatment of organic wastes into a source of electricity. Fuel cell technology, despite its recent popularity as a possible solution for a fossil-fuel free future, is actually quite old. The principle of the fuel cell was discovered by German scientist Christian Friedrich Schönbein in 1838 and published in 1839. The operating principle of a fuel cell is fairly straightforward. It is an electro-chemical energy conversion device that converts the chemical energy from fuel (on the anode side) and oxidant (on the cathode side) directly into electricity. Today, there are many competing types of fuel cells, depending on what kind of fuel and oxidant they use. Many combinations of these are possible. For instance, a hydrogen cell uses hydrogen as fuel and oxygen as oxidant. Other fuels include hydrocarbons and alcohols.

A number of research institutes are developing MFCs. One of them, the Center for Environmental Biotechnology at Arizona State University, explains the three reasons why the MFC shapes up to become a revolutionary tech-nology: "First, it makes the treatment of organic pollutants a direct producer of electricity, not a consumer. Second, it expands fuel-cell technology to use renewable organic materials as a fuel; conventional fuel cells use hydrogen gas, which is today produced from fossil fuels. Furthermore, the MFC can use organic fuels that are wet, the usual condition for wastes and fuel crops. Third, the MFC, by operating at ambient temperature, can double to triple the elec-tricity-capture efficiency over combustion, while eliminating all the air pollu-tion that comes from combustion."[2]

The link from MFC to nanotechnology comes in the shape of CNTs. Researchers in Spain have fabricated MWCNT scaffolds with a microchannel structure in which bacteria can grow. This scaffold structure could be used as electrodes in microbial fuel cells.

"Given that CNTs are also suitable supports for cell growth, one could consider the use of CNT-based electrodes in microbial fuel cells," says Francisco del Monte. "MFCs function on different carbohydrates but also on complex

---

[2] Source: http://www.biodesign.asu.edu/centers/eb/

substrates present in wastewaters. As yet there is limited information available about the energy metabolism and nature of the bacteria using the anode as electron acceptor; few electron transfer mechanisms have been established unequivocally. Nonetheless, the efficient electron transfer between the micro-organism and the anode (*e.g.* microorganisms forming a biofilm on graphite fibers) seems to play a major role in the performance of the fuel cell. To further enlarge the electrode surface exposed to the bacterial growth medium, the design and preparation of 3-D architectures through which bacteria can grow and proliferate is indeed of great help in further improving the performance of this sort of device."

Del Monte, a scientist at the Instituto de Ciencia de Materiales de Madrid, Spain, leads the institute's Bioinspired Materials group. Together with colleagues from the Centro Nacional de Biotecnología, also in Madrid, he describes the group's development of a long-range microchannel-structured MWCNT scaffold that offers a high internal reactive surface easily accessible to bacterial immobilization and proliferation.

"We think that MWCNT scaffolds could offer a self-supported structure with large surface area through which hydrogen-producing bacteria such as *E. coli* can eventually grow and proliferate," says del Monte. "Our first concern was to study the biocompatibility of the MWCNT scaffolds. MWCNTs have been reported to be biocompatible for different eukaryotic cells, but no data existed for bacteria." Del Monte's team found that MWCNT scaffolds did indeed exhibit excellent biocompatibility for immobilization of *E. coli*.

The Spanish researchers tried to grow bacteria on the scaffolds by two different means: by direct soaking in a bacterial culture medium and by the immobilization of nutrient-containing beads before scaffold preparation.

The former approach provided a higher bacterial population, but only in a few layers at the surface of the scaffold, while the latter colonized the whole of the nanostructure.

"Given that full colonization is highly desirable, we are focusing on the improvement of bacterial viability during the scaffold formation process," says del Monte. "We believe that the efficient proliferation of hydrogen-producing bacteria throughout an electron-conducting scaffold like this can form the basis for the potential application of these MWCNT scaffolds as electrodes in MFCs."

Featured scientist: Francisco del Monte
Organization: Bioinspired Materials Group, Instituto de Ciencia de Materiales de Madrid (Spain)
Relevant publication: María C. Gutiérrez, Zaira Y. García-Carvajal, María J. Hortigüela, Luis Yuste, Fernando Rojo, María L. Ferrer, Francisco del Monte: Biocompatible MWCNT scaffolds for immobilization and proliferation of *E. coli*, *J. Mater. Chem.*, **17**, 2992–2995.

13.5 Nanotechnology and Water Treatment

According to the World Water Council, only 30% of all fresh water on our planet is not locked up in ice caps or glaciers (although not for much longer, unfortunately). Of that, some 20% is in areas too remote for humans to access and of the remaining 80% about three-quarters comes at the wrong time and place—in monsoons and floods—and is not always captured for use by people. The remainder is less than 0.6% of all the fresh water on the planet. Expressed another way, if all the Earth's water were stored in a 5 L container, the available fresh water would not quite fill a teaspoon. The problem is that we do not manage this teaspoonful very well. Currently, 600 million people face water scarcity. Depending on future rates of population growth, between 2.7 billion and 3.2 billion people may be living in either water-scarce or water-stressed conditions by 2025.

Fresh water looks as if it could become the oil of the 21st century—scarce, expensive, and the reason for armed conflicts. Nanotechnology could play various roles in resolving issues relating to water shortage and water quality, ranging from areas relevant to water purification, including separation and reactive media for water filtration, to nanomaterials and nanoparticles for use in water bioremediation and disinfection.

"The potential impact areas for nanotechnology in water applications are divided into three categories: treatment and remediation, sensing and detection, and pollution prevention," says Eugene Cloete. "Within the category of treatment and remediation, nanotechnology has the potential to contribute to long-term water quality, availability, and viability of water resources, *e.g.* through the use of advanced filtration materials that enable greater water reuse, recycling, and desalinization. Within the category of sensing and detection, of particular interest is the development of new and enhanced sensors to detect biological and chemical contaminants at very low concentration levels in the environment, including water."

Cloete is head of the Microbiology Department at the University of Pretoria in South Africa and chairperson of the university's School of Biological Sciences. He summarizes the role of nanotechnologies in water treatment.

Nanomaterials and Water Filtration

Membrane processes are considered key components of advanced water purification and desalination technologies and nanomaterials such as CNTs, nanoparticles, and dendrimers[3] are contributing to the development of more efficient and cost-effective water filtration processes.

There are two types of nanotechnology membranes that could be effective: nanostructured filters, where either CNTs or nanocapillary arrays provide the basis for nanofiltration; and nanoreactive membranes, where functionalized nanoparticles aid the filtration process.

---

[3] Dendritic polymers or dendrimers are repeatedly branching molecules.

Advances in macromolecular chemistry such as the synthesis of dendritic polymers have provided opportunities to refine, as well as to develop effective filtration processes for purification of water contaminated by different organic solutes and inorganic anions.

Nanotechnologies for Water Remediation

Many areas, especially in developing countries, are seriously contaminated or damaged with consequent impoverishment of natural resources and serious effects on human health. Remediation of contaminated water—the process of removing, reducing or neutralizing water contaminants that threaten human health and/or ecosystem productivity and integrity—is a field of technology that has attracted much interest recently.

In general, remediation technologies can be grouped into categories using thermal, physicochemical, or biological methods. The various techniques usually work well when applied to a specific type of water pollution, though no readily available treatments have been discovered that could clean all types of pollutants. Because of the complex nature of many polluted waters, it is frequently necessary to apply several techniques to soil from a particular location to reduce the concentrations of pollutants to acceptable levels.

Cloete says that traditional technologies such as solvent extraction, activated carbon adsorption, and common chemical oxidation, although effective, are often costly and time-consuming: "Biological degradation is environmentally friendly and cost-effective; but it is usually time-consuming. Thus, the ability to remove toxic contaminants from these environments to a safe level and to do so rapidly, efficiently, and within reasonable costs is important. Nanotechnology could play an important role in this regard. An active emerging area of research is the development of novel nanomaterials with increased affinity, capacity, and selectivity for heavy metals and other contaminants. The benefits from use of nanomaterials may derive from their enhanced reactivity, surface area, and sequestration characteristics. A variety of nanomaterials are in various stages of research and development, each possessing unique functionalities that are potentially applicable to the remediation of industrial effluents, groundwater, surface water, and drinking water."

Bioactive Nanoparticles for Water Disinfections

There is a growing threat of water-borne infectious diseases, especially in the developing world. This threat is rapidly being exacerbated by demographic explosion, a global trend towards urbanization without adequate infrastructure to provide safe drinking-water, increased water demand by agriculture that draws more and more of the potable water supply, and emerging pollutants and antibiotic-resistant pathogens that contaminate our water resources. No country is immune. Even in OECD countries,[4] the number of outbreaks

---

[4] The Organisation for Economic Co-operation and Development has 30 members, most of which are developed countries.

reported in the last decade demonstrates that transmission of pathogens by drinking-water remains a significant problem. It is estimated that water-borne pathogens cause between 10 million and 20 million deaths a year worldwide.

According to Cloete, nanotechnology may present a reasonable alternative for development of new chlorine-free biocides. Among the most promising antimicrobial nanomaterials are metallic and metal-oxide nanoparticles, especially silver, and titanium dioxide catalysts for photocatalytic disinfections.

As with any other nanotechnology application where there is a possibility that engineered nanoparticles may eventually appear in various environments, the potential human and ecological risk factors associated with this are largely unknown and subject to much debate (see Chapter 15).

His conclusion is that it might be advisable to obtain some definite answers regarding nanoparticle ecotoxicology before we embark on the large-scale use of engineered nanoparticles in water applications. Nevertheless, there is a growing body of research and development that will lead to nanomaterials playing a key role in future water and wastewater treatment.

A specific nanotechnology water application is described in the next section.

Featured scientist: Eugene Cloete
Organization: Microbiology Department, University of Pretoria (South
  Africa)
Relevant publication: J. Theron, J. A. Walker, T. E. Cloete:
  Nanotechnology and water treatment: applications and emerging
  opportunities, *Crit. Rev. Microbiol.*, **34**, 43–69.

13.6   Removing Heavy Metals from Water

Numerous nanomaterials are in various stages of research and development, each possessing unique functionalities that are potentially applicable to the remediation of polluted industrial wastewater, groundwater, surface water, and drinking-water. The main goal of most of this research is to develop cheap and environmentally friendly materials for removal of heavy metals from water. One example is a novel, low-cost magnetic sorbent material for the removal of heavy metal ions from water, developed by scientists in China, who coated iron oxide magnetic nanoparticles (Fe_3O_4, magnetite) with humic acid (HA). The coating greatly enhanced material stability and heavy metal removal efficiency of the nanoparticles.

"Magnetic nanoparticles have been used for this purpose because these materials with the adsorbed heavy metals can be easily recovered by utilizing magnetic separation," explains Gui-Bin Jiang. "However, bare magnetite nanoparticles in aqueous systems are highly susceptible to air oxidation and are easily aggregated, resulting in reduced saturation magnetization and adsorption capacity for metals." In his research, Jiang and his group describe the development of a novel material—HA-coated Fe_3O_4 magnetic nanoparticles—to resolve these problems.

Jiang, a professor at the Research Center for Eco-Environmental Sciences of the State Key Laboratory of Environmental Chemistry and Ecotoxicology, Chinese Academy of Sciences in Beijing, together with his colleagues Jing-Fu Liu and Zong-Shan Zhao, found that coating Fe_3O_4 magnetic nanoparticles with humic acid can (1) greatly enhance the stability of dispersed nanoparticles by preventing their aggregation; (2) maintain the saturation magnetization by avoiding their oxidation; and (3) enlarge the adsorption capacity for some heavy metals by making use of the abundant carboxylic acid and phenolic hydroxyl functional groups of HA to complex with heavy metal ions.

"We were aware of recent research which indicated that HA shows a high affinity for Fe_3O_4 nanoparticles, and that sorption of HA on the Fe_3O_4 nanoparticles enhances the stability of nanodispersions by preventing their aggregation," he says. "It is also well known that HA, which is abundant in natural aqueous systems, has a skeleton of alkyl and aromatic units that are attached to carboxylic acid, phenolic hydroxyl, and quinone functional groups. These functional groups have high complex capacity with heavy metal ions, and HA has therefore been used to remove heavy metal ions from water."

In previous research it was also found that the adsorption capacity for metal ions with the complexes of HA and iron oxides was larger than that with the respective iron oxides and HA alone. The Chinese researchers therefore hypothesized that by coating Fe_3O_4 magnetic nanoparticles with HA they could develop a very effective sorbent material for the removal of heavy metals from water.

For removal of metals from fresh water, 50 mg of the coated iron oxide nanoparticles—prepared by a co-precipitation procedure with cheap and environmentally friendly iron salts and HA—was added to 100 mL of water. Then the magnetic Fe_3O_4/HA with sorbed heavy metals was separated from the mixture with a hand-held magnet.

To test the effect of time on the leaching of Fe_3O_4/HA components and the sorbed heavy metals, Jiang and his colleagues re-suspended Fe_3O_4/HA laden with heavy metals in deionized water. They found that the leaching of both the sorbed heavy metals and the material components was negligible.

"We also found that the removal of heavy metals by Fe_3O_4/HA was not affected by environmentally relevant parameters such as pH, salinity, and dissolved organic matter" says Jiang. "Furthermore, our proposed procedure was very efficient as the sorption equilibrium was reached in less than 20 min, and the Fe_3O_4/HA with adsorbed heavy metals can be simply recovered from water with magnetic separations at very low magnetic field gradients within a few minutes."

This research is a good example of the ways nanotechnology can be used to take substances that are abundant in nature (such as iron oxide and HA) and use them to synthesize novel, highly effective adsorbent materials that are inexpensive and have no adverse effect on the environment. The result will not only be more effective and efficient water treatment processes but also lower overall costs.

> *Featured scientist:* Gui-Bin Jiang
> *Organization:* Research Center for Eco-Environmental Sciences of the
>   State Key Laboratory of Environmental Chemistry and
>   Ecotoxicology, Chinese Academy of Sciences, Beijing (China)
> *Relevant publication:* Jing-fu Liu, Zong-shan Zhao, Gui-bin Jiang:
>   Coating Fe_3O_4 magnetic nanoparticles with humic acid for high
>   efficient removal of heavy metals in water, *Environ. Sci. Technol.*, **42**,
>   6949–6954.

13.7   A Solution for Radioactive Waste Cleanup

Radioactive material is toxic because it creates ions—by stripping away electrons from atoms—when it reacts with biological molecules. These ions can form free radicals that damage proteins, membranes, and nucleic acids in the body. Many forms of cancer are thought to be the result of reactions between free radicals and DNA, resulting in mutations that can adversely affect the cell cycle and potentially lead to malignancy.

Nanotechnology has provided numerous constructs that reduce oxidative damage in engineering applications with great efficiency. One research result shows how nanotechnology applications could also help to remediate radioactive contamination at the source, by removing radioactive ions from the environment.

Environmental contamination with radioactive ions that originate from the processing of uranium or the leakage of nuclear reactors is a potentially serious health threat because it can leach into groundwater and contaminate drinking-water supplies for large population areas. The key issue in developing technologies for the removal of radioactive ions from the environment—mainly from wastewater—and their subsequent safe disposal is to devise materials which are able to absorb radioactive ions irreversibly, selectively, efficiently, and in large quantities from contaminated water.

"Clays, zeolites, and other natural inorganic cation-exchange materials have been extensively studied and used in the removal of radioactive ions from water via ion exchange and are subsequently disposed of in a safe way," explains Huai Yong Zhu. "However, synthetic inorganic cation-exchange materials—synthetic micas, g-zirconium phosphate, niobate molecular sieves, and titanate—have been found to be far superior to natural materials in terms of selectivity for the removal of radioactive cations from water. Radioactive cations are preferentially exchanged with sodium ions or protons in the synthetic material. More importantly, a structural collapse of the exchange materials occurs after the ion exchange proceeds to a certain extent, thereby forming a stable solid with the radioactive cations being permanently trapped inside. Hence, the immobilized radioactive cations can safely be disposed of."

Zhu, an associate professor in the School of Physical and Chemical Sciences at the Queensland University of Technology in Brisbane, Australia, points out that this phenomenon—that the uptake of large, radioactive cations eventually triggers the trapping of the cations—in itself represents a desirable property for any material to be used in decontamination of water from radioactive cations.

"Generally, ion-exchange materials exhibiting a layered structure are less stable than those with 3-D crystal structures and the collapse of the layers can take place under moderate conditions," says Zhu. "Then again, it has also been found that nanoparticles of inorganic solids readily react with other species or are quickly converted to other crystal phases under moderate conditions and thus are substantially less stable than the corresponding bulk material." Based on this, Zhu and his colleagues focused their search for potential candidates for intelligent absorbents on nanoparticles of inorganic ion-exchange materials with a layered structure.

"The novelty of our project is that the adsorption of bivalent toxic radioactive cations by the nanofibers finally induces structure collapse and deformation of the nanofibers, which permanently locks in the toxic radioactive cations," Zhu explains. "The permanent entrapment prevents the radioactive cations being released from the adsorbents and assures that they can be safely disposed of. Furthermore, the titanate nanofiber we used can selectively remove the radioactive ions in the presence of abundant competing ions."

It has been know for years that titanate solids possess a layered structure and exchangeable sodium ions. Titanate materials are stable to radiation, chemicals, and thermal as well as mechanical stress, so that they make an ideal carrier for radioactive ions.

According to Zhu, titanate nanofibers have a much larger capacity to take up the bivalent toxic radioactive cations, and they can do it much faster than other materials. "Our most important finding is that the nanofibers can trap the toxic radioactive cations permanently". This makes them an ideal absorbent to remove cations from contaminated water while the used sorbents can be disposed of safely without risk of the release of the absorbed cations from the absorbents which could cause secondary contamination.

The scientists also mention that the titanate nanofibers possess a number of additional advantages for practical application in the removal of radioactive cations:

(1) The synthesis of phase-pure titanate nanofibers with a small and uniform particle size can be achieved in large quantities from abundant raw materials, such as titanium dioxide minerals, using simple and cost-effective processes.
(2) Titanate nanofiber sorbents can be readily dispersed into solution because the fibers do not aggregate as much as clays and zeolites.
(3) Moreover, because of their fibril morphology the absorbents can be separated from any liquid after the absorption simply by filtration, sedimentation, or centrifugation.
(4) Absorption of radioactive ions by the fibers is very prompt. The amount absorbed within the first 24 h is approximately 80% of the final equilibrium capacity.

Zhu and his team are already working on developing inorganic adsorbents with metastable structures for irreversible ion exchange—a material that will be useful for removal of toxic ions from water. They are also making efforts to improve the selectivity and increase the capacity of the materials so that they can be used for treatment of radioactive wastes on an industrial scale and produce waste solids suitable for long-term storage and disposal.

> *Featured scientist:* Huai Yong Zhu
> *Organization:* School of Physical and Chemical Sciences, Queensland University of Technology, Brisbane (Australia)
> *Relevant publication:* Dong Jiang Yang, Zhan Feng Zheng, Huai Yong Zhu, Hong Wei Liu, Xue Ping Gao: Titanate nanofibers as intelligent absorbents for the removal of radioactive ions from water, *Adv. Mater. (Weinheim, Ger.)*, **20**, 2777–2781.

13.8   Self-healing Anticorrosion Coatings as an Alternative to Toxic Chromium

Remember the blockbuster movie *Erin Brockovich*? The film is based on a real-world legal case that revolved around hexavalent chromium, also known as chromium(VI), used by the Pacific Gas and Electric Company (PG&E) to control corrosion in cooling towers in its compressor station in Hinkley, California. Chromium(VI), a natural metal, is known to be toxic and is recognized as a human carcinogen via inhalation. It also is widely used by industry in the manufacture of stainless steel, welding, painting and pigment application, electroplating, and other surface coating processes. PG&E would periodically treat the surface of the cooling coils in its Hinkley station with anticorrosion paint and release the chromium-containing wastewater into the environment, thereby causing severe health problems for the neighboring population.

The huge economic impact of the corrosion of metallic structures is a very important issue for all modern societies. Estimates for the direct and indirect costs of corrosion degradation run to about €200 billion a year in Europe and more than $270 billion a year in the United States. The direct costs are related to including corrosion protection in the design, manufacturing, and construction of structures. Indirect costs are concerned with corrosion-related inspection, maintenance, and repairs.

In spite of its toxicity, chromium(VI) has remained an essential ingredient for corrosion control in the metal finishing industry. But combine the economic impact of corrosion damage, the environmental and health problems caused by chromium(VI), and the increasing regulatory restrictions, and scientists have a huge incentive to develop a new generation of protective coating systems.

The advanced materials that are being developed and used in modern industries require increasingly sophisticated coatings for improved performance and durability. With a degrading environment due to industrial factors, environmental compatibility is an aspect that gains in importance during the design phase of novel materials—and chromium(VI) compounds certainly would not make the list. Furthermore, while conventional anticorrosion coatings are just passive barriers that prevent the interaction of corrosive species with a metal, future nanotechnology based protective coatings will be 'smart', *i.e.* they will provide several functionalities that will in effect result in self-healing capabilities.

The whole concept of 'smart' materials that react on external impact (pH, humidity changes, or distortion of the coating integrity) and repair themselves, has experienced a tremendous boost with the advent of nanotechnology. The nanoscale multilayer structure of a coating, in which the components are integrated and mutually reactive, is an important point in sophisticated and strong corrosion protection.

Researchers in Germany have developed a novel method of multilayer anticorrosion protection including the surface pretreatment by sonication and deposition of polyelectrolytes and inhibitors. This method results in the formation of a smart polymer nanonetwork for environmentally friendly organic inhibitors.

"Our novel coating exhibits very high resistance to corrosion attack, long-term stability in aggressive media, and an environmentally friendly, easy, and economical preparation procedure," says Daria Andreeva. "We have demonstrated the general procedure for a surface that is important for the aircraft industry but it is similarly applicable for many types of surfaces, thus enabling many applications in advanced technologies."

Andreeva is a researcher at the Max Planck Institute of Colloids and Interfaces in Potsdam, Germany. Together with her colleagues, she proposes a novel multilevel protection approach where the protective systems—the 'smart' multilayers—will not only be a barrier to external impacts but also respond to changes in their internal structure and combine different damage prevention and reparation mechanisms in the same system.

The Max Planck scientists started with the assumption that the layer-by-layer (LbL) deposition procedure would be a very effective solution for the preparation of self-healing anticorrosion coatings. The LbL process involves the stepwise electrostatic assembly of oppositely charged species (*e.g.* polyelectrolytes and inhibitors or nanoparticles) on a substrate surface with nanometer-scale precision, and allows the formation of a coating with multiple functionalities.

A novel step in this anticorrosion system is the surface pretreatment of aluminium alloy by intensive sonication in water with an ultrasonic horn. Although the typical aluminium surface is covered by a natural oxide film 3–7 nm thick, this layer is not sufficient to protect against corrosion agents and does not yield good adhesion for subsequent layers of coating.

"The ultrasonic pretreatment is crucial for the formation of a uniform film," says Andreeva. "The surface of ultrasonically pretreated samples exhibits better wettability, adhesion, and chemical bonding with the polymer layers of the subsequent LbL coating. It results in a homogeneous distribution of the polymer film on the aluminium surface." After pretreatment, 5–10 nm thick layers of polyelectrolytes and inhibitor were formed by LbL deposition on the freshly sonicated aluminium alloys.

The scientists were amazed that even the nanometer-thick polyelectrolyte/inhibitor coating provides effective corrosion protection for the aluminium alloy. They explain that the nature and properties of their novel anticorrosion coating simultaneously provides three mechanisms of corrosion protection: (1) passivation of the metal degradation by controlled release of inhibitor; (2) buffering of pH changes at the corrosive area by polyelectrolyte layers; and (3) self-curing of the film defects owing to the mobility of the polyelectrolyte constituents in the LbL assembly.

Since the release of the inhibitor is stimulated by corrosive species and corrosion products, the 'smart' coating enables prolonged self-healing activity.

Because of its versatility, this anticorrosion protection method has a very broad range of potential applications. All components (polyelectrolytes and inhibitors) could be adjusted for a particular surface. The novel coating could be applied in aerospace, automotive, and marine industries and all other areas that suffer from corrosion damage, such as oil and gas pipelines. "Although we concentrated on corrosion, our method could also be more generally applied for self-repairing coatings like antifungal or antifriction applications," Andreeva points out.

One of the practical problems the Max Planck team is working on is the automation of the layer formation procedure in order to allow the scaling up of their technique for industrial applications. Beyond that, they are already looking to introduce other components with other mechanisms of corrosion protection into the system such as the combination of several inhibitors or self-polymerized compounds.

Featured scientist: Daria Andreeva
Organization: Max Planck Institute of Colloids and Interfaces, Potsdam (Germany)
Relevant publication: Daria V. Andreeva, Dmitri Fix, Helmuth Möhwald, Dmitry G. Shchukin: Self-healing anticorrosion coatings based on pH-sensitive polyelectrolyte/inhibitor sandwichlike nanostructures, *Adv. Mater. (Weinheim, Ger.),* **20**, 2789–2794.

Energy—Renewable and Clean

Nanotechnology applications could provide decisive technological break-throughs in the energy sector and have a considerable impact on creating the sustainable energy supply that is required to make the transition from fossil fuels. Possibilities range from gradual short- and medium-term improvements for a more efficient use of conventional and renewable energy sources all the way to completely new, long-term approaches for energy recovery and utilization. With enough political will and funding, nanotechnology could make essential contributions to sustainable energy supply and global climate protection efforts.

Nanotechnology innovations could impact each part of the value-added chain in the energy sector: energy sources, energy conversion, energy distribution, energy storage, and energy usage. The possible applications range from high-duty nanocomposite materials for lighter and more rugged rotor blades for wind and tidal power plants, to efficient photovoltaic systems, fuel cells, and batteries, and energy savings through more efficient lighting sources, better insulation, or better lubricants.

What follows are just a few examples that show nanotechnology's role in future energy systems.

14.1 Catching a Rainbow—Quantum Dots Brighten the Prospects for Solar Energy

Harnessing the power of the Sun to replace the use of fossil fuels holds tremendous promise—solar energy is by far the largest source of renewable energy. One way to do this is through the use of solar, or photovoltaic, cells. Large-scale installations already show the technical feasibility of this technology although the major problem of photovoltaic solar energy, its relative inefficiency, still needs to be overcome to make the cost of electricity produced

RSC Nanoscience and Nanotechnology No. 8
Nano-Society: Pushing the Boundaries of Technology
By Michael Berger
© Michael Berger 2009
Published by the Royal Society of Chemistry, www.rsc.org

by solar cells competitive with that of electricity produced by nuclear and fossil fuels. So far, solar cells that convert sunlight to electric power have been dominated by solid state junction devices made of silicon wafers. Efforts are under way in laboratories worldwide to design ordered assemblies of semi-conductor nanostructures, metal nanoparticles, and carbon nanotubes (CNTs) for the construction of next-generation solar energy conversion devices.

Quantum dots have been identified as an important light-harvesting material for building highly efficient solar cells. Quantum dots are nanoscale semi-conductor structures which, when exposed to light at certain wavelengths, can generate free electrons and create an electrical current. Quantum dot technology represents an exciting field of research in solar energy, yet the actual research results of using them in solar cells have been relatively limited so far.

By combining spectroscopic and photoelectrochemical techniques, researchers have demonstrated size-dependent charge injection from cadmium selenide (CdSe) quantum dots into titanium dioxide nanoparticles and nano-tubes, showing a way to maximize the light absorption of solar cells based on quantum dots. Termed 'rainbow solar cells', these next-generation devices consist of different-sized quantum dots assembled in an orderly fashion. Just as a rainbow displays multiple colors of the visible light spectrum, the rainbow solar cell has the potential to absorb multiple wavelengths of light simulta-neously and convert it to electricity in a very efficient manner.

"One of the most important things we show in our work is that one can use the same material—here, CdSe quantum dots—to collect light across much of the solar spectrum," says Prashant V. Kamat. "Because the Sun emits light at a variety of wavelengths, it is important to be able to collect as many of these wavelengths as possible to maximize solar cell efficiency. CdSe quantum dots have long been known to collect light at multiple wavelengths and have also previously been utilized in solar cells. However, our work is the first that directly compares how different-sized quantum dots, which absorb different wavelengths, perform when incorporated into solar cells."

Kamat, a professor of chemistry and biochemistry at Notre Dame University in Indiana, and his team found that the smallest quantum dots—absorbing the shortest wavelengths of the solar spectrum—perform best because they move electrons through the cell, *i.e.* create current, at the fastest rate.

This work both advances the general understanding of the nature of quan-tum dots as well as provides information on how one can construct a better solar cell. Kamat's findings are threefold:

(1) The ability to tune the photoelectrochemical response and photo-conversion efficiency via size control of CdSe quantum dots. Although Kamat's team employs the same CdSe material as light absorber, size selection in the 2–8 nm region allows them to selectively tune the pho-tocurrent response in the visible region.
(2) An improvement in photoconversion efficiency by facilitating the charge transport through a titanium dioxide nanotube architecture. The Notre

Dame researchers have achieved an incident photon-to-charge carrier conversion efficiency approaching 50% through their CdSe quantum dot/titanium dioxide nanotube architecture.

(3) Laying the foundation for the construction of their rainbow solar cell by identifying the strengths of size-selective quantum dots and the ability to modulate the absorption profile of the solar cell.

Kamat tells us that his group was motivated to conduct this work for two specific reasons. "First, as a research group, we recognize the scientific promise that quantum dots offer and are interested in studying their properties and adding to a rapidly growing body of research which attempts to both better understand their physical behavior and find new uses in an attempt to improve the daily life of an average person.

"Secondly, we are interested in laying the foundations for the best possible solar cell. The entire research group at Notre Dame is proud to be working on a problem—the improvement of solar cells—which society needs solved now more than ever."

According to Kamat, the direct application of this work is to create the most efficient solar cell possible. In order to do this, the researchers need to accomplish three things:

(1) to collect as much of the solar spectrum as possible (accomplished here by using multiple sizes of quantum dots)

(2) to make sure that, once such light is absorbed, it is converted into moving electrons (charge separation)

(3) once the electrons are created, to make sure that they have the ability to move freely throughout the cell (charge transport), which creates electrical current.

"While all three of these steps must be present to create the best possible solar cell, researchers often study each step individually to isolate and improve upon any potential problem, with the hopes of eventually incorporating such improvements into a future solar cell," says Kamat. "One way in which our recent findings improve upon the third condition is by utilizing a tubular instead of a spherical titanium dioxide base. This advanced architecture allows electrons generated by the quantum dots to travel more freely within the cell because a tube provides an elongated direct pathway for electron travel whereas multiple spheres permit electron travel only if the electron "hops" between them, potentially slowing electron transport and hindering overall cell efficiency."

Kamat emphasizes that, although the construction of this type of next-generation solar cell is still in the works, the theoretical framework behind this technology is sound: "Through the use of quantum dots as solar light absorbers, the fabrication of rainbow solar cells, is on the horizon."

Featured scientist: Prashant Kamat
Organization: Department of Chemistry and Biochemistry, University of Notre Dame, Notre Dame, IN (USA)
Relevant publication: Anusorn Kongkanand, Kevin Tvrdy, Kensuke Takechi, Masaru Kuno, Prashant V. Kamat: Quantum dot solar cells. Tuning photoresponse through size and shape control of CdSe-TiO_2 architecture, *J. Am. Chem. Soc.*, **130**, 4007–4015.

14.2   Improving Lithium Ion Battery Performance

Lithium ion batteries seem to be everywhere these days. They power most of the electronic devices we carry around with us—cellphones, laptops, MP3 players, digital cameras, and so on. They get their name from the lithium ion that moves from the anode to the cathode during discharge and from the cathode to the anode during recharging. Because of their good energy-to-weight ratios, lithium batteries are some of the most energetic rechargeable batteries available today.

In terms of weight and size, batteries have become one of the limiting factors in the continuous process of developing smaller and higher-performance electronic devices. This does not only apply to consumer electronics. As with so many other nanotechnology research, the military is a strong driver behind battery R&D. All of the electronic gizmos of a modern soldier—night vision goggles, flashlights, laptops, radios, GPS—are powered by batteries. The backup batteries soldiers are required to carry generally add several kilograms to their basic load and the logistics necessary to supply the troops with sufficient numbers of replacement batteries is costly.

To meet the demand for batteries with higher energy density and improved cycle characteristics, researchers have been making tremendous efforts to develop new electrode materials or design new structures of electrode materials. For instance, metallic tin has recently been widely researched as one of the promising anode materials for next-generation high energy density lithium ion batteries. This new anode material has a high theoretical specific capacity but its practical application is limited by its poor cycling performance. The problem with increasing the performance of lithium batteries is that none of the existing electrode materials alone can deliver all the required performance characteristics including high capacity, higher operating voltage, and long cycle life. Consequently, researchers are trying to optimize available electrode materials by designing new composite structures, often at the nanoscale.

Demonstrating the benefits of directed nanostructure-design of electrode materials, Chinese scientists have prepared tin nanoparticles encapsulated in elastic hollow carbon spheres. This tin-based nanocomposite exhibits a very high specific capacity and excellent cycling performance, and therefore shows great potential as anode material in lithium ion batteries.

"Metallic tin is considered to be a very promising anode material for lithium batteries mainly for three reasons," explains Li-Jun Wan. "First, its theoretical specific capacity is much higher than that of conventional graphite. Secondly, the tin anode has higher operating voltage than graphite, so it is less reactive and the safety of batteries during a rapid charge/discharge cycle could be improved. And thirdly, a significant advantage of metallic tin over graphite is that it does not encounter solvent intercalation—which causes irreversible charge losses—at all. Unfortunately, the biggest challenge for employing metallic tin as an active anode material is that it experiences huge volume variation during the lithium insertion/extraction cycle, which leads to pulverization of the electrode and very rapid capacity decay."

Wan, a professor and director of the Institute of Chemistry at the Chinese Academy of Sciences (CAS) in Beijing, together with members of the Beijing National Laboratory for Molecular Sciences, describes their novel carbon nanocomposite as a promising anode material for high-performance lithium ion batteries: "Not only is this a further example of the directed nanostructure design of electrode materials for lithium ion batteries, but also the strategy could be extended to other anode and cathode materials by using elastic hollow carbon spheres as buffer and container."

The CAS scientists have successfully realized a novel design structure of tin-based anode material to solve the problem of capacity decay. The key to their composite material is nanosized tin particles that are placed into hollow carbon containers, resulting in tin nanoparticles encapsulated with elastic hollow carbon spheres (TNHCs).

The TNHCs were prepared by *in situ* reduction of tin oxide hollow spheres with carbon coating. The tin oxide spheres were first synthesized according to the Stöber method and used as templates to prepare hollow spheres. Next, polycrystalline tin oxide was deposited on silicon dioxide spheres to form uniform shells. Then their cores were etched to obtain hollow spheres. After that, the carbon precursor layers were coated on to the outer surface of the spheres. Finally the product was dried and heat-treated to carbonize the carbon precursor shell, and meanwhile the inner tin oxide shells were reduced to metallic tin by the carbon shells themselves to get the final TNHCs.

These TNHCs, with a uniform diameter of $\sim 500\,nm$, encapsulate multiple tin nanoparticles with a diameter of less than 100 nm in one thin, hollow carbon sphere with a thickness of only about 20 nm. This nanocomposite material is characterized by a tin content of up to 74% by weight (which results in a high theoretical specific capacity of $831\,mA\,h\,g^{-1}$) and a void volume in the carbon shell as high as about 70–80% by volume.

Wan explains that the ratio of tin nanoparticles to void volume (roughly 1:3) and the elasticity of the thin carbon spherical shell efficiently accommodate the volume change of tin nanoparticles due to the lithium–tin alloying/dealloying reactions, and thus prevent the pulverization of the electrode. "As a result, this type of tin-based nanocomposite has very high specific capacity of more than $800\,mA\,h\,g^{-1}$ in the initial 10 cycles, and more than $550\,mA\,h\,g^{-1}$ after the

100th cycle, as well as excellent cycling performance, exhibiting a great potential as anode material in lithium ion batteries."

The researchers also point out that their results successfully demonstrate the power of the strategy of using elastic hollow carbon spheres as buffer and container, and could be extended to other anode and cathode materials.

> *Featured scientist:* Li-Jun Wan
> *Organization:* Institute of Chemistry at the Chinese Academy of
>   Sciences (CAS), Beijing (China)
> *Relevant publication:* Wei-Ming Zhang, Jin-Song Hu, Yu-Guo Guo,
>   Shu-Fa Zheng, Liang-Shu Zhong, Wei-Guo Song, Li-Jun Wan:
>   Tin-nanoparticles encapsulated in elastic hollow carbon spheres
>   for high-performance anode material in lithium ion batteries,
>   *Adv. Mater. (Weinheim, Ger.)*, **20**, 1160–1165.

14.3 Building Better Fuel Cells

We have already looked at one form of fuel cells in Section 13.4. In general, fuel cells have attracted a lot of attention because they provide a potential solution to our addiction to fossil fuels. Energy production from oil, coal, and gas is an extremely polluting, not to mention wasteful, process that consists of the extraction of heat from fuel by burning, conversion of that heat to mechanical energy, and transformation of that mechanical energy into electrical energy. In contrast, fuel cells are electrochemical devices that convert a fuel's chemical energy directly to electrical energy with high efficiency and without combustion.

Despite their modern, high-tech aura, fuel cells actually have been around since the early 19th century. In 1839, William Robert Grove developed the Grove cell, which used zinc and platinum electrodes exposed to two acids and separated by a porous ceramic pot to generate about 12 A of current at about 1.8 V.

An example of how nanotechnologies can improve fuel cell technology has been demonstrated by researchers in India who have developed a carbon nanotube (CNT) composite material which has the potential to make better fuel cell modules due to enhanced electrolyte properties.

"An improvement in the proton conductivity of the electrolyte membrane even by only one order of magnitude could change the performance of fuel cells dramatically," explains Vijayamohanan Pillai. "Currently, membranes based on Nafion (a sulfuric acid in a solid polymer form) are widely used as the proton exchange membrane (PEM) in fuel cells that operate from 60 to 80 °C. Although these state-of-the-art membranes show good proton conductivities from 0.1 to 0.01 Siemens/cm in a humid environment, they have many limitations: (1) dependence on water for conductivity; (2) high methanol permeability; (3) a tendency to disintegrate in the presence of hydroxyl radicals, an intermediate in the cathode reaction; and (4) moderate mechanical and chemical stability."

Pillai, a researcher at the National Chemical Laboratory in Pune, India, and head of its Materials Electrochemistry Group, together with his team has developed a chemical strategy to increase the sulfonic acid content of Nafion membranes by incorporating sulfonic acid functionalized single-walled carbon nanotubes (S-SWCNTs), and has demonstrated the remarkable utility of this composite membrane as electrolyte in PEMFC applications.

Pillai notes that significant efforts by fuel cell researchers go into developing composite materials that aim to increase the water retention capabilities of Nafion so that it does not lose its proton conduction at high temperatures. "In our work, we introduced S-SWCNTs into the Nafion matrix, thereby increasing the number of sulfonic acid groups—which is the key for its conduction—in the membrane". The incorporation of SWCNTs would also help increase the mechanical stability of the composite membranes over that of Nafion membranes, and a further major benefit would be a much needed cost reduction in PEMFC technology.

Pillai explains that the high proton conductivity of Nafion is attributed to a mechanism in which the reorganization of hydrogen bonds plays a key role. "As the extra sulfonic acid moieties are anchored on the surface of our SWCNTs, these could provide more facile hopping of protons, which in turn would help to increase the proton mobility, thus accounting for the observed enhancement in conductivity (the Nafion/S-SWCNT composite membranes showed conductivity almost an order of magnitude higher). In contrast, we found that an unmodified SWCNT/Nafion composite membrane does not show any improvement in proton conductivity."

This work has introduced a new method of designing the polymer electrolytes for electrochemical power sources. It also demonstrates that, apart from a remarkable improvement in proton conductivity, properly designed membranes with appropriate incorporation of functionalized CNTs could help decrease the methanol cross-over without sacrificing the proton conductivity of the membrane.

Pillai cautions that several challenges have to be overcome before these advantages can be commercially exploited, including a rigorous evaluation of the chemical stability and durability of the membranes, lifetime studies of membrane–electrode assemblies, and possible corrosion of the electrode materials due to the higher sulfonic acid content.

Featured scientist: Vijayamohanan Pillai
Organization: Materials Electrochemistry Group, National Chemical Laboratory, Pune (India)
Relevant publication: Ramaiyan Kannan, Bhalchandra A. Kakade, Vijayamohanan K. Pillai: Polymer electrolyte fuel cells using Nafion-based composite membranes with functionalized carbon nanotubes, *Angew. Chem., Int. Ed.*, **47**, 2653–2656.

14.4 Nanotechnology is Key to Improving Fuel Cell Performance

Modern fuel cells have the potential to revolutionize transportation. Like battery electric vehicles, fuel cell vehicles are propelled by electric motors. But whereas battery electric vehicles use electricity from an external source and store it in a battery, fuel cells on board a vehicle are electrochemical devices that convert a fuel's chemical energy directly to electrical energy with high efficiency and without combustion.

One of the leading fuel cell technologies developed, in particular for transportation applications, is the proton exchange membrane (PEM) fuel cell, also known as polymer electrolyte membrane fuel cell—both resulting in the same acronym, PEMFC. They are powered by the electrochemical oxidation reaction of hydrogen and by the electroreduction of the oxygen contained in air.

These fuel cells run at relatively low temperature (less than 100 °C) and therefore need catalysts to generate useful currents at high potential, especially at the electrode where oxygen is reduced (the cathode of the fuel cell). Presently, platinum-based electrocatalysts are the most widely used in PEM fuel cell prototypes. However, platinum is expensive and its price is highly volatile. This creates one of the major barriers preventing commercialization of PEMFCs—the lack of suitable materials to make them affordable.

According to the U.S. Department of Energy (DOE), the system cost for automotive fuel cells has gone from \$275/kW in 2002 to \$95/kW in 2008 and is projected to be \$60/kW in 2009. The target is \$30 by 2015. The estimated cost for a gasoline engine is about \$30 per kilowatt (depending on the price of gasoline).

Although nanotechnology promises cheap bipolar materials using nanocomposites, more efficient nonplatinum electrocatalysts, and thermally stable and more durable membranes to become available in the near future, platinum still remains the workhorse of PEM fuel cells. This is why DOE has also established some performance targets for platinum use in PEM fuel cells for automotive applications. These targets are 0.3 g and 0.2 g of platinum per kW of PEM fuel cell stack for 2010 and 2015, respectively.

"In order to meet these targets, it is imperative to improve the specific activity of platinum-based fuel cell catalysts, especially at the cathode where the oxygen reduction reaction (ORR) is sluggish," Jean-Pol Dodelet tells us. "This may be done by alloying platinum with one or more other metals. However, our work has shown that it is also possible to improve the specific activity of platinum by using platinum nanowires as electrocatalyst for ORR instead of the usual platinum nanoparticles."

Dodelet is a professor at the INRS—Energy, Materials and Telecommunications at the Université du Québec in Canada. Together with his team he has demonstrated a simple room-temperature aqueous phase synthesis of single-crystal nanowires of platinum on the nanospheres of carbon black (a commonly used catalyst support in fuel cells). This use of carbon nanospheres as a substrate provides a cost-effective procedure for growing platinum nanowires

The resulting nanostructures—with the high-surface-area carbon black as the core and the electrocatalytically active platinum nanowires growing radially from the surface of the carbon particles—show enhanced catalytic activity for the ORR compared with a state-of-the-art platinum/carbon catalyst made of platinum nanoparticles.

Although the team concedes that the growth mechanisms of nanowires and branched nanowires of platinum are not yet fully understood, they believe that anisotropic growth, preferentially in the (111) direction, is promoted by the very slow reduction rate at room temperature and the lowest-energy principle. The length of the nanowires can be controlled by adjusting the reduction time of the platinum precursor, and their density on the carbon nanospheres by adjusting the weight ratio for platinum precursor to carbon.

"To the best of our knowledge, this is the first time that platinum nanowires have been used as electrocatalysts at the cathode of PEM fuel cells," says Dodelet. "It has been known for several years that the specific activity for oxygen reduction on platinum in acid medium is face-dependent, the most active faces being the low-index single crystal surfaces: $Pt(110)$, followed by $Pt(111)$ and $Pt(100)$. Platinum nanoparticles usually display all these low-index single surfaces. As it was expected that platinum nanowires would display a smaller number of single crystal surfaces than platinum nanoparticles and would also have fewer structural defects than the particles, we undertook the synthesis of platinum nanowires according to a very facile route using the anisotropic growth of a platinum nanostructure performed in an environmentally friendly aqueous solution without the help of stabilizer or template."

The platinum nanowires at the cathode of a membrane electrode assembly prepared in Dodelet's lab can reach much better performance in fuel cells than a commercial membrane electrode assembly. According to Dodelet, their in-house platinum nanowire catalyst shows a 50% higher mass activity than the commercial cathode. "Quite surprisingly, this improvement occurred in spite of a 50% lower platinum area for the platinum nanowire catalyst. Taking both effects into account, a specific ORR activity of the platinum nanowire catalyst of $275\,\mu A$ per cm^2 of platinum (at 0.9 V) was calculated, which is threefold better than that of the commercial cathode when tested at the same fuel cell test station."

The synthesis of these platinum nanowires is therefore an important step toward the fabrication of nanostructured platinum that would retain the high specific activity of bulk platinum. Consequently, Dodelet expects that platinum nanowires may favorably replace spherical platinum nanoparticles in any electrocatalytic application.

He points out that some particular platinum alloy single-crystal surfaces display higher specific activities than pure platinum in PEM fuel cells. "Therefore our next challenge is to obtain platinum alloy nanowires, especially those containing iron, copper, or nickel, in order to further improve the ORR activity of these catalysts in fuel cells."

> *Featured scientist:* Jean-Pol Dodelet
> *Organization:* INRS—Energy, Materials and Telecommunications,
>   Université du Québec, Varennes (Canada)
> *Relevant publication:* Shuhui Sun, Frédéric Jaouen, Jean-Pol Dodelet:
>   Controlled growth of Pt nanowires on carbon nanospheres and
>   their enhanced performance as electrocatalysts in PEM fuel cells,
>   *Adv. Mater. (Weinheim, Ger.)*, **20**, 3900–3904.

14.5   High-performance, Flexible Hydrogen Sensors

In a possible future hydrogen-based economy, hydrogen sensors will be a critical and widely needed safety component. Sensors will detect leaks from hydrogen-powered cars and fueling stations long before the gas becomes an explosive hazard. But even today, there is a wide range of potential applications for hydrogen sensors, such as sensing hydrogen build-ups in lead–acid storage cells found in most vehicles; detecting hydrogen leaks during petrochemical applications where high-pressure hydrogen is used; detecting impending transformer failure in electric power plants; or monitoring hydrogen build-up in radioactive waste tanks and in plutonium reprocessing. Another example is the space shuttle, which uses a combination of hydrogen and oxygen as fuel for its main engines. Any hydrogen leak could potentially result in a hydrogen fire, which is invisible to the naked eye. Today, the leakage of hydrogen caused by a tiny pinhole in the pipe of a space shuttle could not be easily detected by individual rigid detectors because the locations of pinholes are not predetermined.

The problem with most current hydrogen sensor designs is that they are built on rigid, unbending substrates, and this mechanical rigidity limits their applications. A new type of flexible sensor uses SWCNTs to improve efficiency and reduce cost. In the space shuttle, for example, laminating a dense array of flexible sensors over the whole surface of a pipe could detect any leakage of hydrogen prior to diffusion and alert control units to remedy the malfunction. This use of large-area sensory skins would not significantly increase the overall weight of the Shuttle because of the lightweight nature of these flexible sensors.

"The most exciting contribution of our research is the first-time fabrication of hydrogen sensors with mechanical flexibility and superb sensing performance by using nanostructured materials," says Yugang Sun. "In comparison to previously designed hydrogen sensors, which are rigid and rely on expensive pure palladium for sensing components, our sensors are flexible and use SWCNTs to improve efficiency and reduce cost. The flexibility of our hydrogen sensors is beneficial to application in many systems which require demanding low cost, large area, light weight, mechanical flexibility, and mechanical shock resistance." (See Figure 14.1.)

Sun is a scientist at the Center for Nanoscale Materials at Argonne National Laboratory in Argonne, Illinois. Together with H. Hau Wang from Argonne's

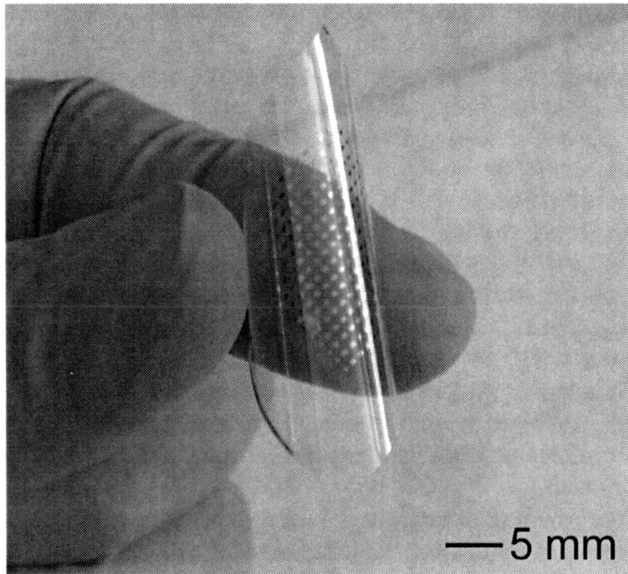

Figure 14.1 The as-fabricated flexible hydrogen sensor (Image: Dr Y. Sun, Argonne National Laboratory)

Materials Science Division, he fabricated the new sensing devices using a two-step process with high- and low-temperature steps. First, at around 900 °C, Sun and Hau grew SWCNTs on a silicon substrate using chemical vapor deposition. Then, they transferred the SWCNTs onto a plastic substrate at temperatures less than 150 °C using a technique called *dry transfer printing*.

It has been shown previously that nanotube networks grown through chemical vapor deposition (CVD) have enhanced sensing capability for hydrogen when the SWCNTs are decorated with palladium nanoparticles via electron beam evaporation. One problem with this method is the high-temperature step ($\sim$900 °C) involved in the growth of SWCNTs, which is not compatible with plastic substrates that can only withstand temperatures of less than 300 °C. A solution became available with the dry transfer printing process for transferring CVD nanotubes on to plastic substrates where device fabrications can be processed at relatively low temperatures of less than 100 °C.

With the separation of high-temperature and low-temperature steps, the fabrication of flexible thin-film transistors on plastic sheets with the use of CVD SWCNTs became possible. Sun and Hau have combined the dry transfer printing technique and modification of SWCNTs with palladium nanoparticles to prepare high-performance hydrogen sensors with excellent mechanical flexibility on plastic substrates.

This precise process is what allows the film of nanotubes to form on the plastic, after which the palladium nanoparticles can be deposited on the SWCNTs to make the sensors. The palladium nanoparticles play an important role in increasing the interaction between hydrogen and the SWCNTs to enhance the change of resistance of the device when it is exposed to hydrogen molecules.

"The driving force for us to develop low-cost, high-performance hydrogen sensors was the DOE Energy Hydrogen Program, which is making progress towards the goal of a 2015 commercialization decision," says Sun. "Fast and precise detection of leakage of hydrogen is critical to ensure the safe use of hydrogen technologies. Therefore, hydrogen sensors represent one of the core technologies in the program. As one of the largest national laboratories of DOE, Argonne has the opportunity to set the research pace in this field."

These flexible hydrogen sensors, although not yet optimized, can detect hydrogen with concentrations as low as 30 ppm in air at room temperature. The sensors show a change of 75% in their resistance when exposed to hydrogen at a concentration of 0.05% in air. The devices can detect the presence of 1% hydrogen at room temperature in 3 s. The devices still perform with as much effectiveness even after bending and relaxing 2000 times, with a bending radius of $\sim$7.5 mm.

The technique described by the Argonne scientists provides the possibility of a versatile route to fabricate flexible sensors for various gases. According to Sun, next steps for the scientists will include scaling up the fabrication to produce large-area arrays of hydrogen sensors, *i.e.* sensory skins, as well as integrating the sensory skins with flexible electronic control circuits.

Featured scientist: Yugang Sun
Organization: Center for Nanoscale Materials, Argonne National
    Laboratory in Argonne, IL (USA)
Relevant publication: Y. Sun, H. H. Wang: High-performance, flexible
    hydrogen sensors that use carbon nanotubes decorated
    with palladium nanoparticles, *Adv. Mater. (Weinheim, Ger.)*, **19**,
    2818–2823.

Nanotoxicology—Assessing the Risks

New technology, whether it is a novel cancer treatment or an innovative approach to making a new material, almost always comes with risk. Nanotechnologies are no exception. Certain nanofabrication techniques employ toxic chemicals, the production of carbon nanotubes results in dangerous by-products, and the big question as to what degree certain engineered nanoparticles could be harmful to humans and the environment has not yet been answered. The potentially adverse health effects of fine and ultrafine particles have been studied for decades, but nanoparticles are not just smaller versions of certain particles—they are very different from their everyday counterparts with regard to their physical properties and catalytic activities. Their adverse effects cannot simply be derived from the known toxicity of the macroscale material.

To complicate things further, in talking about nanoparticles it is important to remember that a powder or liquid containing nanoparticles is almost never monodisperse, but will contain a range of particle sizes. This complicates the experimental analysis, as larger nanoparticles might have different properties than smaller ones. Also, nanoparticles show a tendency to aggregate and such aggregates often behave differently from individual nanoparticles.

There are four entry routes for nanoparticles into the body: they can be inhaled, swallowed, absorbed through skin, or deliberately injected during medical procedures. Once within the body they are highly mobile and in some instances can even cross the blood–brain barrier. How these nanoparticles behave *in vivo* is one of the big issues that needs to be resolved.

Not enough data exists to know for sure if nanoparticles could have undesirable effects on the environment. Two areas are relevant here:

(1) In free form, nanoparticles can be released in the air or water during production (or production accidents) or as waste by-product

RSC Nanoscience and Nanotechnology No. 8
Nano-Society: Pushing the Boundaries of Technology
By Michael Berger
© Michael Berger 2009
Published by the Royal Society of Chemistry, www.rsc.org

of production, and ultimately accumulate in the soil, water, or plant life.

(2) In fixed form, where they are part of a manufactured substance or product, they will ultimately have to be recycled or disposed of as waste.

We do not yet know if certain nanoparticles will constitute a completely new class of nonbiodegradable pollutant. If they do, we also do not yet know how such pollutants could be removed from air or water because most traditional filters are not suitable for such tasks (their pores are too big to trap nanoparticles).

To properly assess the health hazards of engineered nanoparticles the whole life cycle of these particles needs to be evaluated, including their fabrication, storage and distribution, application and potential abuse, and disposal. The impact on humans or the environment may vary at different stages of the life cycle.

Regulatory bodies such as the U.S. Environmental Protection Agency (EPA) and Food and Drug Administration (FDA) or the Health and Consumer Protection Directorate of the European Commission have started to deal with the potential risks posed by nanoparticles. So far, neither engineered nanoparticles nor the products and materials that contain them are subject to any special regulation regarding production, handling, or labeling. The Material Safety Data Sheets that must be issued for certain materials often do not differentiate between bulk and nanoscale size of the material in question.

Studies of the health impact of airborne particles are the closest thing we have to a tool for assessing potential health risks from free nanoparticles. These studies have generally shown that the smaller particles get, the more toxic they become. This is due in part to the fact that, given the same mass per volume, the dose in terms of particle numbers increases as particle size decreases.

Looking at all available data, it must be concluded that current risk assessment methodologies are not suited to the hazards associated with nanoparticles; in particular, existing toxicological and ecotoxicological methods are not up to the task. Exposure evaluation (dose) needs to be expressed as quantity of nanoparticles and/or surface area rather than simply mass; equipment for routine detecting and measuring nanoparticles in air, water, or soil is inadequate; and very little is known about physiological responses to nanoparticles.

Nanomaterials are hugely variable in their nature. They are not a uniform class of materials, and attempts to regulate or legislate solely on the basis of particle size (1–100 nm in one or more dimensions) or how they are made are misguided. It is the functionality of nanomaterials, *i.e.* what they do and how they behave, that matters and this should form the basis of governance and regulation.

15.1   Risk Assessment could Benefit from Nanoparticle Categorization Framework

On some days you see articles in the media that slam certain nanomaterials, and nanotechnology in general, as inherently risky because of a new scientific paper that reports alarming levels of toxicity of a particular nanoparticle. On other days you find research papers that report no apparent risk whatsoever with the same nanomaterials. This has become quite a familiar pattern for regular readers of nanoscience and nanotechnology literature. The cosmetics industry, for instance, claims its products and the nanoparticles in them are perfectly safe; environmental groups contend that nanomaterials are inherently risky and that rigorous testing is needed before large-scale commercialization of nano-products can happen; some even call for a complete moratorium on nanomaterial production. The result is a situation that is confusing not only for lay people but also for industry, researchers, and the journalists who cover the field.

The fact is that every new technology is inherently risky. Plenty of people are being injured or killed every year by electricity, cars, or chemicals, just to name a few. In order to reap the benefits of a new technology and make it acceptable to society there has to be a general perception that the risks are fully understood, can be managed, and it is clear who is responsible for what. All of that is currently missing in nanotechnology. Although the speed and scope of nanotechnology risk research, and the emerging field of nanotoxicology, is picking up, a lot of this work is stand-alone research that is not being coordinated within a larger framework.

As an emerging science, nanotoxicology is expanding the boundaries of traditional toxicology from a testing and auxiliary science to a new discipline where toxicological knowledge of nanomaterials can be put to constructive use in therapeutics as well as the development of new and better biocompatible materials. Until now, though, no one has been able to pinpoint which properties determine or influence the inherent hazards of nanoparticles.

A group of Danish scientists proposes a categorization framework that enables scientists and regulators to systematically identify the various categories of nanomaterials.

"We believe that, in order to reap the many benefits offered by nanotechnology, it is necessary that environmental and human health risks are considered at an early stage in product development," says Steffen Foss Hansen. "However, before this can be done, there is a need for clarification of terminology. The current literature addressing the potential hazards of nanomaterials shows a strong tendency to use the terms 'nanotechnology' and 'nanomaterials' as synonyms for 'nanoparticles'. The hazards related to nanotechnology/nanomaterials have so far predominantly been documented for specific nanoparticles, mainly titanium dioxide and carbon-based nanoparticles. However, the physical, chemical, and biological properties of various nanomaterials differ quite substantially from those of specific nanoparticles, as do the expected routes of exposure, making it necessary to differentiate nanomaterials in order to identify the potential hazards and risks they pose."

Hansen, a researcher in the Department of Environmental Engineering at Technical University of Denmark (DTU), together with colleagues, has developed a framework that can be applied to a proposed hazard identification approach and is aimed at identifying causality between inherent physical and chemical properties and observed adverse effects reported in the literature.

The DTU scientists tested the workability of their proposed procedure using nanoparticles as an illustrative case study. "We generated a database noting the reported inherent physical and chemical properties of the nanoparticles tested and the main effects observed," Hansen explains. "428 studies were noted in the database, reporting on a total of 965 nanoparticles. We found that, although a limited number of studies have been reported on ecotoxicity, more than 120 have been reported on mammalian toxicity and 270 on cytotoxicity. In general there was a lack of characterization of the nanoparticles studied and it was not possible to link specific properties of nanoparticles to the observed effects. Our study shows that future research strategies must have a strong focus on characterization of the nanoparticles tested."

Hansen and his colleagues suggest that nanomaterials should be categorized depending on the location of the nanoscale structure in the system. This leads to a division of nanomaterials into three main categories, which then can be further divided into subcategories: (1) materials that are nanostructured in the bulk; (2) materials that have nanostructure on the surface; and (3) materials that contain nanostructured particles.

Hansen points out that it is possible for a system to consist of nanostructured elements belonging to different categories in this framework. He gives the example of catalysts used to remove nitrogen oxide from car exhaust: "The chemical reaction that removes nitrogen oxide is catalyzed by platinum and ruthenium nanoparticles 2–3 nm in size. These nanoparticles are bound to the surface of a support material. At the same time, the support material is a nanoporous material consisting mostly of $G\text{-}Al_2O_3$ (70–85%) and other oxides such as cerium oxide or lanthanum oxide. A similar analysis would apply to fuel cells, for example."

A major benefit of the proposed categorization framework is that it provides a tool for dividing nanosystems into identifiable parts and thereby facilitates evaluations of relevant exposure routes or analysis of effect.

Other aspects that need to be considered in assessing the toxicity of nanomaterials are their physical and chemical properties. Today, which properties determine or influence the inherent hazards of nanoparticles is still an open question.

"After an initial literature review, and considering the information needed in order to describe a nanomaterial from a physical and chemical perspective when estimating the hazard of nanomaterials, we propose the following nine properties as being important," says Hansen: "(1) chemical composition, (2) size, (3) shape, (4) crystal structure, (5) surface area, (6) surface chemistry, (7) surface charge, (8) solubility, and (9) adhesion, defined as the force by which the nanoparticle and its components are held together."

The Danish researchers then went ahead and combined their categorization framework with this list of properties to construct a hazard identification scheme.

"Clearly, not all the properties apply to all the different categories," says Hansen. "We distinguish three cases: (1) the property is relevant in determining the hazard; (2) the property does not apply to the category; and (3) the property can be determined, but it is not relevant in order to determine the hazard of the nanomaterial of that category, because it is unlikely that the surface charge of nanoparticles suspended in a solid plays any role in the overall toxicity given the limited exposure potential."

Having constructed their model, the researchers then went on to test its workability by reviewing the literature on the potential hazards of nanomaterials belonging to two of the proposed categories: nanoparticles suspended in liquids and airborne nanoparticles. Altogether, they identified 428 relevant studies which, in total, reported the observed adverse effects of 965 tested nanoparticles of various compositions. Hansen and his colleagues then analyzed each of the 428 studies in order to identify which inherent physical and chemical properties were reported and the main effects observed. Then they filled the information into their hazard identification scheme.

"In general, there is a lack of characterization of the nanoparticles tested in the identified studies and this hampers the potential for identifying causality between observed hazards and specific physical and chemical properties," explains Hansen. "Furthermore, it is evident from our work that the information provided is all over the map, making it impossible to analyze the studies systematically for properties of the nanoparticles that are important for the observed effects."

Although the lack of characterization is troublesome, it is hardly surprising, as nanotoxicology is a very new field. Hansen points out that a true understanding of the hazardous properties that materials begin to exhibit at the nanoscale requires a level of interdisciplinary research that has not yet been achieved.

"In order to conduct and interpret scientific studies on the hazardous properties of nanomaterials that are relevant for future risk assessment of nanotechnology-based compounds and products, we need strong interdisciplinary collaborations between (eco)toxicologists, and nanoscientists such as physicists, chemists, and material engineers."

Featured scientist: Steffen Foss Hansen
Organization: DTU Environment, Technical University of Denmark, Lyngby (Denmark)
Relevant publication: Steffen Foss Hansen, Britt H. Larsen, Stig I. Olsen, Anders Baun: Categorization framework to aid hazard identification of nanomaterials, *Nanotoxicology*, **1**, 243–250.

15.2 Comparing Apples with Oranges—the Problem of Nanotube Risk Assessment

Here is a specific example that underlines the conclusions made by Hansen and his colleagues. Despite their name, CNTs are not made of 100% carbon. Depending on which of the various synthesis techniques is used in their production, CNTs have variable chemistries and physical properties resulting from their different metal catalysts or amorphous carbon coatings. As a result, they may contain large percentages of metal and carbonaceous impurities which will have different environmental and toxicological impacts. In early toxicological studies, researchers obtained contradictory results: in some studies nanotubes were toxic; in others, they were not. The apparent contradictions were actually a result of the materials that the researchers were using, not appreciating that 'CNTs' are really 'CNTs + metal + amorphous impurities'. Ignoring these impurities prohibits scientists from fully understanding the material's electronic character, environmental transport, transformation, and ecotoxicology.

Scientists have begun to acknowledge that the identity of these impurities and co-products is critical to the toxicology and chemical behavior of CNTs. However, the chemical compositions of these fractions are not well defined and there have been no concerted efforts to identify and compile this information— without which accurate environmental risk assessments for specific CNT stocks is not possible.

To address these needs, a group of researchers measured the elemental, molecular, and stable carbon isotope compositions of commercially available single-walled carbon nanotubes (SWCNTs) produced by 10 companies in the United States, giving a true picture of their diversity and chemical complexity. This diversity and complexity is extremely important from both fate and toxicity perspectives.

"Our specific goals were (1) to identify metal catalysts and aromatic hydrocarbons that would be released with and affect the properties of SWCNTs, (2) to seek compositional data suited to pursuing environmental exposure modeling of SWCNTs, and (3) to find properties that would be helpful for detecting, and perhaps apportioning the sources of, SWCNTs in environmental matrices," Desirée Plata tells us. "If we are going to predict the toxicities of nanomaterials, we need to know what they contain and understand how those components vary—*e.g.* are they always 15% nickel and 5% yttrium? Or do they all have hydrocarbon contaminants that may desorb in the environment?"

Plata, a joint program graduate student at MIT and the Woods Hole Oceanographic Institution and her mentors, chemists Phil Gschwend and Chris Reddy, found that the 10 different CNT samples they tested had vastly different compositions.

This study is the first time that anyone has explored the use of carbon isotopes or metal ratios to track CNTs in the environment. Both analytical methods can be used to detect nanotubes in bulk samples (*e.g.* in sediments or aerosols), allowing for high-throughput quantification of CNTs in complex matrices.

The results show that the metals associated with CNTs are available for reactions with the outside world. "Many people suspected that they would not present a true danger, as they would not be free to react with or travel to the environment," says Plata. "Since that is not the case, we need to adjust the way we account for nanotube toxicity, reactivity, and potential environmental effects."

Plata and her collaborators think that the most concerning problem is the reactivity of the metal catalysts that travel with the CNTs. There are many approaches to try to minimize this effect, and probably the most effective way, until we know more, will be to embed the materials inside impermeable layers such as polymer matrices (which are used for instance in consumer products such as CNT-reinforced golf balls or tennis rackets).

"If a manufacturer chooses to use CNTs in clothes, sunscreens, water-filtration devices, or permeable reactive barriers (to treat groundwater), they may be exposing the public and the environment to an unintended risk," says Plata.

This of course is the challenge that regulators are facing today: you cannot regulate a toxin, or tell if it is sitting in your back yard, if you don't know how to find it in the first place or, even if you find it, you don't know exactly what its effects are. Rather than modeling the risk of a generic, *i.e.* oversimplified SWCNT, researchers need to develop nanomaterial risk assessment methods that take into account the actual diversity of these products and their interaction with the environment. This might lead to mathematical models relating certain CNT parameters to various degrees of toxicity.

Plata says that the research community is moving towards being able to track these diverse chemicals, which will help develop sound analytical techniques. This will also enable manufacturers to weigh the material-specific risk assessments and to design synthetic processes to achieve environmental objectives while simultaneously considering performance and manufacturing cost.

An interesting side result from this research is that the unique metal ratios can be used to 'fingerprint' CNTs. The researchers note that, for example, in Houston, Texas, there are several CNT manufacturers. If there were a release of CNTs to the environment, it would be possible to tell which manufacturer was responsible for the release based on the metal content of the nanotubes. The city of Houston could then identify a 'responsible party' and ask them to assist with the clean-up.

The good news from research like this one of course is that it is taking place before a real problem pops up. This represents a big paradigm shift from the way some sectors of industry and society used to operate (and indeed still do), *i.e.* pollute first, then worry about it later when it becomes a problem. If the emerging nanotechnology-based companies act responsibly (and smartly), they will fully embrace being asked challenging product safety questions and they will proactively support finding all the required scientific answers so that they can take an integral part in the design of new industrial processes and materials.

Plata says that, in a way, this is what environmental champions have been demanding since the 1960s. "Rachel Carson wanted people to use DDT in a smart, discriminating way. She was against the ubiquitous distribution of poorly understood chemicals. Essentially, we're calling for the same type of action: use these chemicals, but use them in a smart way from start to finish. The old adage, 'It's easier to beg forgiveness than ask permission' doesn't apply to Mother Nature and it doesn't apply to public health. We need to be proactive about preventing future environmental catastrophes, and we have the means to do it."

In previous work, Plata and colleagues found that the process of nanotube manufacturing produced emissions of at least 15 aromatic hydrocarbons, including four different kinds of toxic polycyclic aromatic hydrocarbons (PAHs) similar to those found in cigarette smoke and automobile exhaust emissions. They also found that the process was largely inefficient: much of the raw carbon went unconsumed and was vented into the atmosphere. The researchers are currently working with materials scientists and industry to mitigate these effects.

Featured scientist: Desirée Plata
Organization: Woods Hole Oceanographic Institution, Woods Hole, MA (USA)
Relevant publication: D. L. Plata, P. M. Gschwend, C. M. Reddy: Industrially synthesized single-walled CNTs: compositional data for users, environmental risk assessments, and source apportionment, *Nanotechnology*, **19**, 185706.

15.3 Biodegradation of Carbon Nanotubes could Mitigate Potential Toxic Effects

The toxicity issues surrounding CNTs are highly relevant for two reasons. Firstly, as more and more products containing CNTs come to market, there is a chance that free CNTs will be released during their life cycles, most likely during production or disposal, and find their way through the environment into the human body. Secondly, and much more pertinent with regard to potential health risks, is the use of CNTs in biological and medical settings. The interesting structural, chemical, electrical, and optical properties of CNTs are explored by numerous nanomedicine research groups around the world with the goal of drastically improving performance and efficacy of biological detection, imaging, and therapeutic applications. In many of the envisaged applications, CNTs would be deliberately injected or implanted into the body.

"While it has been shown that CNTs can indeed act as a means for drug delivery, negative effects such as unusual and robust inflammatory response, oxidative stress and formation of free radicals, and the accumulation of peroxidative products have also been found as a result of CNTs and their accumulated aggregates," Alexander Star points out. "As a possible solution, we have provided compelling evidence of the biodegradation of CNTs by horseradish peroxidase and hydrogen peroxide over the course of several weeks. This marks a promising possibility for nanotubes to be degraded by horseradish peroxidase in environmentally relevant settings."

Star, an assistant professor in advanced functional materials, nanosensors, physical organic chemistry at the University of Pittsburgh, together with Valerian Kagan and collaborators from the Departments of Chemistry and Environmental and Occupational Health at the University of Pittsburgh, have demonstrated the natural biodegradation of SWCNTs through enzymatic catalysis.

Researchers have been experimenting with ways to alleviate the potential negative side effects of CNTs either by functionalizing them to make them more biocompatible or by degrading them after their use. So far, methods for degrading nanotubes, or 'cutting' them, have involved the use of a harsh solvent consisting of sulfuric acid and high concentrations of hydrogen peroxide.

"When dealing with environmental issues it is important not to introduce any contaminants harsher than what is being cleaned," says Star. "Our method provides a mild, natural approach for the safe removal of CNT material."

In their work, the University of Pittsburgh scientists show the natural biodegradation of SWCNTs through enzymatic catalysis. "By incubating CNTs with the common enzyme horseradish peroxidase and low levels of hydrogen peroxide under static conditions, these nanomaterials are oxidized," says Brett Allen, a chemistry PhD candidate in Star's lab. "The formation of a highly oxidizing intermediate from this enzyme, known as Compound I, facilitates this biodegradation process. These results mark promising possibilities for nanotubes to be degraded in environmentally relevant settings."

In their 16 week experiment, the researchers started to observe a substantial decrease in the average nanotube length and the appearance of globular material after 8 weeks. By the end of the 16 week incubation period, they found that it had become difficult to account for any nanotube structure at all. Allen says that examination of the samples at 12 weeks already revealed that the bulk of nanotubes were no longer present, and globular material had amassed.

These findings could lead to the development of immobilized horseradish peroxidase/hydrogen peroxide mixtures into a chemical spill kit to clean up CNTs in the environment, thus mitigating CNT toxicity.

"It is tempting to speculate that other plant and animal peroxidases may also be effective in oxidative degradation of CNTs," says Star. "If so, enhancement of these catalytic biodegradation pathways may be instrumental in avoiding their cytotoxicity in drug delivery, gene silencing, and tumor imaging."

He and Allen point out that, with further insight into this type of biodegradation process, it will be possible to engineer better, more efficient drug delivery platforms where patients need not worry about the injection of materials that could possibly accumulate, causing cytotoxic effects.

"Our team is investigating the effects and degradation of CNTs using other relevant peroxidases," Star explains the team's next steps. "Furthermore, we need to understand the products that are formed. Although it appears that nanotubes are degraded, we have still not identified these degradation products."

Featured scientists: Alexander Star, Brett Allen

Organization: Department of Chemistry, University of Pittsburgh, Pittsburgh, PA (USA)

Relevant publication: Brett L. Allen, Padmakar D. Kichambare, Pingping Gou, Irina I. Vlasova, Alexander A. Kapralov, Nagarjun Konduru, Valerian E. Kagan, Alexander Star: Biodegradation of single-walled carbon nanotubes through enzymatic catalysis, *Nano Lett.*, **8**, 3899–3903.

15.4 Nanoparticle Penetration of Human Skin—a Double-Edged Sword

Engineered nanoparticles are at the forefront of the rapidly developing field of nanomedicine. As we have seen, their unique size-dependent properties, of which optical and magnetic effects are the most used for biological applications, make them suitable for a wide range of biomedical applications such as cell labeling and targeting, tissue engineering, drug delivery and targeting, magnetic resonance imaging (MRI), probing of DNA structure, tumor destruction via hyperthermia, and detection and analysis of biomolecules such as proteins or pathogens. Many of these applications can also be tailored to target skin to help in the early diagnosis of a skin disease, which then could also be treated via nanocarriers. In addition, a tissue engineering approach could be useful for skin wound healing therapies and the magnetic properties of nanoparticles might help in directing and localizing these agents in a particular layer of the skin where their action is desired. Unfortunately, if nanoparticles are able to penetrate layers of skin for therapeutic purposes, they might equally be able to penetrate skin unintentionally. This raises the question of whether people who are exposed to such nanomaterials could accidentally be contaminated and thus exposed to a potential local and/or systemic health risk.

Researchers in Italy have begun to systematically evaluate both risks and applications of nanoparticle skin absorption. They have investigated whether superficially modified iron-based nanoparticles, not designed for skin absorption but having dimensions compatible with skin penetration routes, are able to penetrate and perhaps permeate the skin. Their results, which showed profound skin penetration by these nanoparticles, opens up two main directions of investigation: nanomaterial toxicological risk assessment and awareness, and potential exploitation of nanomaterials as carriers for drug delivery into and through the skin.

"So far, the skin, which is our first defense against the environment, has been considered an unlikely path of entry for engineered nanoparticles," says Biancamaria Baroli, a researcher at the Department of Pharmacy at the University of Cagliari. "Yet these conclusions arise from investigations carried out with much bigger particulate vehicles than the ones we used (<10 nm). Even bacteria and viruses are bigger; for instance, an adenovirus particle is around 150 nm. We think that it is important to ask whether such small materials could penetrate and permeate the skin."

In their work, Baroli and her collaborators applied two different stabilized nanoparticle dispersions to human skin samples. The results of this study showed that nanoparticles were able to passively penetrate the skin and reach the deepest layers of the stratum corneum (the outermost layer of the skin) and hair follicle and, occasionally, reach the living cells of the epidermis. Yet, nanoparticles were unable to permeate the skin. From a therapeutic point of view, these results represent a breakthrough in skin penetration because it is early evidence that rigid nanoparticles have the ability to passively reach the viable epidermis through the lipid matrix of the stratum corneum.

"Our findings are in agreement with previously published results," says Baroli. "Although more experiments are needed to help us completely understand the penetration mechanism, this study represents a proof of principle and provides a major breakthrough in the study of skin absorption, which allows us to envisage potential toxicological risks as well as advance nanoparticle biomedical applications.

"In fact, it is now possible to foresee synthesized particles that have been designed specifically to target the skin—as we are currently doing—both to understand the penetration mechanism and to study whether the amounts that penetrate could be of any use in biomedical applications. In addition, from a nanotoxicological point of view, one can ask 'how much is enough' to trigger toxicological responses."

Unfortunately, researchers cannot yet answer this last question, as nanotoxicological risk assessment investigations have only recently been undertaken and the cutaneous route of exposure has not yet received great attention. Baroli points out that her team's commitment is rather to evaluate both the risks and the applications of nanoparticle skin absorption, hoping that their work will help to increase awareness and safety of a technology with great therapeutic potential.

"We are now investigating nanoparticles designed and synthesized for skin applications with the intention of better understanding penetration mechanisms and finding simple but accurate methods for quantifying the amount of particles entering the skin," says Baroli. "A better understanding of penetration mechanisms and our ability to find methods to quantify particles in the skin may help to develop new nanoparticle-based formulations but also prevent accidental contamination and health risks."

Featured scientist: Biancamaria Baroli
Organization: Department of Pharmacy, University of Cagliari (Italy)
Relevant publication: Biancamaria Baroli, Maria Grazia Ennas, Felice Loffredo, Michela Isola, Raimondo Pinna, M. Arturo López-Quintela: Penetration of metallic nanoparticles in human full-thickness skin, *J. Invest. Dermatol.*, **127**, 1701–1712.

15.5 Nanotechnology's Complicated Risk–Benefit Dichotomy

Adding yet another twist to the emerging debate about the potential risks of nanomaterials, researchers have demonstrated how difficult it is to map out the health effects of nanoparticles. They have shown that, even if a certain nanoparticle does not appear toxic in itself, the interaction between this nanoparticle and other common compounds in the human body may cause serious problems to cell function. On one hand, this effect could be used to great advantage for killing cancer cells. On the other hand, unfortunately, it is unknown at present whether the same effect could also be observed with healthy cells. Since the number of possible combinations of nanoparticles and various biomolecules is immense, it is practically impossible to research them systematically. This example of the risk–benefit dichotomy of nanotechnology just shows how thin the line is between promising applications such as effective cancer killers and the unknown risks posed by unintentional effects of exactly the same applications.

"Our recent work with fullerenes provides two core findings," says Emppu Salonen, a scientist at Helsinki University of Technology in Finland. "We found that fullerenes, which are inherently insoluble in water, can be efficiently solubilized by gallic acid, a phenolic acid which is ubiquitous in plants and can be found for instance in tea, grapes, oak bark, and cosmetic products as an antioxidant. Furthermore, when exposed to gallic-acid-solubilized fullerenes, human tumor cells were shown to contract rapidly within tens of minutes and subsequently die."

Salonen explains the significance of these two findings: "The first finding touches two important topics. First, there is intensive ongoing research aimed at finding good solubilizing agents for carbon nanomaterials. Second, while the volume of nanoparticles produced and the number of consumer products containing nanoparticles are both increasing rapidly, we still do not know much about their environmental and biological effects. We do not know, for example, how nanoparticles might be discharged into the environment, whether they could be solubilized in soil and natural waters, and whether they could enter the food chain. To us, that is quite a lot of important unknowns."

According to Salonen, the second finding has an interesting twist. Fullerenes are envisioned to be used in nanomedicine as drug delivery agents or even as the drugs themselves, for instance in therapies to fight cancer, HIV, Alzheimer's and Parkinson's diseases.

"Fullerenes are excellent antioxidants and have already shown great promise in *in vitro* experiments. Gallic acid is abundant in substances like tea, red wines, or walnuts. What our work shows is that, when acting together, these *a priori* beneficial compounds induce a fast and dramatic death of tumor cells. What we do not know at present, though, is whether the same effect could also be observed with healthy cells."

Salonen worked with the research groups of Ilpo Vattulainen at Tampere University of Technology in Finland and Pu-Chun Ke at Clemson University in South Carolina. The main motivation for this study was to understand the fate of fullerenes discharged into the environment since, with the increasing industrial-scale production and envisioned application of fullerenes in consumer products, their eventual entry into the environment is unavoidable.

"The fact that nanoparticles are so small and highly reactive and adaptive makes determining their environmental impact much more challenging as compared to other, larger-size pollutants common today," says Salonen. "What we endorse—like many other researchers, governments, and nongovernmental organizations around the world—is proactive work on determining the biological and environmental effects of nanomaterials in general."

The Finnish–U.S. research team combined results both from atomistic molecular dynamics simulations conducted by Salonen and Vattulainen at their labs in Finland, and from experimental observations by Ke's lab at Clemson University.

The scientists used gallic acid as their solubilizing agent. They found that cells exposed to fullerenes or gallic acid separately did not show any loss of viability. However, the interplay of fullerenes and gallic acid combined resulted in dramatic cell death observed both as clear contraction of the cell plasma membranes in fluorescent measurements as well as directly in cell viability assays that provide information on the metabolism of cells.

Basically, the study tried to answer two questions: (1) can fullerenes discharged into the environment be solubilized by some naturally abundant compounds (answer: yes), and (2) what are the possible biological effects of such fullerene-solubilizing agent aggregates (answer: apart from the observed contraction of cell membranes, mostly unknown).

One area the scientists are already working on is determining the effects of fullerene–gallic acid aggregates on cell plasma membranes, in order to provide an explanation for the cell death observed here. The particularly pressing question of how fullerenes solubilized by gallic acid can enter the food chain and eventually interact with animal cells remains unanswered, although Ke's group has already carried out experimental work on this topic and also with other carbon nanomaterials and with other solubilizing agents. Notwithstanding the potential negative effects, this research shows that gallic acid seems to be a very good compound for rendering fullerenes water soluble.

Salonen says that it should be noted that the cell lines which were exposed to fullerene–gallic acid aggregates in this study were human tumor cells (HT-29 cell line). "But while it sounds very promising that fullerene–gallic acid aggregates were so effective in killing tumor cells, we still have no idea whether these nanoaggregates could also induce similar effects with healthy animal cells and bacteria".

Salonen and his collaborators point out that the fact that gallic acid is such an efficient solubilizing agent for fullerenes raises an important question. Since gallic acid shares structural similarity with many components of natural organic matter—a heterogeneous distribution of organic compounds found in soil and natural waters—it is valid to ask whether different chemical species in natural organic matter can also solubilize carbon nanomaterials. If so, what are the implications for environmental transport, bioavailability, and physiological effects?

"Understanding these effects in all their complexity is a daunting task for future research," says Salonen. "Only by understanding the fate of nanomaterials in biological organisms and in the environment can we guarantee a safe and transparent development of nanotechnology."

> *Featured scientist:* Emppu Salonen
> *Organization:* Laboratory of Physics, Helsinki University of
>   Technology (Finland)
> *Relevant publication:* Emppu Salonen, Sijie Lin, Michelle L. Reid,
>   Marcus Allegood, Xi Wang, Apparao M. Rao, Ilpo Vattulainen,
>   Pu Chun Ke: Real-time translocation of fullerene reveals cell
>   contraction, *Small*, **4**, 1986–1992.

15.6   Toxic Nanotechnology—a Problem that Could Result in Surprising Benefits

The fight against infections is as old as civilization. Silver, for instance, was already recognized in ancient Greece and Rome for its infection-fighting properties and it has a long and intriguing history as an antibiotic in human health care. Modern pharmaceutical companies developed powerful antibiotics as an apparently high-tech solution to get nasty microbes under control. In the 1950s, penicillin was so successful that the U.S. Surgeon General at the time, William H. Stewart, declared it was "time to close the book on infectious diseases, declare the war against pestilence won". These days, the U.S. Centers for Disease Control and Prevention (CDC) estimates that the infections acquired in hospitals alone affect approximately 2 million people annually. In the USA, between 44 000 and 98 000 people die every year from infections they picked up in hospitals. As our antibiotics become more and more ineffective, researchers have begun to re-evaluate old antimicrobial substances such as silver. Antimicrobial nanosilver applications have become a very popular early commercial nanotechnology product. Researchers have also made a first step to add CNTs to their microbe-killing arsenal.

Contrary to previous conventional thinking, scientists now suspect that there might be bacterial components to many diseases and disorders, from mental disorders to cancer and even obesity. What makes this a scary prospect is that, in an alarming trend, bacteria and other microorganisms that cause infections are becoming remarkably resilient and can evolve quickly to survive drugs meant to kill or weaken them. This resistance is due largely to the increasing use of antibiotics. Just consider the fact that by some estimates 70% of antibiotics used in the United States (about 12 million kg) are given to farm animals—not to treat disease, but to promote slightly faster growth and to compensate for crowded and insanitary living conditions. These antibiotics find their way through our food into our bodies. Antibiotics have even been found in breast milk of women who didn't take them.

The debate to what degree and under what circumstances CNTs are toxic is far from over. Researchers have now taken an interesting approach: if CNTs can kill human cells under certain conditions, can't this toxicity be used to develop antimicrobial applications?

A team of researchers led by Menachem Elimelech, the chair of the Department of Chemical Engineering and the director of the Environmental

Engineering Program at Yale University, has firmly established that SWCNTs can pierce bacterial cell walls.

Specifically, by using highly purified, pristine SWCNTs with a narrow diameter distribution, they demonstrated that direct cell contact with SWCNTs can cause severe membrane damage and subsequent cell inactivation of *E. coli* bacteria. In contrast, the silver nanoparticles used in antimicrobial applications damage bacterial cells by destroying the enzymes that transport cell nutrient and weakening the cell membrane or cell wall and cytoplasm. Unfortunately, in practical applications, the pure silver nanoparticles are unstable with respect to agglomeration. In most cases, this aggregation leads to the loss of the properties associated with the nanoscale of metallic particles.

"This is the first work to demonstrate antibacterial effects of SWCNTs," says Elimelech. "We took great care in designing the experiments so the results are more conclusive and we prepared pristine, well-characterized SWCNTs with a narrow size distribution."

Elimelech points out that previous studies of the toxicity of SWCNTs used commercial and/or poorly characterized SWCNTs. As we have seen, impurities can play a role in toxicity, and impurities of commercial SWCNTs are quite significant.

"Although our results show that the *E. coli* undergo severe membrane damage and subsequent loss of viability due to SWCNTs, very little information is currently available with regard to the cytotoxic mechanisms of SWCNTs," says Elimelech. "Previous studies, mainly focusing on mammalian cells, have proposed three principal cytotoxic mechanisms: oxidative stress, metal toxicity, and physical piercing. In our study, we postulate that SWCNT aggregates caused irrecoverable damage to the *E. coli* bacteria by physical damage to the outer membrane of the cells, causing the release of intracellular content."

There are two ways to look at these research findings: the negative one of course has to do with the fact that this is a further demonstration of the biotoxicity of CNTs. On the other hand, any discovery of new ways to make antimicrobial materials and applications is welcome news. While it is unlikely that we will see CNT-based microbe-killing medicine replacing ineffective antibiotics, the potential use of antimicrobial surface coatings is huge, ranging from medicine, where medical device infection is associated with significant healthcare costs, to the construction industry and the food packaging industry.

Studies such as this one will also help us learn more about the toxicity of carbon nanomaterials, so that material engineers can design them with appropriate physical and chemical properties that reduce or eliminate toxicity.

Featured scientist: Menachem Elimelech
Organization: Department of Chemical Engineering, Yale University, New Haven, CT (USA)
Relevant publication: Seoktae Kang, Mathieu Pinault, Lisa D. Pfefferle, Menachem Elimelech: Single-walled carbon nanotubes exhibit strong antimicrobial activity, *Langmuir*, **23**, 8670–8673.

15.7 Analyzing Biocompatibility of Nanomedical Applications with Blood

Any drug intended for systemic administration and all medical devices which will contact blood (*e.g.* oxygenators, tubing, catheters, artificial hearts) must undergo thorough biocompatibility testing. These tests include an *in vitro* assay to determine the material's potential to damage red blood cells (erythrocytes). Hemolysis, the abnormal breakdown of red blood cells either in the blood vessels (intravascular hemolysis) or elsewhere in the body (extravascular hemolysis), can lead to anemia or other pathological conditions. In the pharmaceutical industry, hematocompatibility testing is harmonized through the use of internationally recognized standard protocols. ASTM F756—Standard Practice for Assessment of Hemolytic Properties of Materials—is a widely used standard for blood-damage testing. Another standard, ISO 10993–4, recommends investigating damage to red blood cells as a way to study a material's compatibility with blood.

Nanotechnology-based medical devices and drug carriers are emerging as alternatives to conventional small-molecule drugs, and *in vitro* evaluation of their biocompatibility with blood components is a necessary part of early preclinical development. Many research papers have reported nanoparticle hemolytic properties but, so far, no *in vitro* hemolysis protocol has been available that is specific to nanoparticles.

Work by the Nanotechnology Characterization Laboratory in the United States describes *in vitro* assays to study nanoparticle hemolytic properties, identifies nanoparticle interferences with these *in vitro* tests and provides the first comprehensive insight to potential sources of this interference, demonstrates the usefulness of including nanoparticle-only controls, and illustrates the importance of physicochemical characterization of nanoparticle formulations and visually monitoring test samples to avoid false-positive or false-negative results.

"Since biomedical nanoparticle engineering is a rapidly growing field, we realized that having a protocol designed specifically to work with nanomaterials would be useful to many parties," says Marina A. Dobrovolskaia. "We started from an existing standard (ASTM 756–00) for evaluating medical devices. The first task in adapting this to nanoparticles was to minimize material requirements, because biomedical nanoparticles are often expensive and complicated to produce and are not available in gram quantities. The next task was to test various nanoparticles representing different classes of materials and to identify whether or not our method was generally applicable. The main finding of our study is that nanoparticles have unique physicochemical properties that can lead to a host of interference issues with traditional *in vitro* tests."

Dobrovolskaia is an immunologist for the Nanotechnology Characterization Laboratory (NCL), a formal collaboration between the National Cancer Institute (NCI), the U.S. Food and Drug Administration (FDA), and the National Institute of Standards and Technology (NIST). The NCL was established to accelerate the transition of basic nanotechnology research into clinical applications to effectively treat and diagnose cancer. The NCL performs preclinical characterization of nanoparticles intended for cancer therapeutics and diagnostics. It is a free resource available to investigators from academia, industry, and government laboratories. The NCL staff includes scientists from the fields of chemistry, physics, immunology, cell biology, and toxicology, who are now familiar with a wide variety of nanoparticle types, such as dendrimers, liposomes, gold colloids, fullerenes, and polymers. One of the NCL's objectives is to establish a standardized assay cascade which would aid researchers and regulatory agencies in understanding nanoparticle properties that affect biocompatibility.

Dobrovolskaia, together with several NCL colleagues and Scott E. McNeil, director of the NCL, describes validation of an *in vitro* method designed to analyze nanoparticle potential to damage red blood cells and the use of this method to study a variety of nanoparticles. Specifically, their study describes approaches to identify nanoparticle/erythrocyte interferences, when they occur, and how to resolve them to get accurate results (*i.e.* to avoid false-positive or false-negative results).

The NCL team's assay leverages the ASTM F-756–00 standard for analysis of hemolytic properties of medical devices. "We scaled this standard practice to a 96-well plate format assay and conducted a 1 month validation aimed at determining its reproducibility, precision, and accuracy, as well as qualification of negative and positive nanoparticle-relevant controls," explains Dobrovolskaia. "We subsequently used our assay to analyze various types of nanomaterials including polymers, gold nanoshells, nanoliposomes, nanoemulsions, fullerene derivatives, gold colloids, and dendrimers. This second phase was conducted over a 2 year period and included identification and resolution of nanoparticle interference with the assay, in addition to evaluation of reproducibility, precision, accuracy, and control qualification."

"Several previous studies evaluating nanoparticle hemolytic properties *in vitro* have appeared in the literature," says Dobrovolskaia. "However, all these tests were conducted using different methods, and even when similar approaches were used, factors such as plasma anticoagulant, blood incubation times, centrifugal forces, or assay detection wavelength varied from study to study. This made it difficult to compare results." She points out that, most importantly, none of these earlier studies included special controls to identify nanoparticle interference, and therefore it was really impossible to conclude what nanoparticle properties were responsible for the hemolysis—*i.e.* did it depend on size, charge, or surface groups, or was it due to the particle absorbance?

The problem was therefore to develop a hemolysis assay applicable to a wide variety of nanoparticles, so that data can be compared between the many laboratories testing nanoparticle biocompatibility. Consequently, the NCL scientists specifically designed their assay to be applicable to various nanoparticles. That way, it could be used by different investigators working with different nanoparticles, and allow comparison of the results. The protocol they describe is one of a set of assays developed specifically for use with nano-particles and available to the public through the NCL website.[1]

When used in early preclinical drug development, this particular method may help differentiate strong candidates from those with properties that need further tuning before being used in *in vivo* applications. If studies from numerous laboratories on a wide variety of nanoparticles are conducted using the same method, it will increase confidence in the quality of data and conclusions drawn from that data regarding nanoparticle biocompatibility.

Hemolysis is not the only traditional biocompatibility test that nanoparticles interfere with. Nanoparticles are intricate, often delicate systems with unique properties, and their characterization is challenging. Dobrovolskaia mentions that, for instance, many nanoparticles have catalytic properties and can enhance assays that rely on enzymatic reactions, generating false-positive results. "Assays routine to the preclinical characterization of conventional pharmaceuticals, such as the *Limulus amebocyte* lysate (LAL) test for detection of endotoxin contamination, may yield spurious results when applied to nanoparticle samples. Multifunctional nanoparticles have to be characterized quite rigorously, as there are multiple components that must work in concert to achieve functionality."

She adds that a thorough characterization of a nanoparticle-based therapeutic includes evaluation of physicochemical properties, sterility and pyrogenicity assessment, biodistribution (ADME, *i.e.* absorption, distribution, metabolism, and excretion) and toxicity characterization—which includes both *in vitro* tests and *in vivo* animal studies. The NCL has tailored each of these tiers of a rational characterization cascade so that they are relevant to nanoparticles.

Featured scientist: Marina A. Dobrovolskaia
Organization: National Cancer Institute, Nanotechnology
 Characterization Laboratory, Frederick, MD (USA)
Relevant publication: Marina A. Dobrovolskaia, Jeffrey D. Clogston,
 Barry W. Neun, Jennifer B. Hall, Anil K. Patri, Scott E. McNeil:
 Method for Analysis of Nanoparticle Hemolytic Properties in Vitro,
 Nano Lett., **8**, 2180–2187.

---

[1] http://ncl.cancer.gov/working_assay-cascade.asp

PART IV:
A WORLD OF NOVEL
APPLICATIONS AWAITS

CHAPTER 16
Designing Tomorrow's Computer

The success of the semiconductor industry has been due in large part to its ability to continuously increase the complexity, and therefore the processing power, of integrated circuits at a given manufacturing cost. Moore's law states that the number of transistors in a computer chip doubles every 2 years, whilst the cost of making the chip remains the same, because of the miniaturization of the components.

In order to produce the next generation of computer chips it is necessary to keep on shrinking the size of the components on the chip. The miniaturization upon which Moore's law rests has been achieved through advances in the photolithographic process used to pattern the components on to the silicon wafer. A beam of light is projected through a shadow-casting reticule and the light pattern is then directed on to a silicon wafer coated with a photo-chemically sensitive material, known as a *resist*. The solubility of the resist is modified by exposure to the light, allowing specific areas of the resist film to be removed, whilst other areas remain as a mask, so that the silicon wafer can be selectively etched, metallized, or doped.

For many years it has been predicted that the end of photolithography is approaching, and that further miniaturization will require next-generation lithography techniques, such as extreme ultraviolet (EUV) lithography. However, photolithography has proved remarkably resilient, and still continues to improve. Nevertheless, researchers are busily working on the foundations for next-generation processors and technologies.

16.1　Carbon Nanotubes could Buy more Time for Moore's Law

The 'state of the art' in semiconductor fabrication is a rapidly moving target, from 90 nm processes to 65 nm and 45 nm generations and currently 32 nm

RSC Nanoscience and Nanotechnology No. 8
Nano-Society: Pushing the Boundaries of Technology
By Michael Berger

Published by the Royal Society of Chemistry, www.rsc.org

processor technology. The results of these efforts are processors where a billion or more transistor-based circuits are integrated into a single chip. One of the increasingly difficult problems that chip designers are facing is that the high density of components packed on a chip makes interconnections increasingly difficult. In order to be able to continue the trend predicted by Moore's law, at least for a few more years, researchers are now turning to alternative materials for transistors and interconnects, and one of the prime candidates for this job is single-walled carbon nanotubes (SWCNTs). However, one of the biggest limitations of conventional carbon nanotube (CNT) device fabrication techniques today is the inability to scale up the processes to fabricate a large number of devices on a single chip.

Researchers in Germany have demonstrated the directed and precise assembly of single-nanotube devices with an integration density of several million devices per square centimetre, using a novel aspect of nanotube dielectrophoresis. This development is a big step towards commercial realization of CNT-based electronic devices and their integration into the existing silicon-based processor technologies. (See Figure 16.1.)

"The fundamental issue of CNT device fabrication remains the biggest challenge for effective commercialization of nanotube electronics," explains Ralph Krupke. "For CNT electronics to become a reality, it has to be possible to scale up the fabrication technique to fabricate a very large number of such devices on a single chip, simultaneously and reproducibly, with each one individually accessible for electronic transport. Conventional nanotube growth and device fabrication techniques using chemical vapor deposition or spin-casting are unable to achieve this, because of a lack of precise control over positioning and orientation of the nanotubes."

"Since these nanotubes are usually grown at temperatures greater than 500 °C and show no growth selectivity between metallic and semiconducting

Figure 16.1   The figure on the left shows a schematic of an ultra-large-scale array of SWCNT devices fabricated by dielectrophoretic deposition from an aqueous solution. The scanning electron micrograph on the right is a zoom-in to one region of the array showing each electrode pair bridged by an individual carbon nanotube in a self-limiting mechanism. (Image: Dr A. Vijayaraghavan and Dr R. Krupke, Forschungszentrum Karlsruhe)

types, they cannot be directly integrated into silicon-based microfabrication," adds Aravind Vijayaraghavan. "Because of the difficulties in handling and manipulating these nanoscale objects at the individual level, various attempts to assemble them into functional devices have met with limited success. In the ideal case, it should be possible to position an individual nanotube at a predefined location and orientation, forming robust, low-resistance, ohmic contacts to two metallic leads. Furthermore, it should be possible to do this at a scalable integration density with each nanotube forming an individually addressable device."

Krupke and Vijayaraghavan are scientists at the Institute of Nanotechnology (INT) at the Forschungszentrum Karlsruhe in Germany. Together with colleagues from the INT and the University of Karlsruhe they have reported a novel aspect of dielectrophoretic deposition of CNTs, where the dielectrophoretic force field changes upon nanotube deposition and thereby self-limits the directed assembly to a single nanotube or nanotube bundle at predefined locations.

In 2003, the group demonstrated that it is possible to deposit CNT bundles from an aqueous solution using a process called *dielectrophoresis*, which uses nonhomogeneous alternating electric fields to move and assemble nanoscale objects.

"Since then, we have made tremendous advances in understanding the dynamics of a CNT moving in such an electric field," says Vijayaraghavan. "The required nonhomogeneous electric fields are generated by two opposing needle-shaped electrodes with a microscopic gap between their tips. We have discovered the mechanism that allows for a self-limiting deposition of CNTs to one per electrode pair. This happens because the first CNT that is deposited in the gap markedly changes the electric field distribution around it, leading to a repulsion of subsequent CNTs that attempt to enter the region of the gap."

The researchers in Karlsruhe have also developed and optimized the use of capacitatively coupled electrodes, which enables them to reduce their dimensions and increase the density of electrode pairs that can be incorporated on a chip.

"Together, this allows us to fabricate separately addressable, individual SWCNT devices at an integration density comparable to ultra-large-scale integration," says Krupke. "This is three to four orders of magnitude greater than what has been possible so far with any other technique."

This technique is very versatile. It is compatible with SWCNTs from any source that are suitably dispersed in an aqueous surfactant solution. SWCNTs separated on the basis of their length, diameter, or even chirality can be readily assembled into large-scale functional arrays using this technique. The process is fully compatible with post-processing techniques and current microelectronics fabrication technologies, requires no high-temperature steps or chemical modification of the substrate or the CNT, and is a one-step process that can be carried out under ambient conditions.

This achievement takes CNT electronic devices a big step closer to integrating with microelectronics and expanding their scope for commercial viability. On a laboratory scale, it now allows for the fabrication of a large number of devices with identical CNT source and deposition conditions, to perform truly statistical measurements of CNT properties like electronic transport or Raman mapping.

16.2   A Race to the Bottom

As mentioned above, the semiconductor industry is on its way to commercial
32 nm processor technology and beyond. The day may be near when transistors
will reach the limits of miniaturization at atomic levels, putting an end to
currently used fabrication technologies. Apart from the issues of interconnect
density and heat dissipation, which some researchers hope to address with
CNT-based applications, there is the fundamental issue of quantum mechanics
that will kick in once chip design gets down to ~ 4 nm. This is where semiconductor dimensions become so small that quantum effects would dominate
the circuit behavior. Computer designers usually regard this as a bad thing
because it might allow electrons to leak to places where they are not wanted. In
particular, the tunneling of electrons and holes—so-called *quantum tunneling*—
will become too great for the transistor to perform reliable operations. The
result would be that the two states of the switch could become indistinguishable.
Quantum effects can, however, also be beneficial. A group of researchers has
shown that a single bit of data might be stored on, and again retrieved from, a
single atom. Just don't expect this in your computer anytime soon.

At the heart of this 'atomic storage' lies a phenomenon known as *ballistic
anisotropic magnetoresistance* (BAMR), which was predicted by theorists in
2005[1] and has now been demonstrated in the lab.

Magnetoresistance is the property, displayed by all metallic magnetic materials, of changing the value of their electrical resistance when an external
magnetic field is applied to them. The *anisotropic magnetoresistance* (AMR)
effect arises because conduction electrons have more frequent collisions when
they move parallel to the magnetization in the material than when they move
perpendicular to it. Discovered in 1856, AMR is the basis for many modern
data-storage devices. Over the past 30 years, several new forms of magnetoresistance have been found and one of them, *giant magnetoresistance*, gave
birth to the emerging field of *spintronics* (*i.e.* spin-based electronics), a term that
is increasingly used instead of magnetoelectronics.

Physicists theorized that the most recent form of magnetoresistance, BAMR,
would become evident when a magnetized metal wire a few atoms across was

---

[1] J. Velev, R. F. Sabirianov, S. S. Jaswal, and E. Y. Tsymbal: Ballistic anisotropic magnetoresistance, *Phys. Rev. Lett.*, **94**, 127203.

placed in a second magnetic field. The atoms of the wire would be magnetized in the direction of the field. That direction could be used to encode a bit of data. Any electron passing along such a wire should be able to travel ballistically (*i.e.* without being slowed down by bumping into any atoms in the wire, just as a bullet travels down the barrel of a gun). Crucially for data storage, this free flow means that the spins of the electrons in question would be able to align themselves with those of the data-storing atoms, giving a clean signal. In effect, this is a digital version of the magnetoresistance effect.

Andrei Sokolov, a research assistant professor in the Department of Physics and Astronomy of the University of Nebraska, and Bernard Doudin, a professor in the Metallic Materials Department at the Université Louis Pasteur in Strasbourg, France, have reported the first experimental evidence for BAMR by observing a stepwise variation in the ballistic conductance of cobalt nano-contacts as the direction of an applied magnetic field is varied.

The research team used standard lithographic techniques to make two planar gold electrodes separated by a gap of ~ 100 nm and bonded to a silicon chip. A cobalt film was then electrodeposited to close the gap between the electrodes in order to have the current flow through a constriction that tapers down to the size of a single atom. As electrons passed through the film, the signal from the atoms was indeed detected.

"Our results show that BAMR can be either positive or negative, and exhibits symmetric and asymmetric angular dependences, consistent with theoretical predictions," says Sokolov. "Our findings unambiguously demonstrate a new magnetoresistance phenomenon specific to quantum ballistic transport. Measurements performed on magnetically saturated atomic-size contacts indicate that a small change in the magnetization direction can cause a large change in conductance, of amplitude, and of sign depending on the local atomic configuration. This phenomenon is fundamentally different from known magnetoresistive effects related to the relative reorientation of magnetic moments within the sample."

Because of the possibility of controlling the quantized conductance by applied magnetic fields, the researchers believe that BAMR may be appealing for future generations of ultra-small electronic devices, such as magnetic read heads, quantum switches, and logic circuits.

Featured scientists: (a) Andrei Sokolov, (b) Bernard Doudin
Organizations: (a) Department of Physics and Astronomy of the University of Nebraska, Lincoln, NE (USA); (b) Metallic Materials Department, Université Louis Pasteur, Strasbourg (France)
Relevant publication: Andrei Sokolov, Chunjuan Zhang, Evgeny Y. Tsymbal, Jody Redepenning, Bernard Doudin: Quantized magnetoresistance in atomic-size contacts, *Nat. Nanotechnol.*, **2**, 171–175.

16.3 To the Rescue of Overheating Computer Chips

Power and heat have become the biggest issues for chip manufacturers and companies integrating these chips into everyday devices such as cellphones and laptops. The computing power of today's computer chips is provided mostly by operations switching at ever higher frequency. This physically induced power dissipation represents the limiting factor for a further increase of the capability of integrated circuits. Heat dissipation of the latest processors has become a widely discussed issue. By the end of the decade a computer chip might be as hot as a rocket nozzle. As the electronics industry continues to churn out smaller and slimmer portable devices, manufacturers have been challenged to find new ways to combat the persistent problem of thermal management. New research suggests that the integration of CNTs into electronic devices as heat sinks might provide a solution.

As heat is becoming one of the most critical issues in computer and semi-conductor design, the electronics industry faces growing thermal distress in six fundamental ways:

(1) Shortened product lifetimes—every increase of 10 °C in operating temperature cuts product lifetimes in half.
(2) Increased operating costs. Keeping devices cool increasingly requires more and faster-spinning fans, which use more electricity.
(3) Consumer acceptance. More heat requires more cooling fans, which create more noise.
(4) Reduced reliability. There have been several recalls and product failures due to heat.
(5) Degraded performance. RF amplifiers lose as much as 90% of their power through heat, a serious problem affecting battery life on all kinds of wireless devices. With CMOS, achievable performance improvements range from 1% to 3% for every 10 °C lower transistor temperature, depending on the doping characteristics of the chip.
(6) Increased build costs. Ever larger and more complex heat sinks and fans, and even more exotic approaches, are driving up costs.

To reduce high temperatures, today's heat sinks—finned devices made of conductive metal such as aluminum or copper—are attached to the back of the chips to pull thermal energy away from the microprocessor and transfer it into the surrounding air.

Using microfin structures made of aligned multiwalled carbon nanotube (MWCNT) arrays mounted to the back of silicon chips, researchers from Rensselaer Polytechnic Institute and the University of Oulu in Finland have demonstrated that nanotubes can dissipate chip heat as effectively as copper—the best known, but most costly, material for thermal management applications. And the nanotubes are more flexible, more resilient, and 10 times lighter than any other cooling material available.

"We were able to demonstrate a simple and scalable nanotube-on-a-chip assembly, using a unique processing and transfer technique to integrate nanotube structures on the chip," says Robert Vajtai, lab manager at Pulickel M. Ajayan's Carbon Nanomaterials Research Group at Rensselaer Polytechnic Institute. "Once they are attached to the chip, the advantageous mechanical and thermal properties of CNTs can be exploited to remove heat from silicon chip components."

The trend to increasing miniaturization of electronic devices causes a headache for manufacturers. The integrity of materials typically used for cooling structures breaks down when they are reduced to submillimetre sizes. Silicon becomes very brittle and shatters easily, and metallic structures become bendable and weak. CNTs, however, maintain their impressive combination of high strength, low weight, and excellent conductivity. In addition, the CNT heat sinks developed by the Finnish–U.S. research team can be manufactured very cost effectively.

The researchers fabricated thick films consisting of aligned MWCNTs ~ 1.2 mm long with diameters between 10 and 90 nm. After detaching the nanotube layers from the templates, structures of 10×10 fin array blocks were fabricated in the freestanding films by laser-assisted surface patterning. Each cooling element of CNT fin arrays is approximately $1.2 \times 1.0 \times 1.0$ mm^3 in size with a mass of only 0.27 mg.

Compared to a chip with no cooling source, 11% more power was dissipated from the chip mounted with the nanotube cooler. Under forced nitrogen flow, the cooling performance with the fins was improved by 19%.

"These numbers are consistent with the heat dissipated by the best thermal conductors," says Vajtai. "This demonstrates the possibility of a lightweight, solid-state add-on structure for an on-chip thermal management scheme which works without involving heavy metal block and fan or fluid-flow procedures for heat removal which can greatly increase the weight of electronic devices."

The resulting cooling efficiency of the CNT coolers, in terms of power density, is in the range of microcircuit needs (currently around 100 W/cm^2), and they could therefore serve as efficient parts of modern electronic devices.

"In future, we would like to carry out investigations on systems which are closer to real applications in both their size and their geometry," says Vajtai. "We will also try to enhance our method in order to remove a higher percentage of heat and also to explore new paradigms to make the cooler applicable for efficient hot-spot removal."

A large number of parameters could be fine-tuned to improve the efficiency of the CNT coolers: tailoring nanotube structure to obtain higher thermal conductivity; improvement and optimization of the chip–nanotube thermal interface; enlarging the interface surface by using double-sided vertical arrangements; optimizing fin-array geometric spacing; width and height of fins, heat-sink base dimension as well as location; nanotube length and forest density; and improved gas-flow efficiency.

Given the urgency and industrial scale of the problem, there will likely be tough competition among research groups and labs worldwide racing to build the most effective and efficient CNT coolers.

16.4   Telescoping Nanotubes Promise Ultrafast Computer Memory

Nonvolatile random access memory (NVRAM) is the general name used to describe any type of RAM that does not lose its information when the power is turned off. This is in contrast to the most common forms of RAM in use today, DRAM and SRAM, which both require continual power in order to maintain their data. NVRAM is a subgroup of the more general class of nonvolatile memory types, the difference being that NVRAM devices offer random access, as opposed to sequential access like hard disks. The best-known form of NVRAM memory today is flash memory, which is found in a wide variety of consumer electronics, including memory cards, digital music players, digital cameras, and cellphones. One problem with flash memory is its relatively low speed. As chip designers and engineers reach size barriers in downscaling the size of such chips, the research focus shifts towards new types of nanomemory. Molecular-scale memory promises to be low power and high frequency: imagine a computer that boots up immediately on powering up and that writes data directly on to its hard drive, making saving files a thing of the past. Researchers are designing the building blocks for this type of memory device using telescoping CNTs as high-speed, low-power microswitches. The design would allow the use of these binary or three-stage switches to become part of molecular-scale computers.

The advantages that nanostructures such as quantum dots, CNTs, and nanowires offer over their silicon-based predecessors include their tiny size, speed, and density. Several concepts of molecular-scale memory devices have been developed. The main objectives for next-generation memory are that it should be nonvolatile and low power, as well as ensuring high-frequency operation and high resistance to environmental forces.

Researchers have used CNTs to create nanoscale oscillators with frequencies in the gigahertz range. More recently, inner tubes of a MWCNT have been extracted by electrostatic forces and nanotube linear servomotors were realized using the interlayer motion of telescoping MWCNTs. Taking this one step further, researchers have developed a conceptual design for a macroscopically addressable data storage device based on CNTs, which can be utilized both as nonvolatile random access memory and as terabit solid-state storage.

"Our approach is deceptively simple; the design involves inserting one hollow nanotube, closed at both ends, into a slightly larger one, open at both ends, creating a telescoping motion using an electrostatic charge," explains Qing Jiang. "The contact between the nanotube and the electrodes creates a conduction pathway with three possible positions."

Jiang is professor of mechanical and electrical engineering at the University of California, Riverside, where part of his research focuses on the mechanical properties of CNTs and CNT-based devices.

"We have demonstrated that the two telescoped positions of the double-walled CNT, (b) and (c) in Figure 16.2, are both stable, and the switching time from one position to the other is as short as 0.01 ns" Jiang says. "This finding leads to a promising potential to build ultra-fast high-density nonvolatile memory, up to 100 GHz or into the terahertz range. Realization of ultra-fast high-density nonvolatile memory units will bring a broad range of applications in electronics."

The main structural element of this nanomemory is composed of a MWCNT deposited on a metallic electrode. Both the metal electrode and CNT are refined by etching processes to have clean edges of a few nanometers. The proposed nanomemory device is based on a telescoping motion of the nanotubes relative

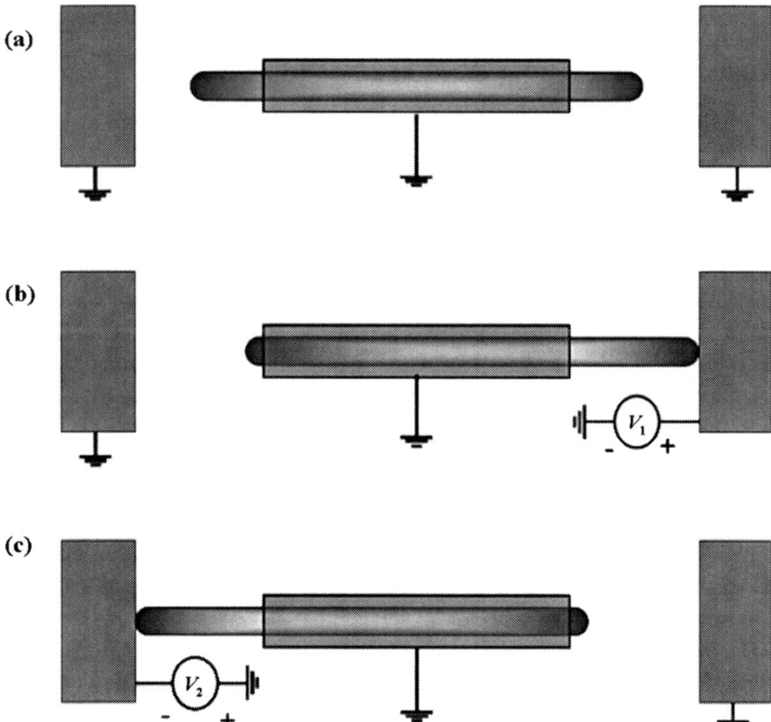

Figure 16.2   A general design of MWCNT-based nonvolatile memory. (a) The initial equilibrium position, (b) the core CNT contacts with the right electrode with V1, and (c) the core CNT contacts with the right electrode with V2. (Graphic: Dr Q. Jiang, UC Riverside)

to each other. The movable core nanotube can slide inside the stationary nanotube by varying the electrostatic forces. This 'telescope' lies between two electrodes, which are neutral when at rest. Negatively charging one of the electrodes and positively charging the core nanotube can overcome the van der Waals force keeping the inner and outer nanotubes together, and make them move toward the oppositely charged electrode. Alternatively, by positively charging the other electrode and negatively charging the core nanotube, the nanotube would slide the other way. High damping would send the core nanotube back to the center.

The potential of such molecular-scale processors has attracted the attention of major corporations such as Hewlett-Packard, IBM, Lucent, Motorola, Siemens and Hitachi, to name a few. Jiang predicts that a functioning prototype of a molecular processor could possibly be demonstrated within a few years.

> *Featured scientist:* Qing Jiang
> *Organization:* Department of Mechanical Engineering, University of California, Riverside, CA (USA)
> *Relevant publication:* Jeong Won Kang, Qing Jiang: Electrostatically telescoping nanotube nonvolatile memory device, *Nanotechnology*, **18**, 095705.

16.5 Next-generation Computer Memory made of Graphene

Taking advantage of the conducting properties of graphene, James M. Tour, Chao professor of chemistry, professor of computer science, and professor of mechanical engineering and materials science at Rice University in Texas, and his team describe a way of how graphene memory could potentially be used as a new type of memory that could significantly exceed the performance of current state-of-the-art flash memory technology.

"What distinguishes graphene from other next-generation memories, in particular phase-change materials, is its vastly higher on/off power ratio—the amount of current a circuit holds when it's on, as opposed to off," Tour explains. "While the on/off ratios of phase-change materials are generally in the 10–100 region, with graphene it can be as high as a whopping million-to-one or even more—we have seen as high as 10 million."

The transistors currently used in computer chips have on/off ratios of 10 000–100 000, but they are three-terminal devices.[2] Two-terminal memories, based for instance on nanowires or CNTs, are actively being researched for future computer applications. Their structure makes 3-D memory practical as the materials can be stacked, multiplying a chip's capacity with every layer.

---

[2] In a three-terminal device, the electric current or voltage between two of the terminals is controlled by applying an electric current or voltage to the third terminal.

The main challenges so far have been a requirement for large-scale fabrication and reliable and large on/off ratios.

The recent research results coming out of Tour's lab show the possibility of building next-generation memory devices with vast amounts of memory using nanocables with a silicon dioxide core and a shell of stacked sheets of grapheme. The team experimented with three different nanocable configurations: two-layer graphite/silicon dioxide and three-layer graphite/silicon dioxide/silicon and graphite/silicon dioxide/silicon carbide. Tour points out that graphene memory would increase the amount of storage in a two-dimensional array by a factor of 5, as individual graphitic thin film sheets could be made as thin as 5–10 nm, compared to the 45 nm circuitry in today's flash memory chips.

When the scientists tested their nanocable devices in the lab, they observed not only excellent endurance—the nanocables showed no degradation after more than 1000 write–read–erase–read cycles—but also no discernable data loss even after weeks of testing, sometimes under harsh temperature conditions that ranged from $-70\,°C$ to $+200\,°C$.

"Interestingly, the devices were also stable to ionizing radiation," says Tour. "In our experiments, we irradiated the unbiased devices with a high dose (>20 Mrad) of 8 keV X-rays, which was significantly higher than the typical failure level for memory devices relying on charge storage, such as flash memories. Thus, nanocable memories demonstrate exceptional stability and are promising for radiation-stable electronics."

The Rice team claims that the new switches are faster than their lab's testing systems can measure (1 µs), although they observed that the write/erase time affected the on/off ratio: "The longer the time, the higher the on/off ratio," says Tour. "When the pulse time is decreased from 100 µs to 1 µs, the on/off ratio in graphite/silicon dioxide nanocable devices decreases to about 100."

Tour's lab is already working on developing scalable manufacturing techniques for their nanocable device. According to Tour, it is possible to deposit a layer of graphene on silicon or another substrate by chemical vapor deposition. "This means that currently used industrial device fabrication processes could be used to grow this type of graphitic thin film *in situ*."

An important aspect of Tour's work is the fabrication atop a silicon wafer in a manufacturable-looking structure. This is essential because nanowires are hard to scale into manufacturable structures since they are tough to place where you want them. Tour's technique shows a way of how to build these future device structures.

Featured scientist: James M. Tour
Organization: Smalley Institute for Nanoscale Science and Technology, Rice University, Houston, TX (USA)
Relevant publication: Yubao Li, Alexander Sinitskii, James M. Tour: Electronic two-terminal bistable graphitic memories, *Nat. Mater.*, **7**, 966–971.

CHAPTER 17

Electronics Beyond Silicon

17.1 Textile Transistors to Create Truly Wearable Electronics

If current research is an indicator, wearable electronics will go far beyond just very small electronic devices. Not only will such devices be embedded on textile substrates, but an electronics device or system could become the fabric itself. Electronic textiles (*e-textiles*) will allow the design and production of a new generation of garments with distributed sensors and electronic functions. Such e-textiles will have the revolutionary ability to sense, act, store, emit, and move (think biomedical monitoring functions or new human–machine interfaces) while leveraging an existing low-cost textile manufacturing infrastructure. So far, only a few steps have been taken toward new architectural possibilities of realizing circuit topologies that can be implemented with textile techniques. An example is an *organic field effect transistor* (OFET), fully compatible with textile processing techniques, that has been developed by researchers in Italy.

Annalisa Bonfiglio's lab at the University of Cagliari is working on the assembly of electronic devices and circuits on textile substrates. This can be done either by a top-down approach, by assembling devices to transfer on to textile substrates, and a bottom-up approach of assembling an electronic fabric starting from electronically functionalized textile basic components.

Top-Down Approach

Organic semiconductors (polymers and oligomers), having the electrical properties of semiconductors and the mechanical properties of plastics, are good candidates for realizing flexible transistors, suitable for transferring on to unconventional substrates such as textiles.

RSC Nanoscience and Nanotechnology No. 8
Nano-Society: Pushing the Boundaries of Technology
By Michael Berger
© Michael Berger 2009
Published by the Royal Society of Chemistry, www.rsc.org

Fiber integration issues are, however, very challenging. Patterning in particular is a significant concern. Although fiber transistors could be fabricated using conventional lithography, these would have limited scalability to large-volume textile processing. What is required is an e-textile technology that facilitates the fabrication of fiber transistors in a textile-compatible, highly scalable manner.

Previously, working with a completely flexible and transparent polyester film, Bonfiglio's lab has studied a 'transistor in a fiber' realized by gluing this film on to a textile ribbon, in order to obtain a flexible yarn that could be employed in a textile process. Here the polyester film is the insulator layer of the field-effect transistor (FET) structure as well as the mechanical support of the whole structure.

Bottom-up Approach

Bonfiglio and her colleagues focused on the possibility of building an OFET with a nonplanar geometry.

"In particular, we have demonstrated the possibility of obtaining a cylindrical organic thin film transistor that, because of the materials used and the dimensions, can be used in textile processes such as weaving or knitting," she says.

The cylindrical OFETs have been obtained starting from a metallic fiber used in textile processes. The metal core of the yarn, covered with a thin polyimide layer, is the gate of the structure. A top-contact device was obtained by depositing a layer of organic semiconductor followed by the deposition of source and drain top contacts, made by metals or conductive polymers, deposited by evaporation or soft lithography. This transistor has shown very interesting performances, with typical values of electronic parameters—mobility, threshold, I_{on}/I_{off} ratio—very similar to those of planar devices.

Bonfiglio explains the potential of this research: "The possibility of making a transistor in yarn form paves the way to an entire new group of applications and offers the possibility of leveraging existing textile technology for building electronic circuits. For example, with weaving technology (the simplest way to make a fabric) one can build a matrix of crossing yarns: if each of these yarns carries a series of transistors, each node of the textile matrix can be singularly addressed and it becomes possible to read or write the matrix one 'node-pixel' at a time. In this way, a flexible and wearable electronic platform can be produced."

Bonfiglio has also been coordinating a project called ProeTEX,[1] funded by the European Union (EU), that focuses on textile-based micro- and nano-technologies to develop textile and fiber-based integrated smart wearables for emergency disaster intervention personnel with the aim of improving their safety, coordination, and efficiency, and additional systems for injured civilians aimed at optimizing their survival management.

---

[1] http://www.proetex.org/

Previously, Bonfiglio's Eolab was involved in two projects that explored the feasibility of textile transistors: a 2003 Information Societies—Future and Emerging Technologies (IST-FET) program, also EU-funded, called ARIANNE—Feasibility study of y**AR**ns and fabr**I**cs with **ANN**exed Electronic functions; and FIRB, funded by the Italian Bureau of Scientific Research, a project for the development of technologies for implementation of electronic components and devices on textile substrates.

> *Featured scientist:* Annalisa Bonfiglio
> *Organization:* Eolab, Department of Electrical and Electronic
>   Engineering, University of Cagliari (Italy)
> *Relevant publication:* Maurizio Maccioni, Emanuele Orgiu, Piero
>   Cosseddu, Simone Locci, Annalisa Bonfiglio: Towards the textile
>   transistor: Assembly and characterization of an organic field
>   effect transistor with a cylindrical geometry, *Appl. Phys. Lett.*, **89**,
>   143515.

17.2 Gutenberg + Nanotechnology = Printable Electronics

Nanoelectronics devices are often made by integrating dissimilar classes of semiconductors and various other disparate materials into one heterogeneous single system. The two primary modes of combining these materials— mechanical bonding and epitaxial growth processes—place stringent require- ments on the ultimate scale or constituent materials of circuits. With mechanical bonding, there is a limited ability to scale to large areas (*i.e.* larger than the wafers) or to more than a few stacking layers; incompatibility with unusual materials (such as nanostructured materials) and/or low-temperature materials and substrates; challenging fabrication and alignment for the through-wafer electrical interconnects; demanding requirements for planar bonding surfaces; and bowing and cracking that can occur from mechanical strains generated by differential thermal expansion and contraction of disparate materials. Epitaxy avoids some of these problems but places severe restrictions on the quality and type of materials that can be grown. Using a process akin to the printing press, researchers have managed to bypass the need for epitaxial growth or wafer bonding to integrate wide-ranging classes of dissimilar semi- conducting nanomaterials on to substrates for the purpose of constructing heterogeneous, 3-D electronics.

Printed semiconductor nanomaterials provide new approaches to 3-D het- erogeneously integrated systems that could be important in various fields of applications such as microfluidic devices with integrated electronics; chemical and biological sensor systems that incorporate unusual materials with con- ventional silicon-based electronics; and photonic and optoelectronic systems that combine light emitters and detectors of compound semiconductors with silicon drive electronics or microelectromechanical structures. Furthermore, the compatibility of this approach with thin, lightweight plastic substrates may create additional opportunities for devices that have unusual form factors or mechanical flexibility as key features.

"Our approach, the combined use of semiconductor nanomaterials and printing techniques, enables high-quality electronics to be formed on diverse substrates, including nonplanar surfaces and thin plastic sheets," John A. Rogers tells us. "These capabilities, in particular the ability to use any mixture of component materials, all lie well outside the range of things that can be achieved with conventional wafer-based approaches to electronics."

Rogers, a professor of chemistry, materials science and engineering at the University of Illinois at Urbana-Champaign, and his research group are trying to develop approaches that enable 'high performance electronics anywhere', *i.e.* they would like to extend semiconductor device technology to substrates other than the semiconductor wafer.

"Our goal is to invent methods that can allow interesting applications which cannot be addressed with conventional technologies, such as flexible displays, large-area solar cells, conformable X-ray imagers, distributed structural and personal health monitors, or curved surface imagers as electronic eyes," says Rogers. "Our belief is that inorganic semiconductor nanomaterials, delivered to a target device substrate using printing techniques, form an attractive way to achieve devices of these types."

Results by Roger's group show how dissimilar single-crystal inorganic semiconductors (such as micro- and nanoscale wires and ribbons of gallium nitride, silicon, and gallium arsenide) can be combined with one another and also with other classes of nanomaterials (such as single-walled carbon nanotubes, SWCNTs) with the use of a scalable and deterministic printing method to yield complex, heterogeneously integrated electronic systems in 2-D or 3-D layouts.

Specifically, the nanoscale semiconductor components are first fabricated, each on its own source substrate, through standard lithographic procedures, with ohmic contacts formed by doping and annealing. These components are then lifted from the source substrate by gentle van der Waals adhesion with an 'inking pad' made of polydimethylsiloxane, and then 'stamped' on to a device substrate such as a sheet of polyimide.

After some additional processing—including deposition and patterning of gate dielectrics, electrodes, and interconnects—the transfer printing and device fabrication steps can be repeated, beginning with spin-coating a new pre-polymer interlayer on top of the previously completed circuit level.

Rogers points out that this fabrication approach has several important features:

(1) All of the processing on the device substrate occurs at low temperatures, thereby avoiding differential thermal expansion and shrinkage effects that can result in unwanted deformations in multilayer stacked systems. This operation also enables the use of low-temperature plastic substrates

and interlayer materials and it helps to ensure that underlying circuit layers are not thermally degraded by the processing of overlying devices.

(2) The method is applicable to broad classes of semiconductor nanomaterials, including emerging materials such as SWCNTs.

(3) The soft stamps enable nondestructive contacts with underlying device layers; these stamps, together with the ultra-thin semiconductor materials, can also tolerate surfaces that have some topography.

(4) The ultra-thin device geometries and interlayers allow easy formation of layer-to-layer electrical interconnects by direct metallization over the device structure. These features overcome many of the disadvantages of conventional approaches.

The researchers demonstrated the capabilities of their printing process by fabricating ultra-thin multilayer stacks of high-performance metal oxide semiconductor field-effect transistors (MOSFETs), high electron mobility transistors, thin-film transistors, photodiodes, and other components that are integrated into device arrays, logic gates, and actively addressable photodetectors on rigid inorganic and flexible plastic substrates.

"We are working to make these kinds of approaches realistic methods for manufacturing," says Rogers. "We focus on the development of tooling to automate the process, and on demonstrator devices such as electronic eye imagers, large area solar cells, and flexible displays that can illustrate the utility of these techniques."

> *Featured scientist:* John A. Rogers
> *Organization:* Materials Science and Engineering, University of Illinois at Urbana-Champaign, Urbana, IL (USA)
> *Relevant publication:* Jong-Hyun Ahn, Hoon-Sik Kim, Keon Jae Lee, Seokwoo Jeon, Seong Jun Kang, Yugang Sun, Ralph G. Nuzzo, John A. Rogers: Heterogeneous Three-Dimensional Electronics by Use of Printed Semiconductor Nanomaterials, *Science*, **314**, 1754–1757.

17.3 Transparent and Flexible Electronics with Nanowire Transistors

Another approach to fabricating flexible electronics includes nanowire transistors. Generally, thin-film transistors (TFTs) and associated circuits are of great interest for applications including displays, large-area electronics, and printed electronics, *e.g.* radio-frequency identification tags (RFIDs). Well-established TFT technologies like amorphous silicon and polysilicon are well suited for many current applications—almost all mobile phone color screens use them—but face challenges in extensions to flexible and transparent applications. In addition, these TFTs have modest carrier mobilities, a measure of the velocity of electrons within the material at a given electric field. The modest mobility corresponds to a modest operating speed for this class of TFTs. Organic TFTs are generally better suited for flexible applications, and can be made transparent. However, mobilities in organic TFTs are also quite low, restricting the speed of operation and requiring relatively large device sizes. Researchers at Purdue University, Northwestern University, and the University of Southern California have developed nanowire TFTs that have significantly higher mobilities than other TFT technologies and therefore offer the potential to operate at much higher speeds. Alternatively, they can be fabricated using much smaller device sizes, which allows higher levels of integration within a given chip area. They also provide compatibility with a variety of substrates, as well as the potential for room-temperature processing, which would allow integration of the devices with a number of other technologies.

"We have demonstrated fully transparent TFTs on both glass and flexible plastic substrates," says David B. Janes. "The TFTs utilize wide-bandgap semiconductor nanowires as the active channels, and transparent conducting oxides for the gate, source, and drain electrodes. The transistors exhibit good performance characteristics, including relatively high on currents (up to 10 µA per nanowire) and high on/off current ratios (required for digital applications in order to achieve low power operation)."

Janes is a professor at the School of Electrical and Computer Engineering, Birck Nanotechnology Center, at Purdue University. Along with colleagues and collaborators from the Institute for Nanoelectronics and Computing at Northwestern University and the Department of Electrical Engineering at the University of Southern California, he has fabricated fully transparent and mechanically flexible nanowire transistors on plastic substrates.

The combination of excellent optical transparency and mechanical flexibility of indium oxide and zinc oxide nanowires, as well as excellent device performance metrics (they have an optical transmission of $\sim 81\%$), make these nanowire transistors an attractive technology for realizing transparent and flexible circuits. Fully transparent nanowire transistors will not only increase aperture ratio efficiency in active matrix arrays, but will also enable low power consumption as well as transparency for future display technologies.

"Earlier reports on semiconductor nanowire transistors have typically employed nontransparent metal electrodes, making the overall structure relatively opaque," explains Tobin J. Marks, Vladimir N. Ipatieff professor of chemistry and professor of materials science and engineering at Northwestern University, and one of Janes's collaborators. "Our study demonstrates that nanowire electronics can be fully transparent as well as flexible while still maintaining high performance levels. This opens the door to entire new technologies for high-performance transparent flexible displays."

The reported nanowire TFTs would be well suited for the drive circuitry in active-matrix displays and would be compatible with pixel technologies such as organic light-emitting diodes (OLEDs). The devices are fabricated using low-temperature processing techniques, which allows integration on to plastics and various other substrates in order to achieve flexibility and ease of packaging.

In particular, there are three broad application areas for this work:

(1) Transparent displays, which are of interest for applications such as heads-up displays on windshields and informational displays on eyeglasses.
(2) Flexible displays. Emerging applications such as 'e-paper' require flexible electronics to be integrated within the pixel array. The demonstration of reliable TFTs on flexible substrates represents an important step toward the required circuitry. Displays suitable for static images have been demonstrated using organic TFTs, but the nanowire transistors demonstrated in this work could operate at much higher speeds and would allow full-motion video.
(3) Transparent/flexible electronics. Applications such as electronic bar-codes, RFID tags, and smart credit cards would be advanced by the availability of relatively high-performance electronics that could be integrated on a variety of substrates. Flexible circuitry would allow integration on curved and nonrigid surfaces. Transparency would allow integration into multilayer packaging in such a way that product information could be seen beneath the electronics.

"The demonstration of individual TFTs with the characteristics we showed represents a first step toward the circuitry required for many novel display and flexible electronics applications," says Janes. "We expect future studies to include demonstrations of analog and digital circuits based on the nanowire TFTs. This will require integration of multiple TFTs into an interconnected circuit, as well as development of appropriate interconnect approaches to provide the specific circuit topology."

Featured scientists: (a) David B. Janes, (b) Tobin J. Marks
Organization: (a) School of Electrical and Computer Engineering, Birck Nanotechnology Center, Purdue University, West Lafayette, IN (USA); (b) Department of Chemistry, Northwestern University, Evanston, IL (USA)
Relevant publication: Sanghyun Ju, Antonio Facchetti, Yi Xuan, Jun Liu, Fumiaki Ishikawa, Peide Ye1, Chongwu Zhou, Tobin J. Marks, David B. Janes: Fabrication of fully transparent nanowire transistors for transparent and flexible electronics, *Nat. Nanotechnol.*, **2**, 378–384.

17.4 Using Quantum Mechanics to Turn Molecules into Transistors

Transistors are the fundamental building blocks of our everyday modern electronics; they are the tiny switches that process the 1s and 0s that make up our digital world. Transistors control the flow of electricity by switching current on or off and by amplifying electrical signals in the circuitry that governs the operation of our computers, cellphones, and any other electronic device you can think of.

The first transistor used in commercial applications was in the Regency TR-1 transistor radio, which went on sale in 1954 for $49.95 (that's over $375 in today's dollars). The first transistors were over 1 cm in diameter, but the smallest transistors today are just 30 nm thick—3 million times smaller. This feat is the equivalent of shrinking the 509 m Taipei 101 Tower, currently the tallest building in the world, to the size of a 1.7 mm grain of rice. The most advanced microprocessors today are packing a whopping 1.9 billion transistors. As we saw earlier, current microprocessor technology is quickly approaching a physical barrier. Switching the current by raising and lowering the electron energy barrier generates heat, which becomes a huge problem as device densities approach the atomic limit. An intriguing—and technologically daunting—alternative would be to exploit the wave nature of the electron, rather than its particle properties, to control current flow on the nanoscale. Such a device, called the *quantum interference effect transistor* (QuIET), has been proposed by researchers in Arizona. This device could be as small as a single benzene molecule, and would produce much less heat than a conventional FET.

Notwithstanding the incredible shrinkage, nanosize transistors work on the same principles as much larger ones: current flows into the base (the gate controller) from one electrode (the collector) and out through another (the emitter). The base switches the current on and off by raising and lowering an electrical potential barrier gate that prevents the flow of electrons. Unfortunately, this type of switching requires a lot of power. As more and more transistors are crammed into the same space, the power density, and with it heat dissipation, becomes an overriding issue.

"Our proposal to use quantum interference is a novel way to control the flow of electrons through a single-molecule device," says David M. Cardamone. "Our work shows that, because they possess exact symmetries one can't achieve in larger systems, molecules are uniquely suited to take advantage of quantum effects needed to build smaller, more efficient electronic devices."

"Our work was originally motivated by curiosity," adds Charles A. Stafford: "How would quantum wave effects modify electron flow through a single-molecule device? We discovered very early, and somewhat fortuitously, that molecular symmetry can lead to unique device properties which hold promise for overcoming important obstacles toward further miniaturization of conventional semiconductor devices."

Stafford, an associate professor of physics at the University of Arizona, and Cardamone, a postdoctoral fellow at Simon Fraser University in Canada, together with Dr Sumit Mazumdar, professor in the Department of Physics at the University of Arizona, explain how the proposed device promises to solve two fundamental problems in nanoscale electronics:

"The first is the problem of heat production. A FET, like the hundreds of millions found in a typical home computer, turns off current flow by building a wall of electric potential energy. Turning this wall on and off causes heating. On the other hand, the QuIET guides the electron waves either forward (on) or back the way they came (off)—a much cooler process.

"The second problem that plagues almost all proposed nanoscale devices is their sensitivity to small perturbations. Most nanoscale devices operate in a narrow resonant range, so that each device on a chip would have to be fine-tuned to its operating point, possibly requiring atomic-precision control of the on-chip environment of each device. Our proposed device does not operate in a narrow resonance, but in a broad valley between resonances, so it is very robust with respect to small electrical perturbations."

On the molecular scale, electrons behave like waves and can interfere with themselves and each other. In wave theory, when one wave passes through another, physicists say the waves interfere. When the crest of one wave passes through, or is superimposed upon, the crest of another wave, the waves interfere *constructively*. However, when the crest of one wave passes through, or is superimposed upon, the trough of another wave, the waves interfere *destructively*. The University of Arizona scientists realized that a current could be switched off in a circuit if it could be constructed so that the electron waves traveling through it cancel each other out through destructive interference.

The QuIET, which right now is only a theoretical construct, exploits this quantum interference stemming from the symmetry of monocyclic aromatic annulenes such as benzene. Because of the exact symmetry possible in molecular devices, the QuIET possesses a perfect mid-gap transmission node, which serves as the off state for the device. By introducing decoherence or elastic scattering from a third lead, the quantum transport through the molecule could be controlled.

"The most obvious application of the QuIET is as part of a smaller, faster, more efficient microprocessor," says Cardamone. "Also, the QuIET doesn't suffer from the environmental limitations of a semiconductor-based field-effect transistor: it could be useful for computation in an aqueous environment, perhaps even *in vivo*."

Although they are incredibly intriguing as a concept, the challenges in realizing single-molecule transistors are enormous. On the practical side, the next challenge is attaching the third lead. Right now, experimentalists can connect two leads to a single small molecule but not yet a third. On the theoretical side, the big challenge is to go beyond a mean-field treatment of electron–electron interactions, which are particularly strong in ultra-small devices, due to their very small capacitance.

Featured scientists: (a) David M. Cardamone, (b) Charles A. Stafford
Organizations: (a) Department of Physics, Simon Fraser University, Burnaby, BC (Canada); (b) Department of Physics, University of Arizona, Tucson, AZ (USA); (b)
Relevant publication: Charles A. Stafford, David M. Cardamone, Sumit Mazumdar: The quantum interference effect transistor, *Nanotechnology*, **18**, 424014.

17.5 DNA Electronics

DNA—the blueprint of life—and electronics seem to be two completely different things but it appears that DNA could offer a solution to many of the hurdles that need to be overcome in the scaling down of electronic circuits beyond a certain point. The reason why DNA could be useful in nanoelectronics for the design of electric circuits is the fact that it actually is the best nanowire in existence—it self-assembles, it self-replicates, and it can adopt various states and conformations. Not surprisingly, performing reliable experiments on a single oligo-DNA molecule is an extremely delicate task, as partly contradictory research reports demonstrate. Different DNA transport experiments have shown that DNA may be insulating, semiconducting, or metallic. Among the numerous factors that could impact the results are the quality of the DNA–electrode interface, the base pair, the charge injection into the molecule, or environmental effects such as humidity or temperature.

A novel CNT-based nanoelectronic platform provides a proof of concept that single DNA molecules can be detected. This detection technique is based on change in electrical conductance upon selective hybridization of the complementary target DNA with the single-stranded probe attached to the system. The single-stranded sequence-specific probe DNA, whose ends are modified with amine groups, is attached between two CNTs/nanowires using dielectrophoresis. This platform can be used for understanding how electrical charge moves through DNA, which could help researchers understand and perhaps develop a technique for reversing the damage done to DNA by oxidation and mutation.

"We are able to use nanoscale electrodes to attach and measure electrical signals through a single DNA molecule," Wonbong Choi tells us. "This is accomplished by suspending the DNA molecule in a nanoscale trench and chemically bonding it to the SWCNT electrodes at its ends. Suspending the DNA between the electrodes eliminates the interaction of the molecule with the chip surface."

Choi, an associate professor and director of the Nanomaterials and Device Laboratory in the Department of Mechanical and Materials Engineering at Florida International University (FIU) in Miami, together with Dr Kalai Mathee from FIU and collaborators from Pohang University of Science and Technology in South Korea and National Institute of Genetics in Japan, also demonstrated the ability to use a SWCNT with the same diameter as a single double helix (~ 2 nm) as an electrode to connect and measure electrical signals through the molecule.

Their findings are significant improvements over the electrical conductivity measurements of DNA demonstrated by other researchers. "Typically, the diameter of the connecting leads is much larger than the size of the DNA molecule," says Choi. "This leads to several DNA molecules becoming attached between connecting leads. Furthermore, because of the dynamic material properties of the DNA molecule itself, several other factors such as the surrounding environment play a major role in studying electrical conductivity in DNA. For instance, the interaction of DNA with the surface on which it is deposited completely changes the charge transport in DNA. All these factors have resulted in controversial results. In the development of our detection platform we are able to take all these issues into account."

Choi explains that there are several unique aspects to his team's approach. "First, we have exploited SWCNT electrodes for anchoring a DNA molecule of compatible diameter (1–2 nm). Second, the application of dielectrophoresis in our system provides controlled manipulation of a DNA molecule. The third important issue is the formation of a covalent bond between each terminus of a DNA molecule and the functionalized end of a SWCNT electrode. Establishment of a strong electronic coupling between the trapped molecule and the nanoelectrodes facilitates charge transport through the system without the Coulomb blockade effect. The fourth important aspect is the presence of a nanotrench between the SWCNT electrodes that eliminates the contribution of the oxide surface to the charge transport through the DNA molecule. The suspended DNA molecule in our system mitigates the problem of compression-induced perturbation of charge transport." The results of several control experiments confirmed that the measured electrical signals did indeed originate from the anchored DNA molecules.

One of the potential applications of this study would be the identification of specific genes based on the hybridization-induced change in electrical signal. The researchers explain that their current detection platform could be used in an application for the electrical detection of several other gene sequences on a single chip, although such simultaneous detection of several different gene sequences using an array of nanoelectrodes is a major challenge, which is currently under investigation.

This platform could also be used in fundamental research studies to understand the properties of DNA at the single-molecule level. Beyond that it has a wide range of applications, for instance a new diagnostic tool that can reveal the presence of disease-related genes.

Featured scientist: Wonbong Choi
Organization: Nanomaterials and Device Laboratory, Department of Mechanical and Materials Engineering, Florida International University, Miami, FL (USA)
Relevant publication: Somenath Roy, Harindra Vedala, Aparna Datta Roy, Do-hyun Kim, Melissa Doud, Kalai Mathee, Hoon-kyu Shin, Nobuo Shimamoto, Viswanath Prasad, Wonbong Choi: Direct electrical measurements on single-molecule genomic DNA using single-walled carbon nanotubes, *Nano Lett.*, **8**, 26–30.

From Supermarket Shelf to Spacecraft

18.1 Nanocomposite Coatings for Superior Dry Self-lubrication

While many dynamic systems are lubricated by fluids of various kinds, high-tech machinery found in many industries like aerospace, clean room equipment, or medical devices, requires the application of dry coatings in order to reduce friction and wear. Lubrication of dynamic surfaces by fluids adds complexity, weight, and cost to the system, which imposes various constraints and limits the performance of these systems. Self-lubricating coatings, *i.e.* coatings which exhibit reduced wear and friction without resorting to tribological fluids, have been known for many years and are in use for a variety of applications. Numerous studies have shown that different nanoparticles, impregnated into metal, polymer, ceramic, and other coatings, can provide these materials with enhanced tribological performance. Prominent among these additives are the fullerene-like (IF) nanoparticles of tungsten disulfide (WS_2) and molybdenum disulfide (MoS_2). It is possible to obtain self-lubrication of hard (*i.e.* dry) coatings such as cobalt by impregnating them with fullerene-like nanoparticles of WS_2. The coating serves as a reservoir of nanoparticles, which are slowly released from the surface and provide easy shear and reduced oxidation of the coating or native metal surface, which is a common phenomenon in tribology (friction and wear) of metallic surfaces.

IF-WS_2 and IF-MoS_2 nanoparticles are known to provide superior tribological action when mixed with oils and greases and when they operate under

RSC Nanoscience and Nanotechnology No. 8
Nano-Society: Pushing the Boundaries of Technology
By Michael Berger
© Michael Berger 2009
Published by the Royal Society of Chemistry, www.rsc.org

harsh tribological conditions. The effect has been attributed to the facile rolling of the nanoparticles; this is why they are also called 'nano ball bearings'. A more thorough analysis, however, shows that this model is too simplistic and does not provide the whole picture. The lubrication mechanism was found to be much more complicated than initially thought. Analysis of the IF nanoparticles following tribological tests revealed that some of the nanoparticles residing at the contact area were deformed, broken, or partially exfoliated. This gradual exfoliation of the nanoparticles leads to the formation of nanosheets of WS_2 which cover the metal surface and provide easy shear and surface protection against wear.

"What was much less known was that, when the nanoparticles are embedded in a coating, they are slowly released and exfoliated and then they basically do the same kind of job," explains Reshef Tenne. "Earlier work on nickel coatings indicated the possibility of having self-lubricating surfaces. Other work showed the potential of such coatings in numerous medical technologies. Our research shows also that very hard coatings like cobalt, which have numerous industrial applications in bearings and machine parts, can be made self-lubricating."

Tenne is director of the Kimmel Center for Nanoscale Science, Drake Family professor of nanotechnology and head of the Department of Materials and Interfaces at the Weizmann Institute of Science in Israel. His group has produced a large body of work on inorganic fullerene-like nanostructures and inorganic nanotubes since the early 1990s. Tenne is most notable for his prediction in 1992, following the discovery of carbon nanotubes (CNTs), that nanoparticles of inorganic compounds with layered structures, such as MoS_2, would not be stable against folding and would also form nanotubes and fullerene-like structures.

In more recent work, Tenne's group impregnated cobalt coatings with IF-WS_2 nanoparticles by depositing them on stainless steel substrates by either electroless or electrodeposition (electroplating) processes. These coatings were found to provide very good tribological behavior under high loads. Furthermore, these new coatings also exhibit magnetic behavior, offering new opportunities for applications of nanocomposite coatings in the automotive, aerospace, and home appliance industries, as well as other fields.

"We assume that the occluded aggregates of the IF-WS_2 nanoparticles are relatively loosely connected to the cobalt matrix and they are gradually released under friction," says Tenne. "Rolling and sliding of these particles around the contact spot and the slow deformation and exfoliation of the nanoparticles provide low friction and wear."

Many machine parts are made with cobalt coatings and there is great interest in industry to apply this technology to improved tribological coatings. Tenne's lab has licensed their technology to a company called ApNano Materials which is working on commercializing this technology. The company has already launched NanoLub™, the world's first commercial solid lubricant based on spherical inorganic nanoparticles.

> *Featured scientist:* Reshef Tenne
> *Organization:* Kimmel Center for Nanoscale Science and Department
> of Materials and Interfaces, Weizmann Institute of Science, Rehovot
> (Israel)
> *Relevant publication:* Hilla Friedman, Orly Eidelman, Yishay Feldman,
> Alexey Moshkovich, Vladislav Perfiliev, Lev Rapoport, Hagai
> Cohen, Alexander Yoffe, Reshef Tenne: Fabrication of self-
> lubricating cobalt coatings on metal surfaces, *Nanotechnology*, **18**,
> 115703.

18.2   Concrete Nanotechnology

Cement is not necessarily a material one would associate with high-tech industries, not to mention nanotechnology. However, it's probably fair to say that our modern society is built on cement. Look around you and you'll find it everywhere—in buildings, roads, bridges, and dams. Early construction cement[1] is probably as old as construction itself. Cement is a mixture of compounds made by burning limestone and clay together at very high temperatures. It is then used, together with water, as binder in a synthetic composite material known as concrete. For concrete to obtain its optimal properties, it needs to harden—and that takes time. For builders, time is money; particularly in industrial settings, time is a major cost issue. Time is also a safety and convenience factor—think about infrastructure repair work on roads and dams, for instance. Cement manufacturers already knew that reducing the particle size of cement results in faster-binding formulations. By taking the ultimate reduction down to the nanoscale, researchers in Switzerland have shown that a one-step preparation of nanoparticulate cement with a conventional Portland cement composition (the most common type) results in greatly increased early reactivity of the cement.

Cement hardening is a surface-dependent process where a reactive starting phase subsequently reacts with water and recrystallizes. (Concrete does not need to dry out in order to harden, as commonly thought.) The time it takes for a sample to achieve a given hardness is related to the amount of surface of the starting material. Imagine a large, solid piece of cement—it would take months for water to slowly enter the block and react with it. If you make the material smaller, the ratio of surface to volume is greater, hence, the reaction is faster. Now, this has been known for over a century.

"Since cements are water sensitive you cannot access them by classical nanomaterials synthesis methods, working with precipitation or sol–gel wet-phase chemistry," explains Wendelin Stark. "We have therefore developed a dry gas-phase process which results in a novel and direct one-step preparation of

---

[1] The word 'cement' goes back to the Romans who used the term *opus caementitium* to describe masonry which resembled concrete and was made from crushed rock with burnt lime as binder.

calcium silicate-based nanoparticles of a typical Portland cement composition by flame spray synthesis. Isothermal calorimetry revealed that the hardening of this new nanocement demonstrated a more than tenfold increase of initial reactivity, with different reaction kinetics from conventionally prepared cements."

Stark, an assistant professor of catalysis at the Institute for Chemical and Bioengineering of the ETH Zürich, emphasizes that at present this nanocement is very porous and less stable than currently used materials. It will need additional improvements in chemical composition and formulation to make this highly reactive material applicable to modern construction work, where load-bearing strength is important.

As a starting point, Stark and his team took a widely used flame aerosol process used in the manufacture of pigments and carbon black (with outputs of over 10 million tonnes per year). "In order to process complex compositions such as Portland cement, we had to do quite a bit of chemistry, but we managed to find very low-cost precursors relying on a by-product from crude oil refining," says Stark. "This cost-effective method allowed us to prepare nanocements with a classical Portland cement composition."

Interestingly for future commercial applications, the process's similarity to existing large-scale production methods facilitates scale-up and transfer to industrial-scale manufacturing and potentially enables the implementation of highly reactive nanoparticles in practical applications.

Conventionally manufactured Portland cement consists of irregularly shaped, micron-sized particles with a broad particle-size distribution ranging from a few microns to over 60 μm. The nanocement contained nanoparticles of various sizes, depending on the temperature of the treatment. On average, the flame-made nanoparticles were up to three orders of magnitude smaller than their conventionally prepared counterparts.

The small size of the particles completely changed the hydration behavior of the cement, suggesting different reaction kinetics while maintaining similar reaction thermodynamics. The reduction in particle size resulted in a drastically increased early reactivity of the cement.

"We found that hardening is way faster. Our materials are solid within a few minutes and reactions are over in less than an hour," says Stark.

Notwithstanding the current unfavorably high porosity of their material, Stark believes that the early reactivity of the nanocement might open up entirely new fields of applications for Portland-type cements, in particular where short hardening times are crucial.

"In its present form, the nanoparticle-derived porous cement could be used in renovation or insulation applications where compressive strength is less critical, or might enhance the hardening behavior of conventional cements if deployed as an admixture," he says. "Further improvements in terms of hardness, however, will be necessary before such materials can be used in load-bearing applications."

> *Featured scientist:* Wendelin Stark
> *Organization:* Institute for Chemical and Bioengineering, ETH Zürich
> (Switzerland)
> *Relevant publication:* S. C. Halim, T. J. Brunner, R. N. Grass,
> M. Bohner, W. J. Stark: Preparation of an ultra fast binding cement
> from calcium silicate-based mixed oxide nanoparticles,
> *Nanotechnology*, **18**, 395701.

18.3   Intelligent Inks—now you see them, now you don't

When they hear the word semiconductor, most people will think about their role in computers. However, semiconductors also absorb light. Some absorb in the visible range, thus appearing colored, *e.g.* gray silicon, and others in the UV, *e.g.* titanium dioxide, thus appearing white (when in microparticulate form) or colorless (when in nanoparticulate form). This light-absorbing feature is used to drive electrons around a circuit in photovoltaic cells, such as the silicon solar cell, but it can also be used to drive chemical reactions at the surface. A good example of the latter is the use of thin (15 nm) titanium dioxide film coatings on self-cleaning glass. When they absorb UV radiation from sunlight, these films can reduce atmospheric oxygen to water and oxidize any organic material on the surface of the glass to its mineral constituents, thereby keeping the surface clean.

Researchers in the UK have used this oxidation feature to develope an irreversible solvent-based blue ink, which upon activation with UV light loses all its color and becomes oxygen sensitive; it will only regain its original color upon exposure to oxygen. A major application area for this oxygen ink is in food packaging where it could be used to detect a modified atmosphere inside food containers.

Oxygen is an essential feature of respiration in most organisms and, as such, it is a key element for life—but it is also the main reason for food spoilage, not only because it can react chemically with food, but also because its presence is essential for the growth of moulds and other aerobic microorganisms.

The food industry has developed a preservation technique used to prolong the shelf life of processed or fresh food by changing the composition of the air surrounding the food in the package. Called *modified atmosphere packaging* (MAP), this technique reduces the oxygen content within the package by completely flushing out the air and replacing it either with nothing (as in vacuum packaging) or with another gas composition containing nitrogen or carbon dioxide. MAP has become widely used and over 50 billion food packages were MAPed in 2007.

Commercial color-indicator oxygen sensors are available, of course, but the reason you still can't find them on your MAPed meat or fish tray is that they require the use of expensive equipment and a trained operator; they are not inks and therefore not readily printed; and they are not easily handled or stored (since they are active all the time and fail after long exposure to oxygen). As a result, they are too expensive as routine indicators for everyday food packaging in supermarkets.

"In our UV-activated oxygen-sensitive ink, upon exposure to a short burst of UV light titania nanoparticles are able to oxidize an organic in the ink and simultaneously reduce a dye that is also present," says Andrew Mills. "This redox dye is highly colored in its original (oxidized) form but colorless in its reduced form. Most importantly, the latter is also oxygen sensitive. So that upon UV activation the highly colored ink bleaches and is oxygen sensitive, only to regain its color upon exposure to air."

Mills, a professor in the Department of Pure and Applied Chemistry, and James Young Chair in Chemistry, at the University of Strathclyde in Glasgow, UK, explains that, for greatest flexibility, a functional ink (some people prefer the term *intelligent ink*) needs to be printable on plastic. This requires a solvent-based rather than a water-based formulation. The researchers have therefore made sure that all the components are solvent (ethanol) compatible.

The result is the first UV-activated color ink that is oxygen sensitive and can be used for qualitative and quantitative analysis. This ink has indefinite storage times—it does not function as oxygen indicator until and unless it is exposed to UV activating light—and it is made from inexpensive components.

The traditional role of ink in food packaging has been that of a passive communicator of information: ingredients, instructions, sell-by dates, and so on. Now, with intelligent inks, this role is changing. The can become reactive, being able to change information based on their response to a change of the substrate they are applied to or the environment that surrounds them.

"I have worked on a wide range of functional inks, not only because they are fun to work with and present great challenges, but also because there is an increasing need for and interest in information, especially if it is inexpensive," Mills describes the motivation for his work. "Everybody would like to be more informed. These functional inks can provide a wealth of information, accessible to most people, especially if color (rather than luminescence) based inks are used, since colors are usually easy to assess."

Mills explains the basic working principles of such an irreversible oxygen indicator, based on the light absorbance feature of semiconductors. Upon irradiation with UVA light, ultra-bandgap illumination of titanium dioxide particles creates electron–hole pairs. The photogenerated holes oxidize the mild sacrificial electron donor glycerol to glyceraldehyde. The remaining photo-generated electrons, *i.e.* titanium dioxide, reduce the redox-sensitive dye, DOx (in Mill's work this is methylene blue), to a reduced form, DRed (leuko-methylene blue) that has a different color from DOx. These key components are encapsulated in a polymer, usually hydroxyethyl cellulose, to create an oxygen-sensitive, UV-activated film.

"All these components are readily dissolved or dispersed in water to create an intelligent, water-based ink for oxygen," says Mills. "The major drawback of this indicator is the hydrophilic, water-soluble nature of the ink, which limits its applications, especially in food packaging, since it is difficult to print such an ink on the mostly hydrophobic plastics used in packaging, such as polypropylene, poly(ethylene terephthalate), or nylon."

In contrast, the work by Mills's team resulted in a colorimetric, UV-activated solvent-based intelligent ink for oxygen that readily wets and prints on to plastics. A major application of this new ink can be found in the quality assurance of MAP packaging. Although an increasing quantity of food is now MAPed, the quality assurance is not 100%. Given the huge number of items, it is not possible to check if every package is properly sealed, and then even after it has been checked it could break in transit or through customer handling on the shelf.

The oxygen indicator inks developed by Mills could be used to provide 100% quality assurance not only to the packager, but also to the retailer and the customer, reassuring them all that the package is intact and the food is safe to eat.

Apart from food packaging, cheap and inexpensive ink sensors could find a wide range of applications. "There are still many analytes that would benefit from routine monitoring via printable inks, such as food spoilage gases, hydrogen peroxide, water, carbon monoxide—the list is long!" says Mills.

Featured scientist: Andrew Mills
Organization: Department of Pure and Applied Chemistry, University of Strathclyde, Glasgow (UK)
Relevant publication: Andrew Mills, David Hazafy: A solvent-based intelligence ink for oxygen, *The Analyst*, **133**, 213–218.

18.4 Radiation Shielding for Spacecraft

Nanotechnology will play an important role in future space missions. Nanosensors, dramatically improved high-performance materials, or highly efficient propulsion systems are just a few examples. One critical area is radiation shielding. NASA says that the risk of exposure to space radiation is the most significant factor limiting humans' ability to participate in long-duration space missions. A lot of research therefore focuses on developing countermeasures to protect astronauts from these risks. To meet the needs for radiation protection as well as other requirements such as low weight and structural stability, spacecraft designers are looking for materials to help them develop multifunctional spacecraft hulls. Advanced nanomaterials such as isotopically enriched boron nanotubes could pave the path to future spacecraft design with nanosensor-integrated hulls that provide effective radiation shielding as well as energy storage.

Space radiation is qualitatively different from the radiation humans encounter on Earth. Once astronauts leave the Earth's protective magnetic field and atmosphere, they become exposed to ionizing radiation in the form of charged atomic particles traveling at close to the speed of light. Highly charged, high-energy particles known as HZE particles pose the greatest risk to humans in space. A long-term exposure to this radiation can lead to DNA damage and cancer.

One of the shielding materials under study is boron 10 (^{10}Bn). Scientists have known since the 1930s about ^{10}Bn's ability to capture neutrons, and they use it as a radiation shield in Geiger counters as well as a shielding layer in nuclear reactors. Boron is a nonmetallic chemical element used in numerous applications, in its elemental form as a dopant in the semiconductor industry and in boron compounds that play important roles as light structural materials, nontoxic insecticides and preservatives, and reagents for chemical synthesis.

"Boron nanotubes (BNTs) have many of the excellent properties of CNTs because they share the same structure," explains Ying Chen. "Compared to CNTs, BNTs have several advantageous properties such as high chemical stability and high resistance to oxidation at high temperatures, and they are a stable wide-bandgap semiconductor. Because of these properties, they can be used for applications at high temperatures or in corrosive environments such as batteries, fuel cells, supercapacitors, or as a solid lubricant in high-speed machines."

Chen, a senior fellow in the Research School of Physical Sciences and Engineering at the Australian National University, together with his colleagues, demonstrates the synthesis of isotopically enriched BNTs for the first time with a high yield and in large quantities using a ball milling/annealing process. Their work reveals the special role of high-energy ball milling in reducing the nitriding temperature, leading to the growth of thin, cylindrical tubes.

Chen says that the specific application of isotopic ^{10}Bn nanotubes includes radiation shielding, multifunctional materials for energy storage, sensing and shell of future spaceships, environmental protection, neutron-related medical applications, and cancer diagnostics and treatment.

"The large-scale production of pure BNTs has been a major problem for potential practical applications," says Chen. "We have developed a ball milling method that can solve this problem. Using our process, we can produce large enough quantities that we have even started to sell BNTs commercially. We have also demonstrated that BNTs with different isotopies, structures, and sizes can be produced."

Chen and colleagues have been refining the ball milling process for preparing BNTs for several years now. It involves grinding down a powder of boron into nanoparticles in a ball mill in which steel balls tumble against each other for hundreds of hours. The role of this high-energy ball milling is to reduce the reaction temperature by creating a chemically reactive structure and to introduce metal catalysts into ^{10}Bn. The fine boron powder is then heated in an atmosphere of nitrogen. This process gives the researchers not only a mass production method but also full control over nanotube size and structure, which is very important for tuning nanotube mechanical and physical properties.

"I have communicated with researchers at NASA about the possible application of our BNTs in space missions," says Chen. "Several years ago they asked me to prepare BNT samples for tests on the space station. We have also discussed the possible use of ^{10}Bn nanotubes. Currently we are conducting radiation tests on the nanotubes at the Australian Nuclear Science and Technology Organization."

The ^{10}Bn nanotubes might find use in various applications where there is a need for strong, lightweight, cost-effective radiation shielding.

"For example," says Chen, "there is a lot of talk about developing fusion energy to feed an energy-hungry world. One of the major challenges in developing fusion energy on a commercial basis is coming up with materials that can provide shielding from the high neutron fluxes produced by the fusion process. BNTs might just fit the bill."

Featured scientist: Ying Chen
Organization: Electronic Materials Engineering, Australian National University, Canberra, ACT (Australia)
Relevant publication: J. Yu, Y. Chen, R. G. Elliman, M. Petravic: Isotopically enriched ^{10}Bn nanotubes, *Adv. Mater.*, **18**, 2157–2160.

18.5 NASA Research into Shape-Shifting Airplanes

NASA is working on revolutionary Morphing program that will literally change the shape of airplanes. The idea is that aircraft of the future will not be built of traditional, multiple, mechanically connected parts and systems; instead, aircraft wing construction will employ fully integrated, nanotechnology-enabled, embedded 'smart' materials and actuators that will enable aircraft wings with unprecedented levels of aerodynamic efficiencies and aircraft control. Able to respond to the constantly varying conditions of flight, sensors will act like the nerves in a bird wing and will measure the pressure over the entire surface of the wing. The response to these measurements will direct actuators that will function like the bird's wing muscles. Just as a bird instinctively uses different feathers on its wings to control its flight, the actuators will change the shape of the aircraft's wings to continually optimize flying conditions. Active flow control effectors will help mitigate adverse aircraft motions when turbulent air conditions are encountered.[2]

NASA's Morphing program makes extensive use of nanotechnology, for example by developing electroactive polymers to improve sensing and actuation. Researchers working in this area have created a novel intrinsic unimorph (*i.e.* consisting of an active and an inactive layer) CNT polymer composite actuator.

---

[2] View a NASA computer animation intended to illustrate the concept of a morphing, or shape-changing, aircraft here: http://www.youtube.com/watch?v=vR3T8mdpdTI

"Our work demonstrates, for the first time, the actuation with CNT polymer composites in a dry state without any electrolytes," explains Cheol Park. "It is an electrostrictive actuation of the composite caused by increased interfacial polarization at the vast interfaces between the nanotubes and the matrix."

Park, a research fellow at the National Institute of Aerospace (NIA) in Hampton, VA, together with his colleague Jin Ho Kang and collaborators from the Advanced Materials and Processing Branch at the NASA Langley Research Center, have developed a novel electroactive SWCNT–polymer composite, an intrinsic unimorph, which can actuate to a large strain (2.6%) at relatively low driving voltages (< 1 MV/m) while maintaining its high performance in mechanical durability, thermal stability, and chemical resistances.

"The fact that this actuator requires very low energy input compared to currently available state-of-the-art actuators and generates much larger strains than those of piezoelectric polymers—poly(vinylidene fluoride) (PVDF) and its copolymers, and piezoceramic lead zirconate titanate (PZT)—is very important for NASA's long-term space exploration missions such as a trip to Mars or lunar habitats," says Park.

He also points out that this intrinsic unimorph actuator does not require adhesive or extraneous inactive layers to generate the large bending actuation. The actuating capabilities of the electroactive polymers are gained from incorporation of SWCNTs into the polymers.

"Our CNT composite forms an intrinsic unimorph actuator during the film processing, which is unprecedented," Park explains. "This is very significant because we do not need any inactive dummy layer and adhesive layer to convert longitudinal to bending strain, which is a prerequisite for most conventional actuators. Therefore, our intrinsic unimorph is easy to make and has no interfacial aging (delamination) problems."

Several CNT-based composites have been reported so far, based on ionic-type actuation with electrolytes. These composites can actuate only at a wet state or in a liquid solution, to provide sufficient mobility of the electrolytes and ions. In contrast, the CNT–polymer composite developed by the NASA/NIA scientists actuates mainly by electrostrictive effect, which is novel and has never been reported before.

Applications of this novel nanocomposite material include lightweight and low-power-consuming actuators for aerospace as well as terrestrial vehicles. Park points out that these high-temperature, flexible actuators are suitable for any complex system in harsh environments. He lists specific application areas from the NASA Morphing program that could benefit from such actuating CNT-polymer composites:

- Sonic fatigue abatement (by active control)
- Noise transmission attenuation (by active control)
- Wing and panel flutter control
- Tail buffet alleviation control
- Airframe surface shape control
- On–off switch
- Valves.

Park and his colleagues caution that the long-term stability of highly concentrated CNT composite actuators during the application of high electric fields needs to be further studied. They note that the response time is not as fast as some leading rapid electrostrictive or piezoelectric actuators since their intrinsic unimorph actuates based on interfacial polarization. Park is confident that this can be overcome with specially designed electrodes or layered structures.

> *Featured scientist:* Cheol Park
> *Organization:* National Institute of Aerospace, Hampton, VA (USA)
> *Relevant publication:* Cheol Park, Jin Ho Kang, Joycelyn S. Harrison, Robert C. Costen, Sharon E. Lowther: Actuating single wall carbon nanotube-polymer composites: intrinsic unimorphs, *Adv. Mater.*, **20**, 2074–2079.

18.6 Portable, Cheap, and Fast Explosives Detector Built with Nanomaterials

Because of the increased use of sophisticated explosive devices in terrorist attacks worldwide, where the amount of metal used in the device is becoming very small, the development of a new approach capable of rapidly and cost-efficiently detecting volatile chemical emissions from explosives is urgently needed. Trained bomb-detecting dogs and physical methods such as gas chromatography coupled to a mass spectrometer, nuclear quadrupole resonance, and electron capture detection as well as electrochemical approaches are highly sensitive and selective, but some of these techniques are expensive and others are not easily made available in a small, low-power package. As a complementary method, chemical sensors provide new approaches to the rapid detection of ultra-small trace analytes from explosives and can be easily incorporated into inexpensive and portable microelectronic devices. Researchers in China have developed a nanocomposite film that shows very fast fluorescence response to trace vapors of explosives such as TNT, DNT, or nitrobenzene.

Professor Guangtao Li from the Key Laboratory of Organic Optoelectronics and Molecular Engineering of the Ministry of Education at Tsinghua University, Beijing, and colleagues have made silica films doped with porphyrins (nitrogen-containing macrocyclic molecules). Even trace levels of explosives such as TNT can cause a fluorescent response in these films.

Current fluorescent sensors for TNT are based upon conducting polymers, which can be unstable and hard to make. According to Li, the films his team has developed can overcome this problem. "We have prepared two kinds of porphyrin-doped mesoporous silica films," says Li. "Interestingly, we found that our mesostructured hybrid films show a high fluorescence quenching sensitivity towards vapor of TNT. In comparison to conjugated-polymer based sensors, the fabrication of these hybrid films is very simple, the materials used are inexpensive, and the trapped organic sensing elements become very stable in the inert silica matrix."

Two key features of these mesostructured films, namely the porous structure and the large surface area, are believed to be principally responsible for the observed remarkable sensing performance. The unique mesoporous structure provides a necessary condition for the facile diffusion of analytes to sensing elements, while the large surface area considerably enhances the interaction sites between analyte molecules and sensing elements, and thereby further improves the detection sensitivity. These nanocomposite films can exhibit different mesostructures—hexagonal and worm-like—and, under optimized conditions, the researchers achieved remarkable fluorescence sensitivity.

Li and his co-workers believe that there are three reasons for the observed superior quenching sensitivity in their nanocomposite films. (1) The unique mesoporous structure provides a necessary condition for the facile diffusion of TNT molecules to the anchored porphyrin entities. (2) Because of the well-known strong tendency to form coordinate bonds between the metalloporphyrin molecule and nitro groups of nitroaromatics, as well as [pi]-stacking between porphyrins and aromatic rings, these interactions provide a strong driving force for fast fluorescence quenching. (3) The adequate energy-level matching between TNT and porphyrin molecules makes the fluorescence quenching very effective.

"Since the preparation is very easy and the materials used are inexpensive, organic sensing elements become stable enough in the inorganic matrix, and the synthesized sensing films are easily incorporated into inexpensive and portable electronic devices, our method should offer a promising alternative to other explosive detection methods," says Li.

He and his team aim to achieve even lower detection limits. "Our goal is the fabrication of chemosensory materials that could have detection performance comparable to that of the historic gold standard—dogs. By using porous materials, we hope that we can develop high-performance, portable chemosensory devices for explosives detection that at least match, if not surpass, the performance of trained bomb dogs."

Featured scientist: Guangtao Li
Organization: Key Laboratory of Organic Optoelectronics and
    Molecular Engineering, Tsinghua University, Beijing (China)
Relevant publication: Shengyang Tao, Guangtao Li, Hesun Zhu:
    Metalloporphyrins as sensing elements for the rapid detection of
    trace TNT vapor, *J. Mater. Chem.*, **16**, 4521–4528.

18.7 Nano-barcodes to Quickly Identify Biological Weapons

The list of possible terror threats is large and, in addition to explosives, counter-terrorism efforts are concerned with uncovering biological weapons. In an effort to detect biological threats quickly and accurately, a number of detection technologies have been developed. This rapid growth and development in biodetection technology has largely been driven by the emergence of new and deadly infectious diseases and the realization of biological warfare as a new means of terrorism. To address the need for portable, multiplex biodetection systems, a number of immunoassays have been developed.

An immunoassay is a biochemical test that measures the level of a substance in a biological liquid. The assay takes advantage of the specific binding of an antigen to its antibody.[3] Physical, chemical, and optical properties that can be tuned to detect a particular bioagent are key to microbead-based immunoassay sensing systems. A unique spectral signature or fingerprint can be tied to each type of bead. Beads can be joined with antibodies to specific biowarfare agents.

A novel biosensing platform uses engineered nanowires as an alternative substrate for immunoassays. Nanowires built from submicron layers of different metals, including gold, silver, and nickel, are able to act as "barcodes" for detecting a variety of pathogens, such as anthrax, smallpox, ricin, and botulinum toxin. This approach could simultaneously identify multiple pathogens via their unique fluorescent characteristics.

"The ability to miniaturize and adapt traditional bench-top immunoassay protocols to a fully automated micro-or nanofluidic chip holds tremendous promise for enabling multiplex, efficient, cost-effective, and accurate pathogen sensing systems for both biodefense and medical applications," says Jeffrey B.-H. Tok, a researcher at Lawrence Livermore National Laboratory (LLNL).

The team, led by LLNL and including researchers from Stanford University, the UC Davis Center for Biophotonics, and Nanoplex Technologies, used multistriped metallic nanowires in a suspended format to rapidly identify sensitive single and multiplex immunoassays that simulated biowarfare agents ranging from anthrax, smallpox, and ricin to botulinum. The entire assay can be performed within 3–4 hours, thus making it feasible to employ it on a rapid diagnostic platform. The core of this portable bioweapon recognition system consists of two parts. Nanowires with a diameter of ~ 250 nm and a length of ~ 6 µm are electrochemically formed and then layered with bands of silver, gold, and nickel to produce patterns that are similar to the ubiquitous barcodes found on consumer products. The other part is an assortment of antibodies, which are essentially glued to the nanowires. Each type of pathogen calls for a unique antibody which is attached to the nanowires for each antibody type with its unique 'barcode'.

The reflection pattern and fluorescence from each stripe sequence can later be clearly recognized, similar to a barcode being scanned.

"In the end, you will have a pool of various striped nanowires, each of which will have a unique antibody assigned to it, which is to detect for that particular pathogen," Tok explains.

To identify pathogens, the barcoded, antibody-carrying nanowires are floated in a neutral liquid called an assay buffer, into which samples of

---

[3] Antibodies are the proteins that the body produces to directly attack, or direct the immune system to attack, cells that have been infected by viruses, bacteria, and other intruders.

suspected pathogens are injected. If an antigen associated with a pathogen meets its corresponding antibody, the two will combine, creating a nanowire/antibody/antigen sandwich that will fluoresce, or glow, under a special light.

According to Tok, an important advantage of the system is that many kinds of barcoded antibodies can be mixed together in the assay buffer liquid, which can be used over and over. "In theory, we could interrogate as many as 100 different striped nanowires in one single snapshot, which makes the analysis very fast."

The system not only applies to biowarfare agents, but could also be used during an outbreak of an infectious disease.

Ongoing work at LLNL is focused on incorporating the assay into a microfluidic device to allow for a portable biosensing system for biological warfare agents. The platform will ultimately enable an affordable and portable multiplex biodetection system for both first responders and clinicians to accurately detect and confirm infectious agents, thus facilitating point-of-care applications.

Featured scientist: Jeffrey B.-H. Tok
Organization: Lawrence Livermore National Laboratory, Livermore, CA (USA)
Relevant publication: Jeffrey B.-H. Tok, Frank Y. S. Chuang, Michael C. Kao, Klint A. Rose, Satinderpall S. Pannu, Michael Y. Sha, Gabriela Chakarova, Sharron G. Penn, George M. Dougherty: Metallic striped nanowires as multiplexed immunoassay platforms for pathogen detection, *Angew. Chem., Int. Ed.*, **45**, 6900–6904.

18.8   Bullets Bouncing Harmlessly off T-shirts

If research undertaken by scientists in Australia is any indication, bullet-proof vests as light as T-shirts could become reality some day.

CNTs have great potential applications in making ballistic-resistance materials. The remarkable properties of CNTs make them an ideal candidate for reinforcing polymers and other materials, and could lead to applications such as lightweight bullet-proof vests or shields for military vehicles and spacecraft. For these applications, there is a need for thinner, lighter, and more flexible materials with dynamic mechanical properties superior to what is currently available. Ongoing research at the University of Sydney explores the energy absorption capacity of SWCNTs under a ballistic impact. CNT-reinforced materials might not only be very effective in stopping ballistic penetration or high speed impact, like Kevlar vests; they might also be able to prevent the blunt force trauma that is still a problem with today's body armor.

"When a bullet strikes body armor, the fibers of these materials absorb and disperse the impact energy to successive layers to prevent the bullet from penetrating," Liangchi Zhang explains. "However, the dissipating forces can still cause a non-penetrating injury, which is known as blunt force trauma. Even when the bullet is stopped by the fabric, the impact and the resulting trauma leave a severe bruise and, at worst, can damage critical organs. Hence the best material for body armor should have a high level of elastic storage energy that will cause the bullet to bounce off or be deflected."

Zhang, a professor at the School of Aerospace, Mechanical and Mechatronic Engineering at the University of Sydney, has been working on the energy absorption capacity of CNTs under ballistic impact. He points out that the challenge in exploring the ballistic resistance capacity of CNTs lies in the understanding of the bullet impact mechanism, involving how the force, energy, momentum, and velocity vary in the time domain at ballistic striking.

"We looked at a 'real' ballistic impingement process," says Zhang: "When impacting a CNT, the speed of a bullet reduces due to its energy loss (the energy is absorbed by the CNT), becomes zero when its forward motion is fully stopped by the CNT, and is bounced back when the CNT releases its stored energy."

Zhang and his team extended their nanoscopic simulation results to a macroscopic estimation to try to get a very rough feel for the possibility of making a better bullet-resistant material based on a fabric of CNT fibers, although this extension involves major assumptions. For example, the scientists assume that on the macroscopic scale, the CNT will have a perfect lattice structure throughout.

For their simulations, the scientists released a 'bullet' (made of nanoscale diamond with the dimensions $35.6 \times 36.6 \times 7.1 \text{ Å}^3$) at a speed varying from 1000 to 3500 m/s. In comparison, the initial velocity of modern rifle bullets is somewhere between 180 and 1500 m/s, depending on gun and bullet type. For a typical over-the-counter gun the speed is less than 1000 m/s.

Based on the properties of CNTs that they obtained in their simulations, the scientists carried out a case study as an example for a potential application.

"We can very roughly estimate the thickness of a possible CNT body armor material composed of several layers of woven CNT yarns if we can assume that the nanoscale properties can be extended simply to a macroscale case," says Zhang. "We feel this is justified because, when a CNT is grown, its diameter is in nanometers but its atomic structure along its length, in micrometres, does not change. In other words, this is completely different from metals where there is a microstructural change from monocrystalline to polycrystalline."

Based on their assumptions, the researchers estimate that body armor 600 μm thick—the same thickness as a very fine T-shirt—woven from six layers of CNT yarn is sufficient to bounce off a .358 inch revolver bullet with a muzzle energy of 320 J.

Zhang also found that nanotubes have excellent resistance to repeated ballistic impacts, which is essential for body armor. He cautions that there is much work ahead, for instance regarding uncertainties in terms of the scale effect and the availability of effective techniques for weaving such fabrics.

Featured scientist: Liangchi Zhang
Organization: School of Aerospace, Mechanical and Mechatronic
  Engineering, University of Sydney (Australia)
Relevant publication: Kausala Mylvaganam, L. C. Zhang: Ballistic
  resistance capacity of carbon nanotubes, *Nanotechnology*, **18**,
  475701.

Subject Index